Introducing Global Issues

FOURTH EDITION

Introducing Global Issues

edited by
Michael T. Snarr
D. Neil Snarr

BOULDER
LONDON

Published in the United States of America in 2008 by
Lynne Rienner Publishers, Inc.
1800 30th Street, Boulder, Colorado 80301
www.rienner.com

and in the United Kingdom by
Lynne Rienner Publishers, Inc.
3 Henrietta Street, Covent Garden, London WC2E 8LU

Library of Congress Cataloging-in-Publication Data
Introducing global issues / edited by Michael T. Snarr, D. Neil Snarr. — 4th ed.
p. cm.
Includes bibliographical references and index.
ISBN 978-1-58826-559-3 (pbk. : alk. paper)
1. World politics—1989– 2. International economic relations. 3. Social history—1970– 4. Ecology. I. Snarr, Michael T. II. Snarr, Neil, 1933–
D860.I62 2008
909.83'1—dc22

2008000945

British Cataloguing in Publication Data
A Cataloguing in Publication record for this book
is available from the British Library.

Printed and bound in the United States of America

Printed on 30% postconsumer recycled paper

The paper used in this publication meets the requirements of the American National Standard for Permanence of Paper for Printed Library Materials Z39.48-1992.

5 4 3 2 1

For Paul Abernathy and Matt Southworth,
who experienced the dark side of war
and have since devoted their lives to peace

Contents

List of Tables and Figures ix
Preface xi

1 Introducing Globalization and Global Issues
Michael T. Snarr 1

Part 1 Conflict and Security

2 Weapons Proliferation and Conflict
Jeffrey S. Lantis 15

3 Nationalism
Lina M. Kassem, Anthony N. Talbott, and Michael T. Snarr 37

4 The Challenge of Human Rights
D. Neil Snarr 57

5 Global Security
Sean Kay 77

Part 2 The Global Economy

6 Free Trade vs. Protectionism: Values and Controversies
Bruce E. Moon 95

7 The Political Economy of Development
Mary Ellen Batiuk 115

8 Poverty in a Global Economy
Don Reeves and Jashinta D'Costa 137

Part 3 Development

9 Population and Migration
Ellen Percy Kraly and Fiona Mulligan 161

10 Women and Development
Elise Boulding and Heather Parker 185

11 Children
George Kent 201

12 Health
Lori Heninger and Kelsey M. Swindler 221

Part 4 The Environment

13 Sustainable Development
Pamela S. Chasek 243

14 Regulating the Atmospheric Commons
Mark Seis 265

15 Conflict over Natural Resources
Jane A. Winzer and Deborah S. Davenport 287

Part 5 Conclusion

16 Future Prospects
Michael T. Snarr 311

List of Acronyms 321
Bibliography 325
The Contributors 343
Index 349
About the Book 361

Tables and Figures

Tables

1.1 Percentage of People Who Use Computers, Selected Countries, 2002–2007 5
1.2 Percentage of People Who Own Cell Phones, Selected Countries, 2002–2007 5
1.3 Advantages and Disadvantages of Globalization 7
2.1 World and Regional Military Expenditures, 1996–2005 18
2.2 Countries Suspected to Possess Chemical or Biological Weapons Capabilities 26
8.1 Poverty Impact of Income Distribution, Selected Countries, 2005 143
8.2 UN Millennium Development Goals 156
9.1 World Population by Geographic Region and for More- and Less-Developed Countries, 1950–2007 166
9.2 Projections of World Population by Geographic Region and for More- and Less-Developed Countries, 2000 vs. 2050 172
9.3 Refugees and Other Persons of Concern to the UNHCR, by Geographic Region of Origin, 2006 177
10.1 Representation of Women in National Parliaments by Region, 1998–2007 194
11.1 Global Annual Child Deaths, 1960–2006 209
12.1 Life Expectancy at Birth, Selected Countries, 2005 223
12.2 Food Security, Selected Countries, 2004 225
12.3 Global Effects of HIV/AIDS, 1996 vs. 2006 231
12.4 Maternal Health, Selected Countries, 2000 236

12.5 Mental Health and Human Rights 238
13.1 Global Spending on Luxury Items Compared with Funding Needed to Meet Selected Basic Needs, 2004 250
14.1 World Carbon Emissions from Fossil Fuel Burning, 1950–2005 267
14.2 Largest Carbon Emissions Producers, 2005 269
14.3 Per Capita Carbon Dioxide Emissions from Consumption and Flaring of Fossil Fuels, Selected Countries, 2005 270

Figures

2.1 Proliferation Matrix 16
2.2 World Nuclear Weapon Status, 2007 24
3.1 Arab-Israeli Wars 47
3.2 The Expansion of Israel 48
3.3 Israeli-Palestinian Peace Attempts 50
4.1 First Generation of Human Rights, UDHR Articles 2–21 60
4.2 Second Generation of Human Rights, UDHR Articles 22–26 63
4.3 Third Generation of Human Rights, UDHR Articles 27–28 64
4.4 UDHR Articles 29–30 65
7.1 Capital Flows in the Triangular Trade, 1500–1800 117
7.2 Western vs. Eastern Spheres of Influence, 1959 125
7.3 DAC Members' Net ODA, 1990–2005, and DAC Secretariat Simulations of Net ODA, 2006–2010 132
8.1 Number and Percentage of Poor People in Developing Countries, 1981–2004 138
8.2 Distribution of Population and Global Income 142
8.3 Number of Undernourished People by Region, 2001–2003 145
8.4 A Virtuous Circle 150
9.1 Ages of Males and Females as Percentage of Population, Spain and Tanzania, 2005 164
9.2 World Population for Development Categories, 1950–2007 165
9.3 World Fertility Rates, 2005–2010 168
9.4 Projected World Population, 1950–2050 170
11.1 Under-Five Mortality Rate, 1990 vs. 2006 210
13.1 UNCED's Agenda 21 249
13.2 The Millennium Round of World Conferences 256
15.1 Natural Resource Matrix 290
15.2 Species Composition of Commercial Harvests in Pacific Salmon Treaty Region, 1990–1994 295
15.3 Jordan River Basin 299

Preface

IN THIS NEW EDITION OF *INTRODUCING GLOBAL ISSUES,* EVERY chapter has been thoroughly updated to cover the most recent developments. In particular, Chapter 5, on global security, has been revised to include a fresh treatment of terrorism, as well as the role of education in increasing global security. Chapter 7 now takes a more historical approach to the development of the global economy, with special focus on capital flows. And Chapter 12, on health, has been revised to cover a broader range of infectious and noninfectious diseases. Ideas for improving this text, and general comments, are welcome at michael_snarr@wilmington.edu.

* * *

We would like to express our appreciation to those who made this book possible. As usual, we are greatly indebted to the scholars and practitioners who have shared their expertise in this volume.

We also thank many students at Wilmington College. Abbey Pratt-Harrington and Deidre Miller both provided excellent research skills for this edition. The students from Wilmington's fall 2007 honors class—Erin Cane, Jerry Collins, Hannah D'Agostino, Emily Donaldson, Sara Douglas, Dylan Givens, Kaitlin Heistan, Laura Kelly, Kristen Matuch, Kimber Owens, Rachel Robinson, Katie Skaleski, Kelsey Stief, Katie Venable, and Jessica Walt—read and offered valuable comments on several draft chapters. Emily also provided valuable research assistance. A very special thanks goes to Katie Venable, who did an outstanding job constructing the bibliography and addressing other important details. Special thanks also to

Nicole Svajlenka, in the Department of Geography at Colgate University, for her work on the map for Chapter 9.

The following people also read parts of the manuscript and made valuable suggestions: Matt Daly, Eileen Kaufman, Bruce Moon, Rodney North, Fiona Ramsey, and Ruth Snarr.

We are grateful to TransFair USA for granting permission to use the Fair Trade Certified™ logo.

We also greatly appreciate the exceptional support we received from Lynne Rienner, Jaime Schwalb, and Karen Williams at Lynne Rienner Publishers.

Finally, our heartfelt appreciation goes to our family for their gracious support—Ruth, Melissa, Madison, Ty, Isaiah, and Elise.

—Michael T. Snarr
D. Neil Snarr

Introducing Global Issues

1

Introducing Globalization and Global Issues

Michael T. Snarr

- Over 200,000 people are added to the world's total population every day (US Census Bureau 2008b).
- People in more than 200 countries and territories have access to the Cable News Network (CNN 2008b).
- In recent years the number of wars and war deaths has declined (HSC 2006).
- In the developing world there are 820 million hungry people. While this number is higher than a decade ago, the percentage of hungry people has declined (FAO 2006a).
- In 1990, 32 percent of the world's population (1.25 billion people) lived on the equivalent of US$1 per day. By 2004 that number had dropped to 19 percent (980 million people) (UN 2007).
- An infant living in Africa is thirteen times more likely to die than one living in Europe or North America (PRB 2004).
- More civilians died in the twentieth century as a result of war than in the four previous centuries combined (Small and Singer 1982).
- Approximately 3,700 square miles of the Amazon rainforest were destroyed between August 2006 and July 2007, the lowest rate since 1988 ("Brazil Says" 2007).
- Dramatic numbers of species are becoming extinct worldwide.
- During the past three decades, life expectancy in the developing world has increased by eight years and illiteracy has decreased by nearly 50 percent (UNDP 2003).
- Nearly 3 million people die each year from AIDS (UN 2007).
- Approximately 27,000 children die every day from preventable diseases (UN 2007).

- In the 1990s, more children died from diarrhea than all the people who died due to armed conflict since World War II (UNDP 2003).
- In poor countries, the percentage of children attending primary schools increased from 80 percent in 1991 to 88 percent in 2005 (UN 2007).

Each of the items above is related to a global issue discussed in this book, and many of them affect the reader. But what is a *global issue*? The term is used here to refer to two types of phenomena. First, there are those issues that are transnational—that is, they cross political boundaries (country borders). These issues affect individuals in more than one country. A clear example is air pollution produced by a factory in the United States and blown into Canada. Second, there are problems and issues that do not necessarily cross borders but affect a large number of individuals throughout the world. Ethnic rivalries and human rights violations, for example, may occur within a single country but have a far wider impact.

For the contributors to this volume, the primary goal is to introduce several of the most pressing global issues and demonstrate how strongly they are interconnected. Since these issues affect each and every one of us, we also hope to motivate the reader to learn more about them.

Is the World Shrinking?

There has been a great deal of discussion in recent years about globalization, which can be defined as "the intensification of economic, political, social, and cultural relations across borders" (Holm and Sørensen 1995: 1). Evidence of globalization is seen regularly in our daily lives. In the United States, grocery stores and shops at the local mall are stocked with items produced abroad. Likewise, hats and T-shirts adorned with the logos of Nike, the Los Angeles Lakers, and the New York Yankees, for example, are easily found outside the United States. In many countries, Britney Spears and other US music groups dominate the radio waves, CNN and MTV dominate television screens, and *Harry Potter* and other Hollywood films dominate the theaters. Are we moving toward a single global culture? In the words of Benjamin Barber, we are being influenced by "the onrush of economic and ecological forces that demand integration and uniformity and that mesmerize the world with fast music, fast computers, and fast food—with MTV, Macintosh, and McDonald's pressing nations into one commercially homogeneous global network: one McWorld tied together by technology, ecology, communication, and commerce" (1992: 53).

For the editors of this book, globalization took on a more personal face in 2007 when we took a group of students to Mexico. As we sat on a bus bound for the pyramids of Teotihuacán, just outside of Mexico City, we met a Canadian named Jag. We learned on the bus ride that Jag was a Hindu from India who lived in Montreal. His job was to assist the newly formed Inuit (Eskimo) government of Nunavet, a new Canadian territory created through negotiations with the Canadian government. Think about it: a Hindu Indian living in French-speaking Montreal, assisting the Inuit government, and visiting a pyramid built by the Teotihuacán peoples, while vacationing in Mexico City—now that's globalization!

Technology is perhaps the most visible aspect of globalization, and in many ways is its driving force. Communications technology has revolutionized our information systems. CNN reaches hundreds of millions of households in over 200 countries and territories throughout the world. "Computer, television, cable, satellite, laser, fiber-optic, and microchip technologies [are] combining to create a vast interactive communications and information network that can potentially give every person on earth access to every other person, and make every datum, every byte, available to every set of eyes" (Barber 1992: 58). Technology has also aided the increase in international trade and international capital flows, and has enhanced the spread of Western, primarily US, culture.

Thomas Friedman, in his boldly titled bestseller *The World Is Flat,* argues that the world is undergoing its third phase of globalization: "Globalization 3.0 is shrinking the world from a size small to a size tiny and flattening the playing field at the same time" (2005: 10). Whereas in the past globalization was characterized by companies becoming more global, this third phase is unique due to "the newfound power for *individuals* to collaborate and compete globally" (2005: 10; original emphasis). For instance, radiologists in India and Australia interpret CAT-scan images from the United States, telephone operators in India answer calls for major US corporations, and Japanese-speakers at call centers in China serve Japanese customers. Thus the playing field is being leveled and individuals and small companies from all over the world, including poor countries, can now compete in the global economy.

Of course, Earth is not literally shrinking (nor flat), but in light of the rate at which travel and communication speeds have increased, the world has in a sense become smaller. Many scholars assert that we are living in a qualitatively different time, in which humans are interconnected more than ever before: "There is a distinction between the contemporary experience of change and that of earlier generations: never before has change come so rapidly . . . on such a global scale, and with such global visibility" (CGG 1995: 12). Or as Friedman puts it, "There is something about the flattening of the world that is going to be qualitatively different from other such profound changes: the speed and breadth with which it is taking hold. . . . This flattening process is happening at warp speed and directly or indirectly touching a lot more people on the planet at once" (2005: 46).

This seemingly uncritical acceptance of the concept of globalization and a shrinking world is not without its critics, who point out that labor, trade, and capital moved at least as freely, if not more so, during the second half of the nineteenth century than they do now. Take, for example, the following quote, which focuses on the dramatic changes that have taken place in the past three decades to make the world more economically interdependent: "The complexity of modern finance makes New York dependent on London, London upon Paris, Paris upon Berlin, to a greater degree than has ever yet been the case in history. This interdependence is the result of the daily use of those contrivances of civilization . . . the instantaneous dissemination of financial and commercial information . . . and generally the incredible increase in the rapidity of communication" (Angell 1909: 44–45). If this statement were to appear in a newspaper today, no one would give it a second thought. But it was written at the start of the twentieth century—illustrating the belief of some critics that globalization is not a new phenomenon.

Some skeptics caution that, while interdependence and technological advancement have increased in some parts of the world, this is not true for the vast majority of third world countries (the terms "third world," "the South," "the developing world," and "the less-developed countries" are used interchangeably throughout this book in reference to the poorer countries, in contrast to "the first world," "the North," "the developed world," and "the more-developed countries" in reference to the United States, Canada, Western Europe, Japan, Australia, and New Zealand). For example, Hamid Mowlana argues that "global" is not "universal" (1995: 42). Although a small number of people in third world countries may have access to much of the new technology and truly live in the "global village," the large majority of populations in the South do not. Research on access to computer and cell phone usage illustrates this point. Tables 1.1 and 1.2 show findings from a forty-seven-country survey of these indicators of globalization. Table 1.1 clearly shows large differences in the numbers of people who use computers. Further, there is little indication that the gap will close significantly in the near future. Table 1.2 tells a moderately different story for cell phones. The gap between haves and have-nots certainly exists, but it is not as extreme as for computer use, and is closing much more rapidly.

In most of the poorer countries of Africa and Asia, the number of cellular mobile subscribers per 1,000 people is in single digits. In contrast, in many of the developed countries, nearly half of all people use this technology (UNDP 2004). A good example of this contrast was seen in the recent war in Afghanistan. While ultramodern US jets flew above Kabul, many Northern Alliance troops entered the city on horses and bicycles.

Similarly, one can argue that the increased flow of information, a characteristic of globalization, goes primarily in one direction. Even those in the South who have access to television or radio are at a disadvantage. The

Table 1.1 Percentage of People Who Use Computers, Selected Countries, 2002–2007

	Percentage Who Use Computers, 2007	Percentage Change, 2002–2007
South Korea	81	+9
United States	80	+7
Canada	76	+1
Japan	66	+15
Lebanon	61	+11
China	40	+5
Nigeria	37	+14
Russia	36	+17
India	28	+6
Indonesia	11	–1
Pakistan	9	+2
Tanzania	6	+1
Bangladesh	5	–3

Source: Pew Research Center (2007), available at http://pewglobal.org/reports/pdf/258.pdf.

Table 1.2 Percentage of People Who Own Cell Phones, Selected Countries, 2002–2007

	Percentage Who Use Cell Phones, 2007	Percentage Change, 2002–2007
South Korea	97	+4
Japan	86	+13
Lebanon	84	+22
United States	81	+20
China	67	+17
Nigeria	67	+56
Russia	65	+57
Canada	60	+12
India	60	+48
Tanzania	42	+32
Bangladesh	36	+26
Pakistan	34	+29
Indonesia	27	+19

Source: Pew Research Center (2007), available at http://pewglobal.org/reports/pdf/258.pdf.

globalization of communication in the less-developed countries typically is a one-way proposition: the people do not control any of the information; they only receive it. It is also true that, worldwide, the ability to control or generate broadcasts rests in the hands of a tiny minority.

While lack of financial resources is an important impediment to globalization, there are other obstacles. Paradoxically, Benjamin Barber, who argues that we are experiencing global integration via "McDonaldization," asserts we are at the same time experiencing global disintegration. He cites the breakup of the Soviet Union and Yugoslavia, as well as the many other ethnic and national conflicts (see Chapter 3), as evidence of the forces countering globalization. Many subnational groups (groups within nations) desire to govern themselves; others see threats to their religious values and identity and therefore reject the secular nature of globalization (Barber 1992). As a result, Hamid Mowlana argues that globalization "has produced not uniformity, but a yearning for a return to non-secular values. Today, there is a rebirth of revitalized fundamentalism in all the world's major religions, whether Islam, Christianity, Judaism, Shintoism, or Confucianism. At the same time the global homogeneity has reached the airwaves, these religious tenets have reemerged as defining identities" (1995: 46).

None of these criticisms mean that our contemporary world is not different in some important aspects. There is widespread agreement that communications, trade, and capital are moving at unprecedented speed and volume. However, these criticisms do provide an important warning against overstating or making broad generalizations about the processes and effects of globalization.

Is Globalization Good or Bad?

There are some aspects of globalization that most will agree are good (for example, the spread of medical technology) or bad (for example, increased global trade in illegal drugs). Events during the 2001–2002 war in Afghanistan revealed the dramatic contrast between friends and foes of globalization. Due to the Taliban's rejection of many aspects of Western culture, some Afghanis apparently buried their TVs and VCRs in their backyards. When Kabul was captured by the Northern Alliance, it was reported that one Afghani anxiously retrieved his TV and VCR in order to view his copy of *Titanic* (Filkins 2001). Judging whether or not globalization is good, however, is complex.

Table 1.3 identifies three areas that are affected by globalization—political, economic, and cultural—and gives examples of aspects considered positive and negative. A key aspect of political globalization is the weakened ability of the state to control both what crosses its borders and what happens inside them. In other words, globalization can reduce the state's sovereignty (its ability to govern matters within its borders). This can be viewed as good, because undemocratic governments are finding it increasingly difficult to control the flow of information to and from

Table 1.3 Advantages and Disadvantages of Globalization

Effects of Globalization	Advantages	Disadvantages
Political	Weakens power of authoritarian governments	Unwanted external influence difficult to keep out
Economic	Jobs, capital, more choices for consumers	Exploitative; only benefits a few; gap between rich and poor
Cultural	Exposure to other cultures	Cultural imperialism

prodemocracy groups. Satellite dishes, e-mail, and the World Wide Web are three examples of technology that have eroded state sovereignty. But decreased sovereignty also means that the state has difficulty controlling the influx of illegal drugs and unwanted immigrants, including terrorists.

In the realm of economics, increased globalization has given consumers more choices. Also, multinational corporations are creating jobs in poor areas where people never before had such opportunities. Some critics reject these points, arguing that increased foreign investment and trade benefit only a small group of wealthy individuals and that, as a result, the gap between rich and poor grows both within countries and between countries. These critics point out that the combined wealth of the fifteen richest people in the world is more than the gross domestic product (the total goods and services produced in a given year) of sub-Saharan Africa (Parker 2002). Related to this is the argument that many well-paying, blue-collar jobs are moving from the North to the poor countries of Latin America, Africa, and Asia.

At the cultural level, those who view increased cultural contact as positive say that it gives people more opportunities to learn about (and purchase goods from) other cultures. But critics of cultural globalization see things differently. Samuel Huntington (1998) has argued that the shrinking world will bring a "clash of civilizations." In this scenario, clashes will occur among many civilizations, including the largely Christian West against Islam. Other critics are concerned with cultural imperialism, in which dominant groups (primarily wealthy countries) force their culture on others. A primary tool of cultural influence is the North's multibillion-dollar advertising budgets used to influence and to some extent destroy non-Western cultures. The fear of cultural imperialism is certainly a key component in the animosity of some Arabs toward the United States. Other critics are increasingly fearful that more and more national languages will become extinct as foreign languages, especially English, penetrate borders. It is estimated that at least one language is lost every month (Worldwatch Institute 2006). In response to cultural influences, countries like Iran have banned Western music from government radio and television stations in an attempt to stop unwanted outside influences. Even Western countries like France have adopted policies to regulate unwanted foreign cultural influences.

The degree to which cultural values can be "exported" is the subject of some debate. Huntington argues that "drinking Coca-Cola does not make Russians think like Americans any more than eating sushi makes Americans think like Japanese. Throughout human history, fads and material goods have spread from one society to another without significantly altering the basic culture of the recipient society" (1996: 28–29). Similarly, others, such as Hamid Mowlana, argue that globalization brings only superficial change: "McDonald's may be in nearly every country, but in Japan, sushi is served alongside hamburgers. In many countries, hamburgers are not even on the menu" (1995: 46). Thus the global product is often altered to take on a local flavor. The term "glocalization" has combined the words *global* and *local* to describe such hybrid products.

In sum, globalization offers a multitude of advantages to people throughout the world, from greater wealth to more choices in consumer products. At the same time, globalization exposes people to greater vulnerability and insecurity. Our jobs become less secure, diseases travel faster, and traditional family structures are weakened (Kirby 2006). It is left to the reader to determine whether globalization is having a positive or negative effect on the issues discussed in this book. Is globalization enhancing human capacity to deal with a particular problem? Or is it making it more difficult? Of course, each individual's perspective will be influenced by whether he or she evaluates these issues based on self-interest, national interest, a religious view, or a global humanitarian viewpoint. Readers must decide, based on what is most important to them, how to evaluate moral questions of good versus bad. For example, when considering the issue of free trade (Chapter 6), those concerned first and foremost with self-interest will ask, "How does free trade affect me?" For nationalist readers the question will be, "How does free trade affect my country?" For religious readers, the question will be, "How does my religion instruct me on this issue?" Finally, global humanitarians will ask, "What is best for humanity in general?"

Interconnectedness Among Issues

As mentioned above, a primary purpose of this book is to explore the interconnectedness of the various issues discussed here. For example, the chapter on poverty should not be considered separate from the chapter on population, even though these two issues are treated separately. Here are several examples of how issues discussed in this book are interconnected:

- The growth in the world's population (Chapter 9) has been significantly affected, especially in Africa, by the AIDS crisis (Chapter 12).

- Many of the value judgments concerning trade issues (Chapter 6) are intricately linked to human rights issues (Chapter 4).
- Ethnic conflict (Chapter 3) (as well as other types of conflict) often leads to internal migration as well as international population movements (Chapter 9).
- One of the recommendations for reducing poverty (Chapter 8) is to educate women and give them more decisionmaking power over their lives (Chapter 10).

The interconnectedness of these issues is even more extensive than these examples demonstrate. For instance, while an increase in AIDS will affect population growth, the connections do not end there. AIDS epidemics also lead to increased government expenditures, which can lead to increased indebtedness, which will likely lead to more poverty, and so on. Thus, each global issue discussed in this book has multiple consequences, as well as a ripple effect.

Key Players

Of the key players or actors involved in these global issues, the most salient are countries. In the following pages you will continually read about the countries of the world and their efforts to solve these various global issues. Often, countries get together and form international governmental organizations (IGOs). The logic is that by cooperating through an IGO—like the United Nations, the World Bank, or the UN Children's Fund (UNICEF)—countries are better equipped to achieve a common goal, like preventing war or alleviating poverty, that they could not accomplish on their own. The reader will notice that IGOs are also mentioned throughout the book.

Nongovernmental organizations (NGOs) working on global issues are part of what is called civil society. For instance, in recent decades there has been a dramatic increase in NGOs seeking to make the world a better place (NGOs are sometimes referred to as international nongovernmental organizations [INGOs]). NGOs, as their name implies, work outside the government and comprise individual citizens working together on one or more problems. There are many very well known NGOs working on global issues: the Red Cross, Greenpeace, Amnesty International, World Vision, and Doctors Without Borders are just a few of the thousands that exist. Because these NGOs are often made up of highly motivated people in the middle of a war or refugee camp, they can often achieve results that countries cannot. NGOs have become extremely active on all of the issues discussed in this book, and often cooperate with IGOs and individual countries.

Other nongovernmental actors include businesses, often referred to as transnational corporations (TNCs). Nike, Apple, Toyota, and many other

TNCs have gained increasing power in recent years to affect global issues. Many critics complain that, due to their economic strength and global networks, TNCs exercise too much power.

Finally, individuals can have an impact on global issues as well. Several examples of individuals working to resolve various global issues can be found in the following chapters.

Outline of the Book

This book has been organized into five parts. Part 1 focuses on conflict and security issues. It considers some of the primary sources of conflict, such as weapons of mass destruction, nationalism, terrorism, and human rights abuses, and some of the many approaches to establishing and maintaining peace. Part 2 concentrates on economic issues, ranging from international trade and capital flows to one of the major concerns that confronts humanity—poverty. Related political and social issues are also examined. Part 3 deals with development issues, such as population, health, and women and children—issues that tend to plague (but are not confined to) the poorer countries. Part 4 focuses on environmental issues, such as the global commons, sustainable development, global warming, and ozone depletion, and on global attempts to solve them. Part 5 discusses possible future world orders, sources of hope and challenges in the coming decades, and innovative actions that are being taken to make a positive impact on global issues.

Discussion Questions

1. What examples of globalization can you identify in your life?
2. Do you think globalization will continue to increase? If so, in what areas?
3. Do you think globalization has more positive attributes or more negative attributes?
4. Can you think of additional examples of how the global issues discussed in different chapters are interconnected?

Suggested Readings

Barber, Benjamin R. (1996) *Jihad vs. McWorld.* New York: Ballantine.

Congressional Quarterly (2002) *World at Risk: A Global Issues Sourcebook.* Washington, DC.

Friedman, Thomas L. (2000) *The Lexus and the Olive Tree: Understanding Globalization.* New York: Anchor.

——— (2005) *The World Is Flat: A Brief History of the Twenty-First Century.* New York: Farrar, Straus, and Giroux.

Iyer, Pico (2001) *The Global Soul: Jet Lag, Shopping Malls, and the Search for Home.* New York: Vintage.

Jones, Ellis, et al. (2001) *The Better World Handbook: From Good Intentions to Everyday Actions.* Boulder: New Society.

Kennedy, Paul, Dirk Messner, and Franz Nuscheler, eds. (2001) *Global Trends and Global Governance.* Sterling, VA: Pluto.

Kirby, Peadar (2006) *Vulnerability and Violence: The Impact of Globalization.* Ann Arbor, MI: Pluto.

Simmons, P. J., and Chantal de Jonge Oudraat, eds. (2001) *Managing Global Issues: Lessons Learned.* Washington, DC: Carnegie Endowment for International Peace.

United Nations Development Programme (annual) *Human Development Report.* New York: Oxford University Press.

World Bank (2007) *World Development Indicators 2007.* New York: Oxford University Press.

Worldwatch Institute (2006) *Vital Signs 2006–2007: The Trends That Are Shaping Our Future.* New York: Norton.

PART 1

Conflict and Security

2

Weapons Proliferation and Conflict

Jeffrey S. Lantis

THE PROLIFERATION OF WEAPONS AND WEAPONS TECHNOLOGY is one of the most serious challenges to international security today. Recent events such as the Iraq War, confrontations between the international community and Iran and North Korea, and the threat of terrorists using weapons of mass destruction (WMD) are reminders that proliferation challenges are here to stay. These issues also raise questions about the future of the nonproliferation regime and the strategy of deterrence.

Proliferation is not simply a problem for politicians and military leaders. When governments choose to use weapons in conflict, they are exposing both soldiers and civilians to danger. In fact, the proliferation of weapons contributed to higher civilian casualties and greater destruction in the twentieth century than in the previous four centuries combined (Small and Singer 1982). When governments devote funds to establish large armies or WMD systems, they are also choosing to divert resources from other programs like education and healthcare. Whether or not there are imminent security threats, citizens of the world experience the effects of proliferation every day.

Types of Proliferation

Proliferation is best understood as the rapid increase in the number and destructive capability of armaments. As illustrated in Figure 2.1, there are four dimensions of proliferation to consider. Vertical proliferation is the development and stockpiling of armaments in one country. Horizontal proliferation is the spread of weapons or weapons technology across country

Figure 2.1 Proliferation Matrix

	Vertical Proliferation	Horizontal Proliferation
Conventional weapons	Type I	Type II
Weapons of mass destruction	Type III	Type IV

borders. Conventional weapons are those systems that make up the vast majority of all military arsenals—including guns, tanks, most artillery shells and bullets, planes, and ships. Weapons of mass destruction are special weapons that have a devastating effect even when used in small numbers. Nuclear, chemical, and biological systems can be used to kill more indiscriminately than can conventional weapons.

Vertical Proliferation of Conventional Weapons

The buildup of conventional weapons arsenals is the oldest form of proliferation. At first glance, one might also view this as the least threatening of all forms. Vertical conventional proliferation, however, can be a threat to international stability for at least two reasons. First, arms stockpiling provides more weapons for governments and groups to engage in more conflicts. Conventional weapons have become more sophisticated (from automatic rifles to precision-guided munitions) and more destructive (from cannon balls to 2,000-pound bombs). Some believe that vertical conventional proliferation in an unregulated world market may contribute to conflicts. A second danger of conventional arms buildups is the social cost, which often includes serious cuts in social welfare spending by governments whose citizens can ill afford them.

More weapons mean more conflicts. Government programs to stockpile conventional armaments ensure that there are more weapons available to engage in more conflicts. Some experts believe that the simple availability of weapons systems and the development of military strategies increase the chances that a country will go to war. They argue that advances in conventional weaponry and offensive military strategies have contributed to the outbreaks of numerous conflicts, including both world wars and the Vietnam War. In this context, arms buildups are seen as one potential cause of war in the international system (Sagan 1986; Sivard 1991).

Traditionally, conventional arms buildups focus on weapons systems that are considered most effective for the times. In the period leading up to World War I, Germany and Great Britain engaged in a race to build the most powerful warships. Adolf Hitler ordered research and development of surface-to-surface missiles and jet aircraft during World War II. During the Cold War, President Ronald Reagan called for the creation of a 600-ship navy, with an emphasis on strong aircraft carrier battle groups and advanced submarines. More recently, attention has turned to the latest technology of warfare, including stealth planes and ships, unmanned aerial vehicles like the Predator drone, antisatellite weapons, and computer technology, to give advantage to the fighting forces of the twenty-first century.

The relationship between arms buildups and the likelihood of conflict is multiplied by the fact that conventional weapons have become more sophisticated and destructive over the years. "Smart" bombs and precision-guided munitions allow militaries to hit more precise targets from a long distance. Explosive devices triggered remotely by cell phones enable insurgent groups to attack unsuspecting soldiers, such as recently in Iraq and Afghanistan. Shoulder-launched missiles may give a terrorist the ability to shoot down a large aircraft.

Finally, it is important to remember that conventional arms have been used repeatedly in conflict since the end of World War II. From landmines to fighter jets, conventional weapons have been blamed for roughly 50 million deaths around the globe since 1945.

The social costs of arms buildups. Many governments have sizable conventional arsenals. US military expenditures have topped $250 billion annually over the past two decades, with most of these funds used to support troops and conventional weapons. In 2003 the George W. Bush administration authorized a total defense budget of $396 billion to support an active-duty military strength of 1,370,000 soldiers and a force structure comprising ten army divisions, twelve navy aircraft carrier groups, three marine divisions, five army special forces groups, and thirteen air force combat wings. More than 250,000 of these US soldiers were stationed abroad (with large deployments in Iraq, Europe, the Persian Gulf, and Asia). Critics point out that in relative terms, 2001 defense expenditures by the United States were more than six times larger than those of its nearest potential competitor, and more than twenty-three times larger than the combined spending of the seven "rogue states"—the countries identified at the time as the most likely adversaries of the United States: Cuba, Iran, Iraq, Libya, North Korea, Sudan, and Syria (Hellman 2001). By 2005, US military spending reached $478 billion, or 48 percent of total world military expenditures.

Table 2.1 illustrates the broader context of rising global and regional defense expenditures. In 2000, global military expenditures reached $784 billion, or $128 for every person on the planet. North America saw a 49 percent increase in defense spending between 1996 and 2005, followed closely by Africa at 48 percent (SIPRI 2006). By 2005, global defense expenditures exceeded $1 trillion, or $173 for every person on the planet.

Table 2.1 World and Regional Military Expenditures, 1996–2005

	Military Expenditures (US$ billions at constant 2003 values)			Percentage Change, 1996–2005
	1996	2000	2005	
Africa	8.6	11.1	12.7	+48
Asia	116.0	126.0	157.0	+36
Central America	3.3	3.6	3.2	–2
Europe	236.0	243.0	256.0	+8
Middle East	39.0	51.5	63.0	+61
North America	328.0	332.0	489.0	+49
South America	15.7	17.8	20.6	+31
World	747.0	784.0	1,001.0	+34

Source: Stockholm International Peace Research Institute, *SIPRI Yearbook 2006: Armaments, Disarmament, and International Security* (London: Oxford University Press, 2006).

Note: Data for some countries have been excluded because of lack of information.

Critics charge that there are dangerous social costs of military spending. In 1996, for example, world military expenditures per soldier equaled $31,480, while government expenditures on education were just $899 per student. Health expenditures per person were only $231 around the world, and just $22 in developing countries (Sivard 1996). The United States, which ranked first in the world in military spending in 2003, placed thirtieth on infant mortality rates (behind countries such as Canada, Slovenia, and Cyprus) (UNDP 2003). And when military expenditures rise in developing countries, studies have shown that economic growth declines and government debt increases (Nincic 1982). Rightly or wrongly, it seems that some countries are more concerned with defending their citizens from foreign attack than they are with protecting them from social insecurities at home.

Horizontal Proliferation of Conventional Weapons

A second category of proliferation is the horizontal spread of conventional weapons and weapons technology across country borders. The main route of the spread of conventional weaponry is through arms sales. These are often legitimate transfers of weapons from sellers to buyers in the international arms market. The conventional arms trade is very lucrative, though, and many experts are concerned that corporate greed may be driving the world more rapidly toward the brink of major conflict.

Arms dealers. The conventional arms trade is a very big business. In 2006, just three advanced industrialized countries—the United States,

Russia, and the United Kingdom—were responsible for 71 percent of global arms deals. The United States was the number-one arms exporter in the world, exporting $16.9 billion worth of conventional arms (responsible for 42 percent of the global total). Russia was second with $8.7 billion in sales, and the United Kingdom ranked third with $3.1 billion. Among these deals in 2006 were agreements by the United States to sell Pakistan thirty-six new F-16 fighter aircraft, and by Russia to sell tanks to Algeria and helicopters to Venezuela (Grimmet 2007).

Arms customers. Who are the main customers for all of these weapons? US defense contractors sell most weapons to their allies in the developing world. The Middle East and Asia have been the largest markets for arms sales in recent decades. For example, US defense contractors sell heavy weapons, equipment, and technology to Saudi Arabia, Egypt, Israel, Turkey, Oman, the United Arab Emirates, and other countries. In Asia, allies such as Taiwan, South Korea, India, and Japan purchase large numbers of conventional weapons systems.

Arms are not always sold to countries considered traditional allies, however. In the 1980s the People's Republic of China spent more than $400 million on US weapons. These arms deals were stopped only after the Tiananmen Square massacre of prodemocracy activists in 1989. Iraq's former leader, Saddam Hussein, built a massive conventional arsenal through the international arms market in the 1980s. Arms sales to Iraq by friends and allies came back to haunt the United States, however, during the Gulf War (1991) and the Iraq War (2003). Indeed, the sale of conventional weapons raises real concern about the potential for "deadly returns" on arms sales (Laurance 1992).

Legal and illegal arms transfers also have contributed to civil wars around the world in the past decade, including conflicts in Sierra Leone, Afghanistan, Algeria, Sudan, Colombia, Sri Lanka, Chechnya, and the Congo. Sadly, many of the casualties in these conflicts have been civilians. Landmines and unexploded ordnance produce some 15,000–20,000 injuries and deaths per year. Children are especially at risk. According to the United Nations Development Programme (UNDP), more than 2 million children were killed and another 4.5 million were disabled in civil wars and conflicts between 1987 and 1997 (Klare 1999).

Vertical Proliferation of Weapons of Mass Destruction

The vertical proliferation of weapons of mass destruction is another serious threat to international security. There are several important dimensions of this problem, including types of WMD systems, incentives for states to build nuclear weapons, and patterns of vertical WMD proliferation.

Types of WMD systems. There are three categories of weapons of mass destruction: nuclear, biological, and chemical. These are often examined as a group, but it is important to note that their effects—and potential military applications—are quite different.

Nuclear weapons are devices with tremendous explosive power based on atomic fission or fusion. During World War II, President Franklin D. Roosevelt authorized a five-year, $2 billion secret research program known as the Manhattan Project, which was the first in the world to develop atomic weapons. On August 6, 1945, the United States dropped a 12.5-kiloton atomic bomb on Hiroshima, Japan. This weapon produced an explosive blast equal to that of 12,500 tons of TNT and caused high-pressure waves, flying debris, extreme heat, fires, and radioactive fallout, killing approximately 135,000 people. A second bomb was dropped on Nagasaki on August 9, 1945, killing 65,000. The Japanese government surrendered one day later.

The use of atomic bombs to end World War II was actually the beginning of a very dangerous period of spiraling arms races between the United States and the Soviet Union. The Soviet regime immediately stepped up its atomic research and development program. In 1949 it detonated its first atomic test device. By the 1980s the Soviet Union had accumulated an estimated 27,000 nuclear weapons in its stockpile. Both superpowers also put an emphasis on diversification of their weapons systems. Land-based, intercontinental ballistic missiles (ICBMs) were the symbolic centerpiece of the arsenal. But the superpowers also deployed nuclear weapons on submarines, in long-range bombers, and even in artillery shells and landmines.

Chemical weapons and biological weapons. Chemical weapons work by spreading poisons that incapacitate, injure, or kill through their toxic effects on the body. Chemical agents can be lethal when vaporized and inhaled in very small amounts, or when absorbed into the bloodstream through skin contact. During World War I, mustard-gas attacks killed 91,000 soldiers and injured 1 million in 1917 alone. Chemical weapons were used again during the Iran-Iraq War (1980–1988), killing an estimated 13,000 soldiers (McNaugher 1990). And in 1995, a radical religious cult in Japan used the nerve gas Sarin against civilians in the Tokyo subway system, killing twelve and injuring thousands more.

Chemical weapons are relatively simple and cheap to produce compared with other classes of WMD systems. Any group with some level of expertise could use a mix of chemicals to create dangerous weapons of mass destruction. Not surprisingly, this type of weapon has become popular for terrorist groups. In 2006, intelligence agents in Britain uncovered a plot to use chemicals hidden in drink bottles to build bombs that could blow up commercial aircraft. That same year, insurgents in Iraq began detonating bombs laced with chlorine gas to maximize deaths and injuries in populated areas.

As dangerous as chemical weapons can be, biological agents are actually much more lethal and destructive. Biological agents are microorganisms such as bacteria, viruses, or fungi that can be used to cause illness or kill the intended target. Anthrax is the most common example of a biological agent. Anthrax is a disease-causing bacteria that contains as many as 10 million lethal doses per gram. At the height of the Cold War, both superpowers had significant stockpiles of weaponized anthrax ready for use. In the aftermath of September 11, 2001, small amounts of anthrax were used in terrorist attacks through the US mail system. At this writing, this bioweapons terror case remains unsolved.

Like chemical agents, biological and toxic weapons are relatively easy to construct and have a high potential lethality rate (if left untreated). Any government or group with access to a pharmaceutical manufacturing facility or biological research lab could develop biological weapons. And like the other classes of WMD systems, information about the construction of such biological systems is available in the open scientific literature and on the Internet.

Why build WMD systems? There are two primary reasons why countries build weapons of mass destruction: security and prestige. First, some government leaders believe that the security of their country is at risk without such systems. During the Cold War, the United States and the Soviet Union established large nuclear weapons stockpiles as well as arsenals of chemical and biological weapons. US policy on bioweapons development was finally reversed by President Richard Nixon in the early 1970s, but clandestine Soviet research and development continued for another decade. In the Middle East, Israel is suspected of having more than a hundred nuclear devices for potential use in its own defense. Indeed, the Israeli government secretly threatened to use them against Iraq during the 1991 Gulf War if Israel were to come under chemical or biological weapons attack.

The standoff between India and Pakistan is another prime example of the security imperative. After years of border skirmishes between the countries, India began a secret program to construct an atomic device that might swing the balance of regional power in its favor. In 1974 the Indian government detonated what it termed a "peaceful nuclear explosion"—signaling its capabilities to the world and threatening Pakistani security. For the next twenty-five years, both Pakistan and India secretly developed nuclear weapons in a regional arms race. In May 1998 the Indian government detonated five more underground nuclear explosions, and the Pakistani government responded with six test explosions of its own.

Second, some governments have undertaken WMD research and development programs for reasons of prestige, national pride, or a desire for influence. It became clear to some during the Cold War that the possession of WMD systems—or a spirited drive to attain them—would gain attention for a country or leader. In the early 1990s, North Korea's drive to build a nuclear device drew the attention of the United States. After extensive negotiations, North Korea was offered new nuclear energy reactors in exchange for a promise not to continue its bomb program. When North Korea renounced the deal in 2002, world leaders scrambled to head off its program to develop more weapons. Nevertheless, the North Korean government successfully tested its first nuclear bomb in October 2006.

Some governments pursue the development of WMD arsenals because they believe it will help them gain political influence. This may help to explain clandestine Iraqi government efforts to develop a WMD arsenal from the 1970s to the 1990s. Iraqi leader Saddam Hussein ordered the creation of a secret WMD research and development program and began to acquire nuclear technology and materials from France, Germany, the United States, and other countries in the late 1970s. Hussein believed that a nuclear bomb would contribute to Iraqi prestige, power, and influence in the Middle East. In response, the Israeli government tried to stop Iraq's nuclear program by launching an air strike against Iraq's research reactor at Osiraq in 1981. But the determined Iraqi drive for a WMD arsenal appears to have ended only with the work of the UN's special commission to investigate and dismantle all Iraqi WMD programs following the Gulf War (Steinberg 1994).

Horizontal Proliferation of Weapons of Mass Destruction

The horizontal proliferation of WMD systems represents the final dimension of this challenge to international peace and stability. In fact, the spread of these weapons and vital technology across state borders is often viewed as the most serious of all proliferation threats. President George W. Bush famously warned that the gravest danger we face is that "the world's most dangerous people" will get their hands on "the world's most dangerous weapons."

Nuclear arsenals. The massive buildup of nuclear arsenals by the superpowers was not the only game in town during the Cold War. In fact, while the Soviet Union and United States were stockpiling their weapons, other countries were working to join the nuclear club. Today, the United States, Russia, France, Great Britain, China, India, North Korea, and Pakistan all openly acknowledge possessing stocks of nuclear weapons (see Figure 2.2). In 1952, Great Britain successfully tested an atomic device and eventually built a nuclear arsenal of about 200 weapons. France officially joined the nuclear club in 1960 and built a somewhat larger nuclear arsenal of an estimated 420 weapons. The People's Republic of China detonated its first atomic device in 1964 and built an arsenal of about 300 nuclear weapons during the Cold War (McGwire 1994).

Several developing countries began secret atomic weapons research and development projects after World War II. As noted earlier, states like India, Pakistan, and Israel pursued clandestine WMD programs because of concerns about security and prestige. When the Indian government detonated its first nuclear explosion in 1974, it symbolically ended the monopoly on nuclear systems held by the great powers. India actually obtained nuclear material for its bomb by diverting it from a Canadian-supplied nuclear energy reactor (which had key components originally made in the United States). Most experts believe that India now possesses a stockpile of about fifty unassembled nuclear weapons. In the late 1970s, Pakistani research scientist A. Q. Khan stole classified information on uranium enrichment technology from a European consortium and became a national hero for directing Pakistan's weapons program. Today, Pakistan has dozens of weapons that could be assembled quickly for use. The Israeli nuclear program was a derivative of research and development projects in the United States and, ironically, the Soviet Union. Like India, the Israeli government proved to be quite resourceful in adapting existing technologies to construct its arsenal (Cirincione, Wolfstahl, and Rajkumar 2002).

According to the Carnegie Endowment for International Peace, several countries are considered "high-risk" proliferants. Iran has attempted to develop or acquire nuclear weapons, and North Korea successfully detonated its first nuclear device in October 2006. In 2004 it was revealed that Khan, the Pakistani scientist, had supplied both governments with the designs and technology to produce enriched uranium (Rohde and Sanger 2004). Experts believe that North Korea now has sufficient weapons material for at least a dozen nuclear bombs. World leaders are extremely concerned about the Iranian government's efforts to develop nuclear weapons. After starting a pilot nuclear facility and constructing a large uranium enrichment plant in 2004, Iranian government leaders have defied international pressure to stop what they call a drive for a peaceful nuclear energy program.

Finally, some states have made political decisions to give up their WMD research and development efforts altogether. Included in this group are South Africa, Brazil, Argentina, Libya, and three former Soviet republics: Belarus, Kazakhstan, and Ukraine. The South African government admitted that it had constructed six nuclear devices for self-defense in the 1970s and 1980s, for example. But the government decided to destroy these weapons in 1990—unilaterally removing itself from the nuclear club. Argentina and Brazil renounced past efforts to develop nuclear arsenals.

Figure 2.2 World Nuclear Weapon Status, 2007

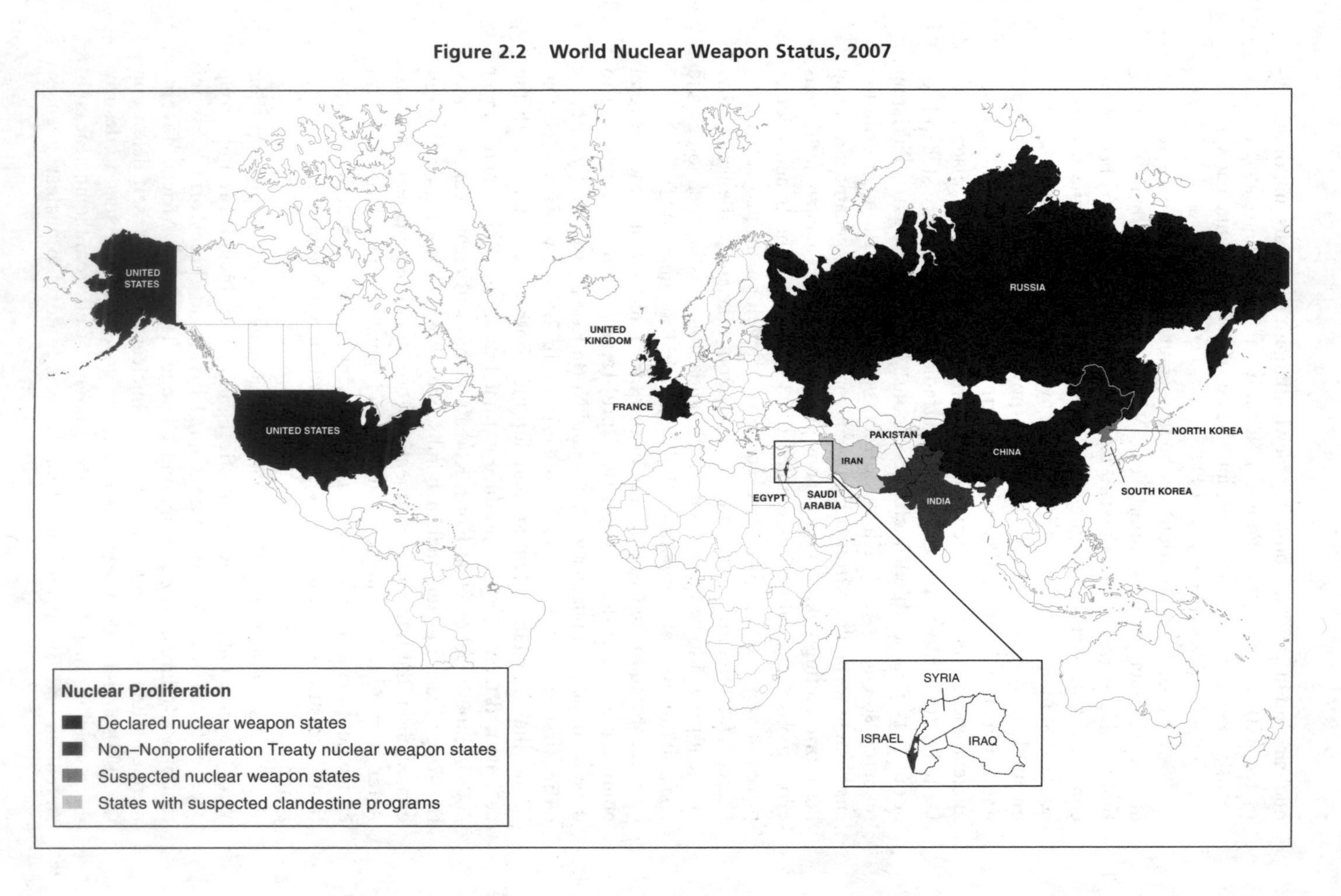

Figure 2.2 continued

Chemical, Biological, and Missile Proliferation

Suspected biological warfare stockpiles[a]	Israel, North Korea, Russia
Suspected biological warfare research programs[b]	China, Egypt, Iran, Syria
Suspected chemical warfare stockpiles[c]	China, Egypt, Iran, Israel, North Korea, Syria
Declared chemical weapons slated for destruction[d]	India, Russia, South Korea, United States
Ballistic missiles with over 1,000-km range	China, France, India, Iran, Israel, North Korea, Pakistan, Russia, Syria, United Kingdom, United States

Notes: a. Country may have offensive biological weapons or agents.

b. Country may have active interest in acquiring the capability to produce biological warfare agents.

c. Country may have some undeclared chemical weapons.

d. Country has declared its chemical weapons and is committed to destroying them under the Chemical Weapons Convention.

Worldwide Nuclear Stockpiles

Country	Total Nuclear Warheads
China	410
France	350
India	75–110
Israel	100–170
Pakistan	50–110
Russia	~16,000
United Kingdom	200
United States	~10,300
Total	~27,600

Missiles with Ranges Exceeding 1,000 km in Six Countries of Proliferation Concern

Country	Missile	Range
India	Agni II	2,000–2,500 km
Iran	Shahab III	1,300 km
Israel	Jericho II	1,500 km
North Korea	No Dong	1,300 km
	Taepo Dong I	1,500–2,000 km
	Taepo Dong II	5,500 km
Pakistan	Ghauri/No Dong	1,300 km
	Ghauri II	1,500–2,000 km
Saudi Arabia	CSS-2	2,600 km

Source: Carnegie Endowment for International Peace (2007), available at http://www.carnegieendowment.org/files/proliferation_status07.pdf.

The three former Soviet republics had about 3,000 strategic nuclear weapons stationed on their soil after the breakup of the Soviet Union. Soon after gaining their independence, however, the three republics agreed to transfer all their nuclear warheads back to Russia in exchange for economic assistance from the United States (McGwire 1994). In December 2003 the Libyan government announced that it would give up its WMD research and development programs. Some experts suggested that Libyan leader Muammar Qaddafi changed course just days before the Iraq War because of the international community's new resolve to fight proliferation.

The spread of chemical and biological weapons. Dozens of countries are suspected to possess chemical and biological weapons (see Table 2.2).

WMD terrorism. The horizontal spread of WMD systems heightens concerns about the possible use of nuclear, chemical, or biological weapons systems in future terrorist attacks. Indeed, the 2001 anthrax attacks in the United States caused panic in many industrialized countries and increased public concerns about other potential threats, such as the use of smallpox or the plague as a terrorist weapon. Terrorist leader Osama bin Laden has called it a "duty" for the Al-Qaida terrorist network to acquire a nuclear bomb to use against the West.

According to the database compiled by the Monterrey Institute's Center for Nonproliferation Studies, some 100 chemical or biological attacks accounted for a total of 103 fatalities and 5,554 injuries in the twen-

Table 2.2 Countries Suspected to Possess Chemical or Biological Weapons Capabilities

Chemical Weapons	Biological Weapons
China	China
Egypt	Egypt
India	India
Iran	Israel
Israel	North Korea
North Korea	Pakistan
Pakistan	Russia
Sudan	Sudan
Syria	Syria
Taiwan	

Sources: E. J. Hogendoorn, "A Chemical Weapons Atlas," *Bulletin of Atomic Scientists* (September–October 1997); "Chemical Weapons in the Middle East," *Arms Control Today* (October 1992); Joseph Cirincione, Jon B. Wolfstahl, and Miriam Rajkumar, *Deadly Arsenals: Tracking Weapons of Mass Destruction* (Washington, DC: Carnegie Endowment for International Peace, 2002).

tieth century. More than two-thirds of all documented incidents occurred outside the United States (Tucker 2000). Many experts believe that the horizontal spread of nuclear weapons, materials, and expertise has increased the likelihood that a group or state will attempt an act of nuclear terrorism in the future. This is of particular concern given the chaos and instability surrounding the nuclear arsenal of the former Soviet Union, and there have been numerous reports of attempts to buy or steal nuclear warheads in that region. Finally, experts are concerned about the possibility that terrorists will combine radioactive materials with conventional explosives to create deadly radiological devices (also known as "dirty bombs"). All of these factors suggest that the threat of WMD terrorism has increased in the modern era.

Global Solutions?

The Nonproliferation Regime

Proliferation is a very complex challenge to international security. World leaders have addressed proliferation threats through a series of regional and global initiatives, and the scope of initiatives to control and limit arms steadily increased from the 1940s to the 1990s. Some experts believe that these efforts have been extremely successful—that there are now even international taboos against the use of chemical, biological, and nuclear weapons (Tannenwald 2005; Paul 2000). However, governments are far from unified in their responses to the threat, and skeptics question the effectiveness of the nonproliferation regime in the twenty-first century.

In some ways, the global nuclear nonproliferation movement began even before the first use of atomic weapons in 1945. Americans involved in the Manhattan Project recognized that such weapons were special and more dangerous than others. President Harry Truman, who had ordered the use of atomic bombs over Hiroshima and Nagasaki, later proposed that all nuclear materials and technology in the world be placed under UN control. While the plan did not receive widespread support, it demonstrated a first step toward global consideration of proliferation problems and set the stage for later progress on the issue. In the 1960s, world leaders agreed to new initiatives, including the Partial Nuclear Test Ban Treaty, which outlawed nuclear tests in the atmosphere, in outer space, and under water. In 1967 the Treaty for the Prohibition of Nuclear Weapons in Latin America created a large nuclear-free zone.

The Nuclear Nonproliferation Treaty. A centerpiece of the global nonproliferation regime, the Nuclear Nonproliferation Treaty was an agreement to halt the spread of nuclear weapons beyond the five declared nuclear powers. Opened for signature in 1968, the treaty had ambitious goals for both vertical and horizontal proliferation. Article 1 of the treaty dictated that no nuclear weapons state (defined by the treaty as a state that detonated a nuclear explosive prior to 1967) would "directly or indirectly" transfer nuclear weapons, explosive devices, or control over these weapons to another party. Article 2 stipulated that a state without nuclear weapons capabilities could not receive, manufacture, obtain assistance for manufacturing, or otherwise try to acquire nuclear weaponry. Another significant element of the treaty was Article 6, which required all nuclear states to pursue general disarmament under strict and effective international control. In many ways, the Nuclear Nonproliferation Treaty represented the crowning achievement of global nonproliferation efforts during the Cold War.

Related nonproliferation initiatives. Several other significant agreements have followed in the spirit of the Nuclear Nonproliferation Treaty. The Biological Weapons Convention of 1972 was the first major effort to gain some control over the world's deadly biological arsenal. More than 140 countries agreed to ban the development and stockpiling of biological agents. In 1993 the Chemical Weapons Convention was opened for signature. This treaty committed all signatories to cease development and stockpiling of chemical weapons. In addition, it included a set of verification procedures somewhat more stringent than those under the Nuclear Nonproliferation Treaty. These procedures supported the rights of a new inspectorate to conduct rigorous investigations and surprise "challenge inspections" of suspected chemical weapons programs in signatory states.

In 1972 the United States and the Soviet Union negotiated their first major arms control treaties. The Strategic Arms Limitation Treaty called for limits on the number of nuclear launch platforms, including missiles and strategic bombers. The Anti–Ballistic Missile Treaty limited the deployment of defensive, ground-based antimissile systems. According to the treaty, each party pledged not to develop, test, or deploy anti–ballistic missile systems or components that were sea-based, space-based, or mobile land-based, beyond allowed limits. Other bilateral agreements led to caps on the number of strategic warheads (the Strategic Arms Reduction Treaties) and even the elimination of an entire class of nuclear weapons (the 1987 Intermediate-Range Nuclear Forces Treaty).

In 1996 the Comprehensive Nuclear Test Ban Treaty, another nonproliferation initiative, was opened for signature. Many UN member states voted to support the treaty and ban all nuclear testing. To become international law, the treaty required the signature and ratification of all forty-four countries known to possess nuclear reactors. However, the governments of both India and Pakistan refused to sign the treaty. India claimed that it wanted a stronger comprehensive treaty that would force nuclear states to comply with Article 6 of the Nuclear Nonproliferation Treaty, and the Pakistani government stated that it would not sign the comprehensive treaty without India's cooperation. Their nuclear tests of May 1998 underscored their continuing resistance to the regime. Other states, including Cuba and Syria, also remain reluctant to sign the treaty. However, the US position was the most surprising of all. The treaty was met by a wave of conservative resistance, charging that the agreement was dangerous because it was not verifiable and would unduly restrict the United States while allowing rogue states to proceed with their nuclear programs. In a move that stunned the world, the Senate rejected the treaty in October 1999.

Controlling Weapons at the Point of Supply

Given serious concerns about the implications of the spread of WMD technology around the world, governments have also tried to limit the supply of critical materials needed to build such weapons. Critics charge that the Nuclear Nonproliferation Treaty was flawed because it did not prevent states from exporting "peaceful" nuclear energy reactors and other types of materials that could potentially be converted for use in the development of WMD programs. To address such concerns, major supplier states agreed in 1976 to establish a "trigger list" of items that could be sold to other countries only under stringent safeguards. In the 1980s, supplier states established the Missile Technology Control Regime, which prohibited the transfer of technology essential for the development of ballistic missile systems.

Like the Nuclear Nonproliferation Treaty, however, supply control efforts have had only marginal success. They helped to limit missile development projects under way in South America and the Middle East, but some twenty countries still have acquired missiles. Nor did these supply controls prevent Iraq, which adapted civilian scientific technology for military use, from progressing toward the development of nuclear weapons. And they did not prevent Iraq from manufacturing and modifying the Scud-B missiles that were used against Israel and Saudi Arabia during the 1991 Gulf War—and that were capable of carrying chemical warheads. Meanwhile, Pakistan developed its own ballistic missile, the Hatf, and acquired about thirty nuclear-capable, medium-range M-11 missiles from China (McNaugher 1990). The North Korean government continues to develop a long-range Taepo Dong missile that may have the potential to reach the continental United States.

On the conventional weapons front, there is growing recognition that arms transfers—even "small arms" like guns and antitank weapons—represent a fundamental threat to international security. Security experts and government leaders have been discussing ways to increase the transparency of the conventional arms trade by making more information on arms transfers available. In 2001 the United Nations estimated that there were more than 500 million small arms in the world, 40–60 percent of which had been acquired illegally through black markets. The concern is that the small arms trade fuels civil wars in developing countries, empowers organized crime, and enables terrible human rights violations (Crossette 2001).

Contemporary Challenges

In the past decade the international community has confronted a number of proliferation challenges, and more have appeared on the horizon. This section outlines three concerns for the twenty-first century: the post–September 11 security environment as illustrated by the Iraq War; responses to drives by Iran and North Korea to develop nuclear weapons; and initiatives that could lead to the future dismantling of the international nonproliferation regime.

Proliferation, Preemption, and Conflict: The Iraq War

The ongoing Iraq War illustrates the link between proliferation and international security. In the aftermath of the September 11, 2001, terrorist attacks and the rapid allied victory against the Taliban in Afghanistan, the George W. Bush administration began to make the case for an invasion of Iraq. One of the primary reasons to attack Iraq, President Bush argued, was to eliminate its clandestine WMD development program. For example, he claimed in 2002:

> The gravest danger to freedom lies at the perilous crossroads of radicalism and technology. When the spread of chemical and biological and nuclear weapons, along with ballistic missile technology—when that occurs, even weak states and small groups could attain a catastrophic power to strike great nations. Our enemies have declared this very intention, and have been caught seeking these terrible weapons. They want the capability to blackmail us, or to harm us, or to harm our friends—and we will oppose them with all our power. (White House 2002b)

In February 2003, Secretary of State Colin Powell gave another high-profile address, to the UN Security Council, outlining Iraqi suspected WMD activity and the purported links between Saddam and Al-Qaida terrorists (Dobbs 2003). In spite of the fact that the United Nations chose not to endorse an invasion of Iraq, President Bush ordered that the war commence. The Iraq War began on March 19, 2003, with US and coalition forces advancing rapidly into the country. In early April, US and coalition forces rushed into Baghdad and forced a general surrender of Iraqi forces.

Iraqis and Americans celebrated the successful "conclusion" of the war. On May 1, 2003, President Bush stood under a "mission accomplished" banner aboard the aircraft carrier USS *Abraham Lincoln* and declared the "end of major combat operations" in Iraq.

But the Iraq War was not yet over. Beginning in the summer of 2003, a major insurgency operation began to claim the lives of coalition soldiers and foreign workers in Iraq. Some of the worst strikes came when suicide bombers and foreign terrorist groups attacked soft targets such as the UN's headquarters in Baghdad, police stations, infrastructure such as electric power stations, and even civilian job fairs. Back in Washington, the Bush administration was forced to reconsider its declaration of victory in Iraq. By 2007, more than 570,000 soldiers had served in Iraq, with a total of more than 1 million tours of duty. The US government was spending $6 billion per month to fund the war (Bilmes and Stiglitz 2006). At the same time, US public support for the war was eroding substantially. The number of Americans who believed that the United States had "done the right thing" in Iraq dropped precipitously from 2003 to 2007. Four years after the start of the Iraq War, the president's approval ratings were at 34 percent—down from a high of 89 percent just days after the terrorist attacks of September 11, 2001 (Stevenson and Elder 2004).

There was another surprising twist in the Iraq story. According to the Bush administration, one of the major reasons for the war was that Saddam Hussein was secretly developing WMD. Soldiers and inspectors who entered the country found no evidence of an active weapons program. Critics began to claim that the war had been waged for illegitimate reasons. Some suggested that conservative Bush advisers had "cooked the books" in favor of the war effort. In particular, they charged that Vice President Dick Cheney and the Defense Department had manipulated the intelligence it was receiving in order to shape American public opinion and policies toward Iraq. Concerned citizens in the United States and around the world began to question the effectiveness of a strategy of preemption for dealing with proliferation threats.

Stopping the Locomotive? Iran and North Korea

Global confrontation with Iran and North Korea over their nuclear programs represents another dimension of the contemporary proliferation challenge. The government of Iran, led by President Mahmoud Ahmadinejad and Grand Ayatollah Ali Khameini, appeared on a collision course with the international community in 2007 over its nuclear ambitions. Following the Iranian Revolution of 1979 and the establishment of a new Islamic fundamentalist regime, Iran acquired nuclear bomb designs and technology through the network established by A. Q. Khan. Russia helped build a nuclear research reactor and a centrifuge plant for uranium enrichment in

Iran, for example (Rohde and Sanger 2004). By 2006 the Iranian government claimed to have made significant strides in its nuclear program.

The international community responded with trade embargoes and sanctions. Iranian nuclear scientists removed the International Atomic Energy Agency (IAEA) at several research laboratories, with the intention of continuing their program unmonitored. In February 2006 the IAEA concluded that Iran was in serious violation of the Nuclear Nonproliferation Treaty. Throughout 2006 and 2007, Iran continued its program while at the same time conducting intermittent negotiations with leaders from the European Union, Russia, the United States, and the IAEA. In 2007, President Ahmadinejad boasted of progress in development of the nuclear fuel cycle in Iran and defied international will to halt the program and open it to international inspectors.

North Korea's drive to build a nuclear device based on an advanced uranium enrichment process drew the attention of the United States and other Western powers in the early 1990s. North Korea is one of the world's few remaining communist dictatorships, currently ruled by President Kim Jong-il. Its nuclear reactors and plutonium reprocessing plant were constructed at Yongbyon, in the country's north. The reactor produces enough plutonium to support the construction of one to two nuclear bombs a year, and this is exactly what the international community suspects that North Korea was doing in the early 1990s. In 1994, tensions with the West came to a boil, and the Bill Clinton administration considered the use of military force to halt North Korea's research progress. Conflict was avoided, however, with the negotiation of an agreed framework whereby North Korea would suspend its nuclear program in exchange for a package of benefits including light-water nuclear reactors, oil shipments, and food aid.

By 2002 it had become clear that North Korea was not abiding by the terms of the agreement. It was revealed that North Korea also acquired designs and technology for a nuclear weapons program through the A. Q. Khan network. In 2003, North Korea formally withdrew from the Nuclear Nonproliferation Treaty. The standoff with North Korea intensified in February 2005 when the government acknowledged that it possessed nuclear weapons and would no longer negotiate with regional actors or the United States regarding the future of its program. Then, on October 9, 2006, North Korea successfully tested its first nuclear bomb. A new round of negotiations in 2007 did yield progress, however, and the North Korean government pledged to stop diverting materials for nuclear bombs in exchange for a package of aid from the West.

These cases represent examples of the challenges of proliferation in the twenty-first century. Optimists believe that international diplomacy may yield solutions such as limited development by Iran and North Korea of peaceful nuclear energy programs in exchange for pledges not to develop weapons. Pessimists warn, however, that these cases are just the tip of the iceberg of the contemporary threat, and that any concessions to rogue states will surely multiply the number of developing countries seeking weapons in the future.

The End of the Nuclear Nonproliferation Regime As We Know It?

US leaders have expressed concern about weapons proliferation for decades, but presidents have taken very different approaches to the problem. President Harry Truman supported a plan to transfer all nuclear technology to UN control. President Dwight Eisenhower's "Atoms for Peace" proposal called for the creation of an international atomic energy agency to pursue peaceful uses of the technology. President John Kennedy negotiated the Partial Nuclear Test Ban Treaty with the Soviet Union soon after the Cuban missile crisis. President Richard Nixon led US negotiations with the Soviet Union that established the first major arms control treaties between the superpowers, the Strategic Arms Limitation Treaty and the Anti–Ballistic Missile Treaty. In 1977, President Jimmy Carter proclaimed that one of his goals would be the complete elimination of nuclear weapons.

In the post–September 11 era, however, the George W. Bush administration has taken an entirely new approach to the proliferation challenge. President Bush and his advisers appear to have little faith in arms control agreements, and are focused instead on ways to enhance US nuclear power. For example, early in his first term, the president announced a plan for a national missile defense system. The Bush administration believed that the United States faced new threats of missile attack—not from Russia or the former Soviet republics—but more likely from the actions of a terrorist group, a "rogue state" like North Korea, or an accidental launch. To build the case for the new system, the president warned: "Deterrence—the promise of massive retaliation against nations—means nothing against shadowy terrorist networks with no nation or citizens to defend. . . . Homeland defense and missile defense are part of stronger security, and they're essential priorities for America" (White House 2002a). Bush announced in December 2001 that the United States would unilaterally withdraw from the Anti–Ballistic Missile Treaty and begin a four-year, $8.3 billion national missile defense research initiative.

National missile defense represents only one dimension of the apparent conversion of US nuclear strategy. Top Bush administration officials have argued that a deterrence policy resting on "mutual assured destruction" is a relic of the Cold War that should be jettisoned as a guiding principle for preserving US security. In the wake of a government study of WMD security (the 2002 *Nuclear Posture Review*), the Bush administration announced that the United States should enhance its ability to launch preemptive strikes (nuclear and nonnuclear) against potentially hostile states. The administration also funded research and development of "mini-nukes" (the Advanced Weapons Concept Initiative) and "bunker-busters" (the Robust Nuclear Earth Penetrator), and a revitalization program for nuclear warheads (Broad, Sanger, and Shanker 2007).

At the same time, however, the Bush administration has also begun to identify countries that it believes will be "responsible partners" in a new global nuclear energy network. On the surface, this means that the United States will encourage countries to develop peaceful nuclear energy programs. Yet experts fear that with some modifications these programs may also fuel clandestine nuclear weapons construction. The 2006 US-India Civil Nuclear Cooperation Agreement is an example of this new approach. The treaty acknowledges India as a declared nuclear power while remaining a nonsignatory of the Nuclear Nonproliferation Treaty. The United States and key allies will sell nuclear fuel and reactor components to India, and the country may continue to build a nuclear arsenal at the same time.

Taken together, these initiatives may indeed represent a reversal of US counterproliferation policies. The United States had steadfastly defended the Nuclear Nonproliferation Treaty for decades as the cornerstone of an international arrangement to curb the spread of weapons of mass destruction. It had refused to sell nuclear materials to nonsignatories of the treaty. Today, the Bush administration is encouraging other governments to join with it to establish a global network of uranium mining, enrichment, export, and waste disposal. A side effect of this network might be the spread of technology that could foster clandestine nuclear weapons programs. At the same time, the United States continues to threaten some would-be proliferants with military strikes. In sum, this paradoxical strategy may have serious implications for regional and international security. Short of a steadfast commitment to the international nonproliferation norm, it may not survive in the twenty-first century.

Conclusion: Prospects for the Future?

The proliferation of weapons is truly a major challenge to global security. One of the original catalysts of global proliferation was the Cold War arms race between the superpowers. But with that conflict a distant memory, many scholars and politicians are taking a new look at incentives for proliferation in the twenty-first century. Optimists say that we may be headed toward a nuclear-free world. They argue that a global build-down in tensions—a reverse proliferation—has occurred since the late 1980s. They cite the Strategic Arms Reduction Treaties between the United States and the Soviet Union (particularly the second treaty's requirement that each country reduce its nuclear arsenal to 3,500 warheads), as well as recent agreements between George W. Bush and Vladimir Putin, as evidence of progress toward a minimal nuclear deterrent relationship.

Pessimists warn, however, that many arms control initiatives are doomed to fail in a proliferating world. New and complex debates have emerged about whether defense and deterrence represent a more effective, pragmatic response to the spread of weapons technology. Some suggest that preparedness for a rogue-state nuclear launch at the United States, or a concentrated toxic terrorist attack, would be the best use of government resources. The September 11 attacks and subsequent international developments appear to have bolstered this pessimistic vision, and the current "war on terror" seems to be one definitive answer to the question of how to respond to proliferation threats. Furthermore, many American conservatives view international nonproliferation agreements with a jaded eye. US government opposition to major arms control treaties, combined with a doctrine of preemption and support for national missile defense, all suggest a new attitude toward proliferation.

Discussion Questions

1. Which of the four types of proliferation do you think represents the most serious threat to international security?
2. Is the proliferation of conventional weapons a challenge that the global community can ever fully meet? Why or why not?
3. Is it possible that weapons proliferation could actually make the international system more stable in the twenty-first century? How might this occur?
4. What are some of the efforts that individual countries and international organizations have undertaken to respond to the proliferation challenge? Which have been most effective, and why?
5. What are some implications of the trade-off between defense spending and social welfare spending?
6. What can governments do to confront the threat of WMD terrorism? How have recent international developments changed the definition of horizontal WMD proliferation?
7. Does national missile defense increase or decrease security in the United States? Do you think deterrence is still an effective strategy for WMD security?
8. Should government leaders use military force to stop would-be proliferants from threatening their security? Why or why not?

Suggested Readings

Allison, Graham (2004) *Nuclear Terrorism: The Ultimate Preventable Catastrophe.* New York: Holt.

Cirincione, Joseph, and Jon B. Wolfstahl (2003) "North Korea and Iran: Test Cases for an Improved Nonproliferation Regime?" *Arms Control Today* (December).

Karp, Aaron (1994) "The Arms Trade Revolution: The Major Impact of Small Arms." *Washington Quarterly* 17 (Autumn).

Lieber, Keir A., and Daryl G. Press (2006) "The End of MAD? The Nuclear Dimension of US Primacy." *International Security* 30 (Spring).

Perkovich, George (2006) "The End of the Nonproliferation Regime?" *Current History* (November).

Pollack, Kenneth M. (2004) "Spies, Lies, and Weapons: What Went Wrong." *Atlantic Monthly* (February).

Sagan, Scott D., and Kenneth N. Waltz (1995) *The Spread of Nuclear Weapons: A Debate.* New York: Norton.

Stockholm International Peace Research Institute (2006) *SIPRI Yearbook 2006: Armaments, Disarmament, and International Security.* London: Oxford University Press.

Tucker, Jonathan B. (2000) *Toxic Terror: Assessing Terrorist Use of Chemical and Biological Weapons.* Cambridge: Massachusetts Institute of Technology Press.

White House (2002) "National Strategy to Combat Weapons of Mass Destruction" (December). Available at http://www.whitehouse.gov/news/releases/2002/12/wmdstrategy.pdf.

3

Nationalism

Lina M. Kassem, Anthony N. Talbott, and Michael T. Snarr

IN LATE FEBRUARY 1986, AS MANY AS 1 MILLION FILIPINOS LEFT their homes in the middle of the night to form a human barricade around the Camp Crame military police base in Manila. The men, women, and children placed themselves in between the tanks of a vengeful dictator and a handful of coup leaders who had attempted to overthrow him. Catholic nuns, schoolchildren, dock workers, attorneys, farmers, business owners, and communist revolutionaries all joined together to stand in defiance of President Ferdinand Marcos. Why? What caused these wildly different people to unite? Was it some shared sense of purpose? Did a common destiny drive them? The answer is yes. The people of the Philippine People Power Revolution shared a common identity: Filipino. They saw themselves as belonging to a community, a community they imagined as encompassing all citizens of their country. They came together to rescue their country. They experienced a strong sense of *nationalism.* But what does this term mean?

Nationalism is a shared sense of identity based on important social distinctions that has the purpose of gaining or keeping control of the group's own destiny. Nationalism arises from many different sources. Shared ethnicity, language, religion, culture, history, and geographical proximity all generate feelings of comradeship and belonging to a certain group. As a result, human beings organize themselves into groups or communities. We are social beings. These communities determine how we interact with others and with whom we interact. They affect our perception of ourselves and of others. We consider other people either to be part of our group or to be outside of our group. Although we may have several identities (daughter or son, mother or father, spouse, club member, student, etc.), our nationality is one of the most important.

People unite into groups in pursuit of certain goals. Often this sense of shared identity becomes political. When the goal is self-determination for the group, the shared identification has become nationalism. In other words, when a group of sports fans identify with one another, but have no political aspirations, this does not constitute nationalism. But when a group of people seek to have political control over a given territory, then it becomes nationalism. Thus national self-determination is the main purpose of nationalism.

The shared identity of nationalism is often called an *imagined community,* because most citizens of a country, despite their strong feelings of fellowship, will never actually meet—let alone get to know—one another. But the feeling of unity remains. To understand nationalism, we must look at the origins of the nation and the state, how they have evolved, and the different shapes that nationalism takes in the world today.

The Evolution of the State

A state (also referred to as a country) is a political unit that has sovereignty over a geographical area. Sovereignty refers to the fact that the state is self-governing; that is, there is no external group or person that has authority over it. It is hard to imagine an alternative, but prior to the seventeenth century, states as we know them did not exist. Prior to the modern state, most people lived under political units referred to as empires. Empires usually included large swaths of land, encompassing many groups or nations. In Europe, between the fall of the Roman Empire and the beginning of the modern system of states, medieval feudalism was largely in place. Under feudalism, power was not as centralized as it was under the Roman Empire. Individual peasants had to answer to local nobles or kings, who in turn were loosely ruled by an empire. Lines of control in medieval Europe were fuzzy at best. As Europe drew closer to the seventeenth century, the Catholic Church's power was increasingly called into question. The Protestant Reformation and secular authorities combined to challenge the Pope's authority. One tragic outcome was the Thirty Years' War (1618–1648), which pitted Protestants against Catholics and destroyed much of Europe.

The Treaty of Westphalia in 1648 ended this devastating war and recognized many independent, secular political units that would become modern states. One key distinguishing characteristic of these states was sovereignty. Unlike the earlier empires, which did not respect the right of other empires to govern their own territory, the new states recognized each other's sovereignty. In short, states were to govern themselves without outside interference. There was no distant secular emperor, king, or religious leader to control them. As a result, states slowly began to develop international law as a way to coordinate or govern relations between them.

The beginning of the modern state system did not happen overnight. It had begun to develop in Europe at least a century before the Treaty of Westphalia, and slowly spread throughout the world in a process that is arguably still continuing today. In fact, most of the countries or states of today are relatively new.

The Evolution of Nationalism

The creation of the modern state system paved the way for nationalism as we know it today. Prior to the spread of nationalism, most people were primarily concerned with local and personal affairs. People knew they were the subjects of powerful kings, queens, and emperors, but there were no serious attempts to foster a sense of common identity. Local rulers and feudal lords governed everyday matters. A vast gulf existed between rulers and ruled. Local people did not participate in government above the most basic levels. People knew only their family and others in the village. There was little sense of belonging to a larger, countrywide community. Slowly this began to change.

Although the roots of nationalism began before the end of the eighteenth century, most scholars point to the French Revolution as the defining moment for nationalism. Influenced by the Enlightenment, writers like John Locke and Jean-Jacques Rousseau argued that the people should govern themselves rather than being governed by kings or queens. Thus the concept of the divine right of kings was gradually replaced with the notion of the will of the people (known as popular sovereignty). The American (1776) and the French (1789) revolutions greatly strengthened the idea that the people had the right to participate in their own governments. Napoleon and the leaders of the French Revolution used this new sense of citizenship to create a psychological bond among the people spread throughout France. For the first time, these varied people began to consider themselves as *Frenchmen.* This new identity led to many changes. People addressed one another as "citizen" rather than "sir," and a new flag was designed. France no longer had to pay mercenaries, but could now motivate an army of Frenchmen to fight for *their* country. United and empowered by nationalism, they did not merely fight for a ruler, they fought for France. Soon after the French Revolution, nationalism began to spread throughout Europe and gradually to the rest of the world.

Since the French Revolution, nationalism has been both a positive and a negative force in the world. It has united and divided peoples. It has brought peace and it has led to war. In one of its more negative forms, nationalism created a strong sense of superiority within Nazi Germany and led to the death of millions of innocent victims. Feelings of superiority and national pride within Europe also fueled the colonization of Africa and other parts of what we now call the South, or the developing world. In the case of Africa, colonization consisted of countries like Portugal, Great Britain, Germany, France, and others forcibly taking over and controlling the continent during the late nineteenth and early twentieth centuries. In what would become known as the "scramble for Africa," the European powers carved up the continent into colonies. These new political units ignored the existing tribal structure. In other words, very diverse groups, even enemies of one another, were forced together into colonies. The European colonial powers were able to take advantage of diverse groups living in Africa in a divide-and-conquer strategy that led to the subjugation of Africa. (Groups that identified with one another were sometimes split into separate units, as happened with the Somalis.)

Conversely, nationalism can often be a very positive force. For certain, nationalism offers many people a sense of belonging and meaning. In addition, it has rallied oppressed people to demand freedom. For example, in places like Africa, nationalism led to anticolonization. The colonized people of Africa and Asia (including the Middle East) eventually overcame their differences long enough to band together and overthrow their colonizers or persuade them to leave. The oppressive tactics of the colonizers ironically helped build a sense of shared identity among the diverse people within each colony that enabled them to build successful anticolonial movements.

Anticolonial nationalism had a dramatic effect on the world. As colonies rejected their colonizers, countries became independent, sovereign states. As a result, the decolonization of the twentieth century led to a dramatic increase in the number of states. To illustrate, consider the following statistics: In 1789 there were 23 countries. By 1900 there were 57 countries. Currently there are approximately 192 countries. These new countries were primarily formed in Africa and Asia, but other countries disintegrated, such as the Soviet Union and Yugoslavia. The new states led to smaller units that more closely resemble nation-states. For example, Croats and Slovenes, who used to live in a more multinational state (Yugoslavia), created more homogeneous countries named Croatia and Slovenia, respectively.

It is also interesting to note that during the short history of nationalism, many have predicted and called for its end or at least a reduction in its influence. After the death of millions in the two world wars, many argued that nationalism was too destructive. To a great extent, the League of Nations and the United Nations were designed to restrain the destructive tendencies of nationalism through international law. During the 1970s, as economic interdependence among countries increased, some argued that national allegiances would be reduced as the world's inhabitants interacted across country borders. At the core of this belief was the idea that, as countries became economically dependent on one another and cultural boundaries were frequently crossed, the strength of people's allegiances to their country would erode. Needless to say, these predictions were premature—nationalism seems to be alive and well today. To confirm nationalism's current importance, one need only look at Iraq and the strong internal divisions between its Kurdish, Sunni Muslim, and Shia Muslim peoples. As the United States attempts to build a single nation out of this divided country, it is gaining appreciation for the strength of nationalism.

In sum, nationalism and the modern state system, although different phenomena, went hand in hand. As states began to form with relatively defined borders, people living within these borders began to identify with one another. This process was aided by the concept of popular sovereignty, since people now had a stake in their government. The natural result was what is called the nation-state.

The Nation-State

Although the term *nation-state* is generally used loosely to mean "country," it is technically defined as a single nation within the boundaries of a single state. Not all modern countries are actually nation-states, but this is the ideal. For example, there exist nations that do not have a common state. One example would be Koreans, who are divided into peoples of North and South Korea. Another example is the Kurds, a group of people who identify with one another but are spread throughout northern Iraq, southern Turkey, and western Iran.

Similarly, there are dozens of examples of multinational states—single states within which multiple nations live. In many African countries, for instance, there are multiple tribes who historically fought one another but who now, since becoming united into sovereign countries in the last half of the twentieth century, live together within the same borders. Canada is another example. Within predominately English-speaking Canada sits the French-speaking province of Quebec. Nationalism in Quebec is so strong that nearly half of its citizens have voted to break away from Canada and create their own country. Also within Canada are several indigenous (native) groups who have resisted considering themselves Canadians. This is not uncommon. Throughout the world, thousands of indigenous groups are part of countries with which they do not identify. As a result, many of the contemporary countries cannot technically be considered nation-states.

Different Perspectives on Nationalism

Nationalism is a very complex subject. Scholars disagree on when it first emerged and whether it is generally good or bad. There are also different types of nationalism.

Civic vs. Ethnic Nationalism

Many scholars divide nationalism itself into two categories: civic and ethnic. This is due to disagreement over the roles of ethnic and political components of nationalism. *Civic nationalism* is associated with the Western experience, and is based on citizenship rather than on ethnic linkages. The nation-state is seen as the core of civic nationalism. Its main role is to promote the principle that a society is united by territoriality, citizenship, and civic rights and legal codes transmitted to all members of the group. All members of this society, regardless of their ethnicity or race, are ideally equal citizens and equal before the law.

By contrast, e*thnic nationalism* is based on ethnicity. Ethnic nationalism draws its ideological bonds from the people and their native history. It relies on elements that are considered purely unique to a group, such as collective memory, common language, values, religion, myth, and symbolism. It is dependent on blood ties, bonds to the land, and native traditions.

To understand nationalism, it is important to consider degrees of inclusiveness and exclusiveness. All nationalism is, by definition, exclusive—it excludes all people who are not members of that nation. In other words, how broadly defined is the nation? How many different subgroups make up the main group of people who compose the nation? History, culture, social-class structure, and form of government are all important. All of these influence how community is imagined and how nationalism is constructed.

Many scholars have treated these two types of nationalism as being diametrically opposed. Civic nationalism, which is more inclusive, is typically seen as the "good" form of nationalism. For instance, in most democracies of the world, an individual is a citizen not because of bloodlines, but because he or she believes in the ideals and symbols of that country and pledges allegiance to the country. Ethnic nationalism is viewed as having more negative characteristics, such as being more exclusive due to its emphasis on ethnic links between people. In other words, if you don't share the common history, language, and other ethnic ties, you are not part of the nation. For example, it would be much easier for a Kurd or Serb to become a citizen in a country that practices civic nationalism (such as the United States, England, or Canada), than to become a member of another national group.

Although many scholars present these two categories of nationalism as being opposites, this is not always the case in practice. For example, the

Philippine People Power Revolution, mentioned in the opening paragraph of the chapter, displayed elements of both civic and ethnic nationalism. Myth, race, religion, and citizenship all combined to unite and empower the people.

Other scholars reject these two categories altogether. They see nationalism mainly in terms of whether it supports or attempts to overthrow an existing government. At least since World War II, every successful war of independence and revolution has been driven by nationalism. During the same period, every government in power has used nationalism to gain the support of its people. Nationalism is a powerful tool of politics that can liberate, oppress, or empower people.

Nationalism, Religion, and Violence

Nationalism and violence often go hand in hand. "United we stand, divided we fall" is the ultimate call to arms of the nationalist. The "us" versus "them" mentality operates in all types and sizes of community—from rival villages arguing over grazing rights, to international coalitions involved in geopolitical disputes. Wars are fought for nationalist reasons. Nationalists overthrow governments. Terrorists attack in the name of nationalism. The potential for violence often increases when the causes of nationalism and religion overlap. This is because nationality and religion are two very powerful forms of identification in the world today. When people think about who they are, many, if not most, think of themselves as Americans or Turks or Thais, and as Christians or Muslims or Buddhists.

Although violence often accompanies nationalism, this is not always the case. Mahatma Gandhi's "Quit India" movement against the British was a remarkably nonviolent effort that relied on anticolonial nationalism. The Philippine People Power Revolution also relied on peaceful means to achieve its goals. Many other, lesser-known examples of nonviolent nationalism also exist.

It should also be mentioned that while religion often intensifies nationalist feelings, it can also be a powerful motivator that transcends or even opposes nationalism. Take, for example, Osama bin Laden. If the traditional notion of nationalism were applied to him, one would expect bin Laden to be a patriotic Saudi Arabian nationalist. Yet this couldn't be further from the truth. Bin Laden's allegiance is not to his country of origin, but rather to Islam. He feels a far greater allegiance to Pakistani, Algerian, Jordanian, and even American Muslims (those who share his interpretation of Islam) than he does to Saudi Arabians who do not share his religious beliefs.

Christianity in the United States is also an interesting case study. American Christians such as Jerry Falwell and others see the United States as a country favored by God and founded on Christian beliefs. To them,

Jesus, US patriotism, and involvement in Iraq go hand in hand. Meanwhile, other Christians are horrified at the meshing of Jesus and US militarism. They focus on the New Testament's commands to "love one's enemy" and "turn the other cheek." In fact, an increasingly vocal group of American Christians argue that followers of Jesus in the United States should reconsider their national allegiances, as they may have more in common with fellow Christians outside the United States (including China and Iraq) than with non-Christians within the United States. Although these radical religious views do not immediately threaten nationalism as a whole, they have raised some interesting issues for people of faith.

In sum, nationalism is a tremendously important political force of our time. From its origins in Europe, it has spread to every corner of the world. Nationalism is also a complicated concept that encompasses a wide range of expressions. It can be inclusive or exclusive, violent or nonviolent. It all depends on the environment in which it develops, on the will of the leaders shaping it, and on how all the people involved imagine it.

Case Study: The Israeli-Palestinian Conflict

The Israeli-Palestinian conflict aptly demonstrates how nationalism works in the contemporary world. Palestinians are united by a strong anticolonial nationalism, while Israelis are motivated by a strong sense of religious nationalism. Self-determination is the guiding principle for both groups. How each of these two communities imagines itself affects how inclusive or exclusive it is. Ethnic nationalism is present for both Palestinians and Israelis. Israel is essentially a Jewish state. In fact, it defines itself as such. Palestine is primarily composed of Arabs. With respect to religion, most Israelis are Jews, although a significant percentage are not practicing Jews. Approximately 20 percent of Israel's citizens are Muslim, Christian, and Druze Arabs (most of whom consider themselves Palestinians). Although they live in Israel, they do not enjoy all the rights of citizenship. The conflict, which has now lasted nearly a century, has been expressed through both violent and nonviolent means. The terrorist tactics used by extremists on both sides are tragic examples of the violent potential of nationalism.

Historical Background

By the beginning of the twentieth century, European colonialists effectively controlled over 85 percent of the world's natural resources as well as its people. Arab nationalism, which is at the center of the Israeli-Palestinian conflict, is a response to this European and foreign intervention. It was an attempt to assert Arab self-determination and independence from colonialism. In Palestine, Arab nationalism developed in response to another form of nationalism, Zionism. Zionists desired to create a homeland for Jews. Mostly born out of the European Jewish experience, Zionism was a response to the violent persecution that Jews suffered at the hands of Europeans. Russian persecution of Jews in the late nineteenth century resulted in the first wave of Jewish emigration to Palestine.

In the late nineteenth century, Theodor Herzl, widely regarded as the father of Zionism, came to the conclusion that the only solution for the plight of the Jews would be the creation of a Jewish homeland. Herzl, a secular assimilated European Jew, considered several locations for a Jewish homeland, including possible locations in East Africa and South America. Other Zionists argued that the more religious Jews would join the Zionist project if the proposed homeland was in the biblical land of Palestine. The idea of reestablishing a Jewish state would also gain the support of Christian Zionists, who advocated restoration of the Jewish homeland based on the belief that the Bible promises it to the Jewish people. On the other hand, the Palestinians have inhabited the land for generations and see themselves as the descendants of the Canaanites, the original inhabitants of Palestine. They saw the new Jewish immigrants as a threat, since the latter did not intend to assimilate into the existing communities but rather to establish a competing claim to the land.

At the outbreak of World War I, the Arabs were under the control of the Ottoman Empire, which was allied with Germany. Britain, hoping to weaken the Ottomans from within, turned to the Arabs. In return for helping the British in the war effort, mostly by revolting against the Ottoman Turks, the Arabs of the region, including Palestine, were promised independence. This promise was articulated through a series of letters in 1915 between Hussein bin Ali, Sharif of Mecca, and Sir Henry McMahon, the British high commissioner in Egypt. The British were also very aware of the strategic importance of the region. Oil had already been discovered in areas of Iraq and Iran, and the Middle East represented an important strategic point on the trade route to India. Motivated by their strategic interest in the area, the British signed a secret agreement with the French that basically turned the entire Middle East into spheres of influence for the two colonial powers. This agreement, negotiated in 1916, became known as the Sykes-Picot Agreement. This agreement contradicted Britain's previous promises to the Arabs, under the Hussein-McMahon letters, and would later also contradict another of Britain's pledges, this time to the Jews in the form of the Balfour Declaration.

In Palestine at the beginning of the twentieth century, Arabs outnumbered Jews—both native and recent arrivals—about ten to one. Despite this, the British decided, for economic and political reasons, to support a Jewish homeland in Palestine. In 1917 the British issued the Balfour

Declaration, which stated: "His Majesty's Government views with favor the establishment in Palestine of a national home for the Jewish people." Many books have been written in an attempt to decipher the true intentions of this declaration; however, it is clear that the British decided that it would serve their interests to support a Jewish homeland in the midst of an Arab world.

The Treaty of Versailles, which ended World War I, gave mandates to France and Britain to divide up the region into client states, under their domination. Palestine, during this mandate period (1919–1947), fell under the direct control of the British. This treaty opened the way for the Jewish National Fund (the Jewish land-purchasing agency of the Zionist movement) to start buying large amounts of land in Palestine. Jewish settlers began building homes on this land. These large land acquisitions by the Jewish National Fund, coupled with the Balfour Declaration and Zionist aspirations, became increasingly threatening to Palestinian landowners and farmers. Palestinians feared that they might lose their land rights and become minorities in their homeland. As a result, fighting between Palestinians and Jewish settlers erupted.

During World War II, the Allied nations, including the United States, refused to open their borders to Jewish refugees fleeing the Holocaust. At the end of the war, when the horror and magnitude of the Holocaust were exposed, the Allies felt a great deal of sympathy for the Jewish people, along with a great deal of guilt. In late 1947, the United Nations decided to "partition" Palestine into two states, one Jewish and one Arab. The UN's partition plan gave 53 percent of the land to the Jews, who accounted for 30 percent of the population, and gave the remaining 47 percent of the land to the Arabs, who accounted for 70 percent of the population and had, until then, owned 92 percent of the land. Immediately after the partition announcement, fighting broke out. Figure 3.1 briefly summarizes this war and three subsequent Arab-Israeli wars.

In the 1947–1948 war and the 1967 war, Israel acquired land beyond what the UN had given it (see Figure 3.2). These land acquisitions are especially important for understanding current land disputes. In 1964, the Palestine Liberation Organization (PLO) was formed by exiled Palestinian nationalists who became disillusioned with the inability of other Arab leaders to liberate Palestine. Yasir Arafat emerged as its national leader, a role he assumed until his death in November 2004. The PLO helped unite the Palestinians both in exile and in the Occupied Territories, and gave voice to their hopes for self-determination and national independence. The PLO has used a combination of diplomatic initiatives as well as armed struggle to gain international recognition for Palestinian national rights.

Figure 3.1 Arab-Israeli Wars

War of 1947–1948. As soon as the United Nations announced the partition plan in November 1947, fighting broke out. Subsequently, Jewish underground organizations (including several terrorists groups), being much more organized than any Palestinian resistance, achieved several strategic victories. Most Arab armies around Palestine were reluctant to intervene. On May 15, 1948, the state of Israel was declared. Over the next few days, armies from several Arab countries invaded Israel. At the end of the war, Israel had acquired close to 80 percent of the Palestine mandate. Jordan annexed what remained of Palestine, the West Bank, and Egypt took control of the Gaza Strip. Of the more than 1 million Palestinians, as many as 800,000 were forced to leave their homes. More than 500 Palestinian villages were either destroyed or depopulated. These Palestinians would become refugees, mostly in neighboring Arab states.

War of 1956. In 1956, Egyptian president Abdel Nasser nationalized the Suez Canal. Israel allied itself with Britain and France and invaded Egypt. The United States asked for and received a cease-fire, with UN peacekeepers maintaining a buffer zone between Israel and Egypt.

War of 1967 (Six Day War). The United Arab Republic (union of Egypt and Syria) asked for the withdrawal of the UN forces from the cease-fire lines. Israel, believing an attack from Egypt was imminent, launched a "preemptive" attack. Israel mobilized and attacked Jordan, capturing the West Bank and East Jerusalem (which had been annexed by Jordan in 1948). Israel also captured the Golan Heights (which was part of Syria), the Sinai Peninsula (which was part of Egypt), and the Gaza Strip (which had been annexed by Egypt in 1948). Following the war, the UN Security Council passed Resolution 242, which requires Israel to withdraw from the West Bank, Gaza, and all other areas it occupied as a result of the 1967 war.

War of 1973 (Yom Kippur War). Egypt and Syria attacked Israel in an attempt to reclaim Syrian and Egyptian territories occupied by Israel. Although the Egyptians were able to make strong advances early in the war, the Israelis rallied and pushed them back. Intervention by the United States and the Soviet Union led to a cease-fire. This war signaled the last major effort by Arab states to liberate Palestinian territory. In 1979, US president Jimmy Carter brokered a peace agreement between Egyptian president Anwar Sadat and Israeli prime minister Menachem Beg7in, which returned the Sinai Peninsula to Egypt. As a result, Egypt agreed to recognize the state of Israel as well as establish full diplomatic relations between the two states. The peace process was based on the acceptance by both parties of UN Security Council Resolution 242, which recognized the legitimacy of the state of Israel in its pre-1967 borders.

Figure 3.2 The Expansion of Israel

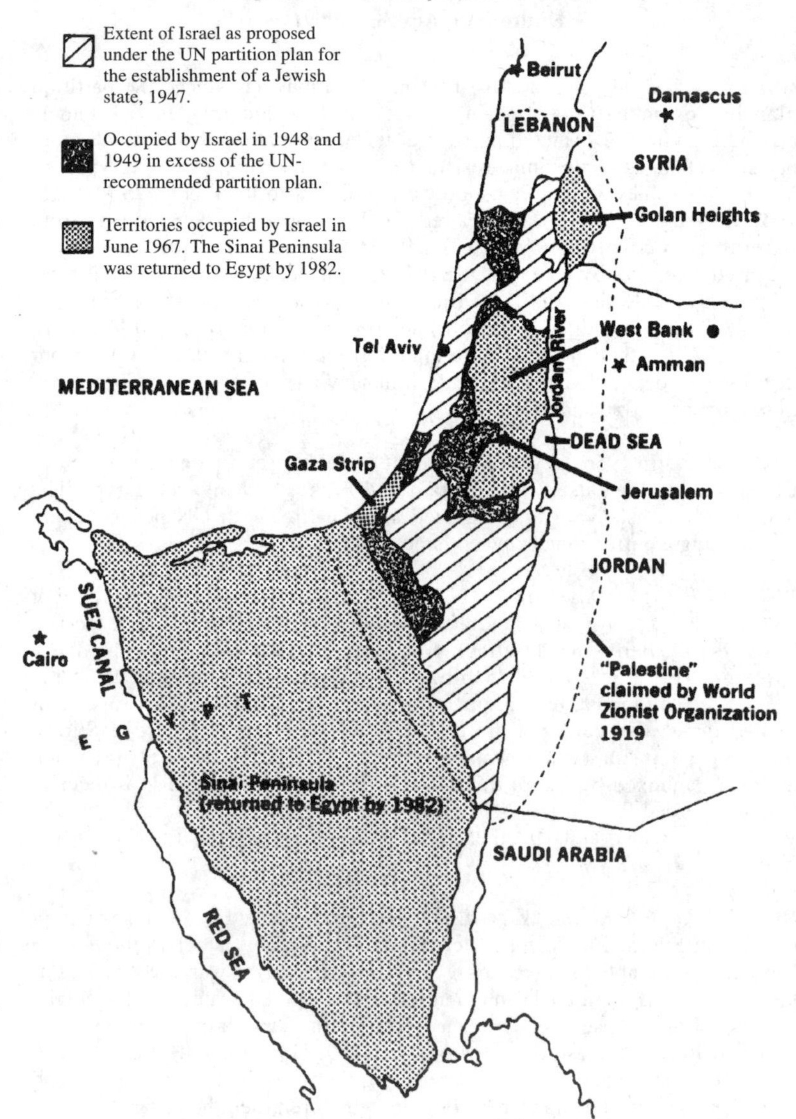

Source: Reprinted from Wayne C. McWilliams and Harry Piotrowski, *The World Since 1945: A History of International Relations,* 6th ed. (Boulder: Lynne Rienner, 2005). © Copyright 2005 Lynne Rienner Publishers.

An important recent development was the first intifada, or uprising (1987–1993). In the West Bank and Gaza Strip, Israeli occupation became increasingly marked by human rights violations, including administrative detentions, land confiscations, and the destruction of Palestinian homes.

Palestinians pointed to the fourth Geneva Convention's prohibition of these tactics. Israeli authorities argued that most of these actions were emergency measures to protect the security of Israeli citizens. Palestinians argued that these measures, such as home demolitions and land expropriations, were methods of acquiring more land for Israeli settlements.

The first intifada began in December 1987, following an incident in which four Palestinians workers were killed in Gaza. As a result, Palestinians took to the streets in protest, which escalated into an all-out revolt. The intifada was symbolized by Palestinian youth throwing rocks at Israeli soldiers, who in turn retaliated with gunfire. The spontaneous uprising was a result of frustration for two decades of Israeli occupation. The demonstrations and the rock-throwing were only a small part of it; the intifada would become a mass mobilization movement of mostly peaceful resistance to the Israeli presence. The intifada and the Israeli response had a wide-ranging effect. For instance, when Israelis shut down Palestinian schools for three years, Palestinian students, teachers, and community leaders organized an alternative education for the students. Palestinians also organized boycotts of Israeli products.

Extremist groups on both sides feared the creation of a lasting peace, which would of course require compromise on both sides. In a dramatic illustration of the determined opposition on the Israeli side, a Jewish member of an extremist group assassinated Israeli prime minister Yitzhak Rabin in Tel Aviv in 1995. Just as Jewish extremists were using violence in an attempt to stop the peace process, Palestinian extremists also stepped up violent opposition in an attempt to destroy whatever was left of Israeli-Palestinian peace talks, known as the Oslo Accords (discussed in Figure 3.3). Members of Palestinian extremist groups carried out suicide bombings in Jerusalem and Tel Aviv. Israelis blamed Arafat, who in 1996 was elected the president in the first ever Palestinian elections. They were frustrated by the inability of his government, the Palestinian Authority, to control violent extremist groups such as Hamas, which has been responsible for many terrorist attacks.

In September 2000, a visit by Israeli defense minister Ariel Sharon to the Temple Mount (the site of the Al-Aqsa Mosque—the third holiest site in Islam) in East Jerusalem sparked widespread demonstrations, which resulted in the killing of six unarmed Palestinian demonstrators. This incident marked the beginning of a second intifada that would become known as the Al-Aqsa intifada, which was marked by increased militarization of the Occupied Territories. Whereas the first intifada was mostly a nonviolent resistance against the occupation, the second was shaped by the excessive use of violence on both sides. The continued military occupation of the West Bank and Gaza Strip facilitated the rise of extremist elements within Palestinian society. Groups such as Hamas increasingly resorted to mass terror attacks against Israeli civilians. Suicide bombers became their weapon of choice. Israelis

Figure 3.3 Israeli-Palestinian Peace Attempts

1988. A key development occurred in 1988 when Yasir Arafat accepted UN Security Council Resolution 242. This resolution called for the withdrawal of Israel from territories seized in the 1967 war, as a basis for any just and lasting peace in the Middle East. This resolution has become the cornerstone of most international efforts to negotiate peace. Also in 1988, Arafat condemned terrorism for the first time and accepted Israel's right to exist.

1993. The Oslo Accords resulted from secret negotiations between Israel and the Palestine Liberation Organization. The basic principle of these negotiations was "land for peace," by which Israel would return land to the Palestinians, who in turn would halt attacks on Israel. As part of the peace accords, the Palestinian Authority was created and given limited home rule. Opposition emerged on both sides, and key issues such as Jerusalem, settlements, and the right of return for Palestinian refugees were left for later negotiations.

2000. Israeli prime minister Ehud Barak met with Arafat at Camp David, Maryland, and President Bill Clinton acted as the moderator. Although Barak was willing to return a larger part of the Occupied Territories (up to 80 percent) than any of his predecessors, he insisted that this portion be divided into several sections. This solution would leave Palestinians with small, unconnected areas of land, surrounded by areas controlled by Israel, making a Palestinian state impossible. The right of return for Palestinian refugees proved to be another major stumbling block.

2003. US president George W. Bush, supporting separate states for Israelis and Palestinians, announced his "roadmap for peace." The roadmap pushed for a stop to the building of Jewish settlements and to Palestinian violence. However, increased violence on both sides led to heightened frustration for Jews and Palestinians, and the US invasion of Iraq diverted attention away from the peace process.

2004. Israeli prime minister Ariel Sharon closed some settlements in the West Bank and all of the settlements in Gaza. However, Israel continued to control the airspace, territorial waters, and land passages of the West Bank and Gaza. Although Israeli public opinion appeared to support this dramatic move, it drew angry protests from within Israel. Palestinian critics argued that the targeted settlements represented only a fraction of Israel's settlements in the West Bank, and that they were already being replaced by additional new settlements in other areas of the West Bank. Most Palestinian and international critics argued that, although settlements in Gaza had been dismantled, Palestinians still lacked real sovereignty.

Palestinians often argue that Israel has never really been interested in Gaza, that it intends to withdraw only from Gaza, but not from most of the West Bank. More recently, Hamas's seizure of control of Gaza could be seen as a major blow to nationalist aspirations of having a viable Palestinian state in the areas occupied by Israel in 1967 (West Bank and Gaza).

began to undertake targeted assassinations of Hamas leaders and activists, often involving the death of innocent Palestinian civilians, including women and children. The Israelis continued to use other means of collective punishment, such as house demolitions, curfews, and mass imprisonment.

In 2002, Israel reoccupied all Palestinian areas it had withdrawn from as part of the Oslo process. Arafat was held responsible for failing to control Hamas and the other extremist groups, and placed under house arrest in his headquarters in Ramallah (in the West Bank). Israel began construction of what it referred to as a "security fence" within the West Bank. It argued that the "fence" (actually a wall that is 403 miles long and a height of 25 feet) was an attempt to protect its citizens from terrorists attacks. Palestinians argued that because the wall was built well into Palestinian territory, it was an attempt to confiscate additional Palestinian lands. Critics also point out that, in many cases, the wall has separated Palestinian communities from their hospitals, schools, and farms. Although in 2004 the International Court of Justice ruled that the wall violated international law, Israel continued to build it.

In November 2004, Arafat died. In January 2005, Mahmoud Abbas was elected to replace him as president of the Palestinian Authority. The election of Mahmoud Abbas, along with the Sharm el-Sheikh summit, brought an end to the second intifada. Some critics argue that whatever glimmer of hope that existed was extinguished by the Hamas win in January 2006. Hamas, which is on the US State Department's list of terrorist organizations, won a majority in the Palestinian Legislative Council elections. Others point to the Hamas victory as a clear sign of Palestinian resentment against a corrupt Palestinian Authority, which for some appears to be taking on the role of "prison guard" for the Israeli occupation, as well as a clear sign of disenchantment with continuous peace proposals that have not materialized, restrictions on their movements, and an illegal wall blocking access to their farmland. Hamas was able to provide badly needed social services and the means for Palestinians to protest their worsening socioeconomic status. The major aid donors to the Palestinian Authority—the United States and several European Union nations—reacted to the election of Hamas by withholding all financial support. The Israelis also tried to tighten the noose around the neck of Hamas. This attempt to starve Hamas out of power had devastating consequences on the entire Palestinian population. Palestinian infighting between Hamas and Fateh (a PLO faction founded by Yasser Arafat), over loyalty to President Mahmoud Abbas, exacerbated the situation in the Occupied Territories. In June 2007, Hamas fighters took control of Gaza, and Mahmoud Abbas dissolved the government and appointed Salam Fayyad as the new prime minister. Palestinians now fear that if the situation is not resolved soon, it could lead to two separate entities, neither of which, alone, could achieve sovereign status.

Despite all the conflict in the region, peace attempts have occurred throughout much of the twentieth century. In fact, attempts to bring both sides to a compromise predate the creation of Israel. Unfortunately, these peace attempts have not been very successful (see Figure 3.3).

A majority of Israelis favor returning land in the West Bank and Gaza to the Palestinians, if it would mean an end to the conflict. And the majority of Palestinians want an end to the occupation of the West Bank and Gaza and the establishment of a viable Palestinian state. However, there are disagreements over how to attain these objectives. No Israeli leadership has ever offered complete withdrawal from the Occupied Territories (based on UN Security Council Resolution 242), and Palestinians argue that anything less will not allow them to establish a viable state.

Major Obstacles to Peace

Finding a lasting resolution of the Arab-Israeli conflict has been proven to be extremely difficult. Some of the main obstacles have been Israeli settlements, Palestinian refugees, East Jerusalem, and terrorism.

Israeli settlements. What to do with Israeli settlements in the disputed territory has been a major stumbling block in Israeli-Palestinian negotiations. Approximately 390,000 Israeli settlers live in the West Bank, controlling 42 percent of the area ("Palestine Fact Sheet" 2004). The increasing size and quantity of the settlements have continued to shrink the area under Palestinian control. The area of land that falls under direct control of the Palestinian Authority in the West Bank is currently less than 7 percent. Palestinians point out that the fourth Geneva Convention prohibits any occupying power from establishing settlements in Occupied Territories. Israel itself is divided on the issue of settlements—some favor moving settlers out of the territories, while others see such a move as betraying the Zionist cause. Even those in favor of moving the settlers acknowledge that it would be extremely difficult politically for the Israeli government to force Jewish settlers from their homes in the Occupied Territories.

Palestinian refugees' right of return. The right of Palestinians to return to their original homes has been a central issue to the larger Israeli-Palestinian conflict. Palestinians complain that Israel, despite international law, has consistently refused to allow Palestinians, who were forced to flee their homes in 1947–1948, to do so. To complicate matters, these Palestinian refugees are often not granted any rights in their host countries. In Lebanon, Palestinians are not allowed to own land or work unless through the United Nations Relief and Works Agency for Palestine Refugees in the Near East (UNRWA). Israel refuses to allow these refugees to return, arguing that it would be a demographic disaster for the Jewish state. If Palestinians were to return to their original homeland in Israel, they would outnumber Israelis, who are already concerned with the high birthrates of Palestinian populations compared to the much lower birthrates of Jewish populations.

East Jerusalem. Christianity, Islam, and Judaism have in common the patriarch Abraham, and all three consider Jerusalem to be holy land. Both the Israelis and the Palestinians lay claim to Jerusalem as their capital. Access to religious sites is a central issue in the conflict over Jerusalem. The Israeli occupation continues to restrict many Palestinians in the occupied West Bank from access to religious sites in occupied East Jerusalem. Security concerns have always been the rationale used by the Israelis for restricting access. Although international law rejects Israel's annexation of predominantly Arab East Jerusalem, which was part of Palestinian-controlled territory prior to the 1967 war, approximately 250,000 Israeli settlers live in occupied East Jerusalem. Palestinians continue to demand that Arab East Jerusalem become the capital of a future Palestinian state, while the Israelis refuse to divide Jerusalem, claiming that it should forever be the capital of the state of Israel.

Terrorism. Israeli officials argue that a major obstacle to peace is the continuous threat of terrorism that the state must endure from extremist groups. Israel has blamed the Palestinian Authority for not prohibiting these groups, such as Hamas, from carrying out their suicide missions. The Palestinian Authority is therefore, Israelis argue, not a legitimate negotiation partner. Palestinians reject the terrorism label and argue that they are engaged in armed struggle for liberation, which is legitimate under international law. They maintain that the occupation of Palestinian lands is the cause of these attacks, and if the occupation of the West Bank and Gaza Strip were to end, then the attacks would end or at least be greatly reduced. Palestinians also argue that the Israeli army engages in state terrorism against Palestinians, which perpetuates the violence. Recent statistics confirm that the number of Israeli civilian deaths caused by armed Palestinian groups is many times fewer than the number of Palestinian civilian deaths caused by the Israeli forces (Amnesty International 2007a; Levy 2007).

■ Nationalism and the Future

Historically speaking, the state system and nationalism are relatively young. However, both have an extremely strong impact on our lives. As in the past few hundred years, nationalism will continue to play an important role in the world by bringing people together and shaping their identities. It will also play an important role where groups seek self-determination (e.g., Chechnya and French-speaking Quebec) and in areas ravaged by war (e.g., the Israeli-Palestinian conflict and the conflict in Iraq).

In light of the power of nationalism and state sovereignty, it will be interesting to see how they fare against the forces of globalization over the next few decades. In a system based on sovereignty, states will continue to seek control over their borders. Yet as the world continues to become more interconnected due to advances in technology, states may find it more difficult to control their borders against external elements such as illegal drugs, weapons, and immigrants. Similarly, globalization may lead citizens to identify with groups in other countries (based on religion, pop culture, etc.) as much as, or even more than, they identify with groups in their own countries. For both nationalism and state sovereignty, changes will likely be gradual.

Discussion Questions

1. What are your primary allegiances? In other words, how do you identify yourself?
2. Will the force of nationalism decline in the future?
3. Has state sovereignty eroded over the past few decades? Will it decline significantly in the future?
4. What do you think would be a fair solution to the conflict between the Israelis and the Palestinians?

Suggested Readings

Anderson, Benedict (1991) *Imagined Communities: Reflections on the Origins and the Spread of Nationalism.* 2nd ed. London: Verso.
Chatterjee, Partha (1993) *The Nation and Its Fragments.* Cambridge: Cambridge University Press.
Finkelstein, Norman G. (1995) *Image and Reality of the Israeli-Palestinian Conflict.* London: Verso.
Gellner, Earnest (1983) *Nationalism.* Ithaca: Cornell University Press.
Hobsbawm, Eric (1992) *Nations and Nationalism Since 1780: Programme, Myth, Reality.* New York: Cambridge University Press.
Hutchinson, John, and Anthony D. Smith, eds. (1996) *Ethnicity.* New York: Oxford University Press.
Khalidi, Rashid, L. Anderson, R. Simon, and M. Muslih, eds. (1991) *The Origins of Arab Nationalism.* New York: Columbia University Press.
Laqueur, Walter (1972) *The History of Zionism.* New York: Schocken.
Morris, Benny (1999) *Righteous Victims.* New York: Knopf.

Pappe, Ilian (1992) *The Making of the Arab-Israeli Conflict, 1947–1951.* New York: Tauris.
Simons, Lewis M. (1987) *Worth Dying For.* New York: Morrow.
Smith, Anthony D. (2001) *Nationalism: Theory, Ideology, History.* Malden, MA: Polity.
Zunes, Stephen (2003) *Tinderbox: US Middle East Policy and the Roots of Terrorism.* Monroe, ME: Common Courage.

4

The Challenge of Human Rights

D. Neil Snarr

THE UNITED NATIONS REFERS TO HUMAN RIGHTS AS "INALIENABLE and inviolable rights of all members of the human family" (UN 1988: 4). According to one scholar: "The very term human rights indicates both their nature and their sources: they are the rights that one has simply because one is human. They are held by all human beings, irrespective of any rights or duties one may (or may not) have as citizens, members of families, workers, or parts of any public or private organization or association. . . . [T]hey are universal rights" (Donnelly 1993: 19). Regardless of one's station in life—their age, color, sex, religious affiliation, wealth (or lack of), or geographic location, among other factors—the intent is clear: all people are to enjoy such rights. This chapter examines these human rights, the controversies that surround them, the efforts to support them, the many forces that inhibit their realization, and some specific case studies.

The Source of Human Rights

Declarations and agreements that contain historic steps toward what we now understand as human rights are numerous and spring from many sources. They come from philosophers, politicians, and religious prophets. They are contained in scripture, speeches, and treaties. The British Magna Carta (1215), the English Bill of Rights (1689), the US Declaration of Independence (1776), the US Constitution (1789), the French Declaration of the Rights of Man and of the Citizen (1789), and the US Bill of Rights (1791) are just a few of the documents that contain references to what we now call human rights. Late in the nineteenth century slave trade was outlawed, and early in the twentieth century slavery itself was outlawed. Later, humanitarian considerations in the conduct of war were agreed upon, and the treatment of workers, prisoners, and women became the subject of international agreements. Because of the genocidal horrors that occurred during World War II against Jews, Gypsies, and other groups in Europe, the world community founded the United Nations and immediately started working on a document that would be called the Universal Declaration of Human Rights (UDHR). Their hope was to avoid such deadly wars and assist poor countries in raising their standard of living.

Today, discussions of human rights generally start in the UN General Assembly, where all member states have representatives. There these rights are debated, publicly scrutinized, and voted on. The passage of a human rights convention in the General Assembly is the easy part; it takes only a majority vote. After the General Assembly approves a convention, it is opened for signature by member states. After a designated number of countries have ratified the convention, it "comes into effect." This process often takes many years, and sometimes a convention is not approved. After a convention comes into effect, the agreeing countries are expected to pass laws, if they do not have them already, to ensure observance and enforcement of the convention. Eventually, it is hoped, all countries will approve the human rights espoused in the convention, and they will become international law.

The Universal Declaration of Human Rights

On December 10, 1948, the UN General Assembly approved the Universal Declaration of Human Rights. Forty-eight countries voted in its favor, including the United States, and there were no votes against the document, although eight countries abstained. Since that time, the UN has approved well over 200 documents that elaborate and expand these rights (such agreements are often referred to as conventions, treaties, or covenants). Since their inception, human rights issues have received tremendous attention. One student of human rights declares, "The recognition of human rights and the weaving of a web of globalization are probably the most important political developments of our lifetimes. Like water carving a canyon, the slow, quiet power of human-rights pressures and aspirations helped bring down the Soviet empire, transform long-suffering Latin America, and construct unprecedented international institutions: The United Nations System" (Brysk 2003: 21).

On the other hand, this focus on human rights has not been universally successful:

> When the United Nations introduced the Universal Declaration of Human Rights in 1948, it was seen by many as a sign of optimism, of the possibil-

> ities of a better world. Yet, over 50 years later, observers recognize that we live in an age when human rights abuses are as prevalent as they ever have been—in some instances more prevalent. The world is littered with examples of violations of basic rights: censorship, discrimination, political imprisonment, torture, slavery, the death penalty, disappearances, genocide, poverty, refugees. The rights of women, children, and other groups in society continue to be ignored in atrocious ways. (O'Byrne 2003: 5)

What rights have been identified as human rights? One way to approach this question is to divide the UDHR of 1948 into three generations, or classes. These three generations have different origins and represent different views of human rights. The UDHR contains thirty articles, the first of which declares: "All human beings are born free and equal in dignity and rights. They are endowed with reason and conscience and should act toward one another in a spirit of brotherhood."

The First Generation of Human Rights

Originating in seventeenth- and eighteenth-century Western ideas, these rights found expression in the revolutions of France, Britain, and the United States. The United States often views these as civil rights, which the US government generally equates with human rights. They also include what are called political rights. It is this first generation of human rights that has received the most attention.

These civil and political rights are contained in Articles 2–21 of the UDHR. They focus on the rights of the individual and emphasize the responsibility of governments to refrain from unjustly interfering in the lives of their citizens (see Figure 4.1). One might think of the civil rights movement in the United States as an example of a group of people gaining the first generation of human rights. It was a struggle for the right to justice: African Americans would no longer be subject to arbitrary arrest, they would be tried before an impartial jury, they would be subject to the same laws as others, and so forth. One might also think of China's controversial stance toward these first-generation rights, which it views as deriving from Western ideas that do not apply to non-Western countries. China has argued that the rights to free assembly and freedom of the press are not universal rights. There has been a long discussion on such points, and it serves to remind the reader that the definition of human rights is often contested.

The Second Generation of Human Rights

The second generation of human rights encompasses social and economic rights. Contained in Articles 22–26 of the UDHR, they stem from the Western socialist tradition. To some degree they are seen as a way to balance what many consider to be the excessive individualism of the first gen-

Figure 4.1 First Generation of Human Rights, UDHR Articles 2–21

2. Everyone is entitled to all the rights and freedoms set forth in this Declaration, without distinction of any kind, such as race, color, sex, language, religion, political or other opinion, nation of social origin, property, birth or other status. Furthermore, no distinction shall be made on the basis of the political, jurisdictional or international status of the country or territory to which a person belongs, whether it be independent, trust, non-self-governing or under any other limitation of sovereignty.
3. Everyone has the right to life, liberty and security of person.
4. No one shall be held in slavery or servitude; slavery and the slave trade shall be prohibited in all their forms.
5. No one shall be subjected to torture or to cruel, inhuman or degrading treatment or punishment.
6. Everyone has the right to recognition everywhere as a person before the law.
7. All are equal before the law and are entitled without any discrimination to equal protection of the law. All are entitled to equal protection against any discrimination in violation of this Declaration and against any incitement to such discrimination.
8. Everyone has the right to an effective remedy by the competent national tribunals for acts violating the fundamental rights granted him by the constitution or by law.
9. No one shall be subjected to arbitrary arrest, detention or exile.
10. Everyone is entitled in full equality to a fair and public hearing by an independent and impartial tribunal, in the determination of his rights and obligations and of any criminal charge against him.
11. (1) Everyone charged with a penal offence has the right to be presumed innocent until proved guilty according to law in a public trial at which he has had all the guarantees necessary for his defense.
    (2) No one shall be held guilty of any penal offence on account of any act or omission which did not constitute a penal offence, under national or international law, at the time when it was committed. Nor shall a heavier penalty be imposed than the one that was applicable at the time the penal offence was committed.
12. No one shall be subjected to arbitrary interference with his privacy, family, home or correspondence, nor to attacks upon his honor and reputation. Everyone has the right to the protection of the law against such interference or attacks.
13. (1) Everyone has the right to freedom of movement and residence within the borders of each State.

continues

Figure 4.1 continued

(2) Everyone has the right to leave any country, including his own, and to return to his country.

14. (1) Everyone has the right to seek and to enjoy in other countries asylum from persecution.
(2) This right may not be invoked in the case of prosecutions genuinely arising from nonpolitical crimes or from acts contrary to the purposes and principles of the United Nations.

15. (1) Everyone has the right to a nationality.
(2) No one shall be arbitrarily deprived of his nationality nor denied the right to change his nationality.

16. (1) Men and women of full age, without any limitation due to race, nationality or religion, have the right to marry and to found a family. They are entitled to equal rights as to marriage, during marriage and at its dissolution.
(2) Marriage shall be entered into only with the free and full consent of the intending spouses.
(3) The family is the natural and fundamental group unit of society and is entitled to protection by society and the State.

17. (1) Everyone has the right to own property alone as well as in association with others.
(2) No one shall be arbitrarily deprived of his property.

18. Everyone has the right to freedom of thought, conscience and religion; this right includes freedom to change his religion or belief, and freedom, either alone or in community with others and in public or private, to manifest his religion or belief in teaching, practice, worship and observance.

19. Everyone has the right to freedom of opinion and expression; this right includes freedom to hold opinions without interference and to seek, receive and impart information and ideas through any media and regardless of frontiers.

20. (1) Everyone has the right to freedom of peaceful assembly and association.
(2) No one may be compelled to belong to an association.

21. (1) Everyone has the right to take part in the government of his country, directly or through freely chosen representatives.
(2) Everyone has the right of equal access to public service in his country.
(3) The will of the people shall be the basis of the authority of government; this will shall be expressed in periodic and genuine elections which shall be by universal and equal suffrage and shall be held by secret vote or by equivalent free voting procedures.

eration of rights and the impact of Western capitalism and imperialism. They focus on social equality and the responsibility of the government to its citizens, rather than on protection of citizens from their government as the first generation of rights does (see Figure 4.2). Second-generation rights necessitate a proactive government acting on behalf of its citizens. They establish an acceptable standard of living for all—that is, a minimal level of equality. Everyone should have, for instance, access to healthcare and education, and receive equal pay for equal work.

The Third Generation of Human Rights

The third generation of rights encompasses solidarity rights—those rights whose realization requires the cooperation of all countries—as expressed in Articles 27–28 of the UDHR (see Figure 4.3). These articles claim rights for poorer countries of the South, or the third world. In 1948, when the UDHR was approved by the United Nations, most of the world's population was represented by the colonial powers, such as Britain, France, Portugal, and Holland, and could not represent themselves. Most of Africa and much of Asia were under the thumb of these European powers. The interests of these peoples were hardly considered by the colonial powers in writing the UDHR. They currently constitute over 80 percent of the world's population, but receive a very small portion of its benefits. The third generation of rights represents a hope, or even a demand, for the global redistribution of opportunity and well-being.

These rights do not have the status of the rights of the first and second generation, and are in the process of being developed. Richard Claude and Burns Weston say the following about them:

> [They] appear so far to embrace six claimed rights. . . . Three of these reflect the emergence of Third World nationalism and its demand for a global redistribution of power, wealth, and other important values: the right to political, economic, social, and cultural self-determination; the right to economic and social development; and the right to participate in and benefit from "the common heritage of mankind" (shared earth-space resources; scientific, technical, and other information and progress; and cultural traditions, sites, and monuments). The other three third-generation rights—the right to peace, the right to a healthy and balanced environment, and the right to humanitarian disaster relief—suggest the impotence or inefficiency of the nation-state in certain critical respects. (1992: 19–20)

As an example of the implementation of the third-generation solidarity rights, in December 1986 the UN General Assembly adopted the Declaration on the Right to Development. As Winston Langley notes: "The Declaration confirms the view of the international community that the right to development is an inalienable human right 'by virtue of which every

Figure 4.2 Second Generation of Human Rights, UDHR Articles 22–26

22. Everyone, as a member of society, has the right to social security and is entitled to realization, through national efforts and international cooperation and in accordance with the organization and resources of each State, of the economic, social and cultural rights indispensable for his dignity and the free development of his personality.
23. (1) Everyone has the right to work, to free choice of employment, to just and favorable conditions of work and to protection against unemployment.
(2) Everyone, without any discrimination, has the right to equal pay for equal work.
(3) Everyone who works has the right to just and favorable remuneration ensuring for himself and his family an existence worthy of human dignity, and supplemented, if necessary, by other means of social protection.
(4) Everyone has the right to form and to join trade unions for the protection of his interests.
24. Everyone has the right to rest and leisure, including reasonable limitation of working hours and periodic holidays with pay.
25. (1) Everyone has the right to a standard of living adequate for the health and well-being of himself and of his family, including food, clothing, housing and medical care and necessary social services, and the right to security in the event of unemployment, sickness, disability, widowhood, old age and other lack of livelihood in circumstances beyond his control.
(2) Motherhood and childhood are entitled to special care and assistance. All children, whether born in or out of wedlock, shall enjoy the same social protection.
26. (1) Everyone has the right to education. Education shall be free, at least in the elementary and fundamental stages. Elementary education shall be compulsory. Technical and professional education shall be made generally available and higher education shall be equally accessible to all on the basis of merit.
(2) Education shall be directed to the full development of the human personality and to the strengthening of respect for human rights and fundamental freedoms. It shall promote understanding, tolerance and friendship among all nations, racial or religious groups, and shall further the activities of the United Nations for the maintenance of peace.
(3) Parents have a prior right to choose the kind of education that shall be given to their children.

human person and all peoples are entitled to participate in, contribute to and enjoy economic, social, cultural and political development, in which all human rights and fundamental freedoms can be fully realized'" (1996: 361). This agreement is important, as are many others with similar intent, but the wealthy countries of the world are reluctant to share too much of their advantages and wealth.

Figure 4.3 Third Generation of Human Rights, UDHR Articles 27–28

27. (1) Everyone has the right freely to participate in the cultural life of the community, to enjoy the arts and to share in scientific advancement and its benefits.
    (2) Everyone has the right to protection of the moral and material interests resulting from any scientific, literary or artistic production of which he is the author.
28. Everyone is entitled to a social and international order in which the rights and freedoms set forth in this Declaration can be fully realized.

An example is a case that has recently come before the World Trade Organization (WTO), which regulates trade between countries and is committed to free trade. The crux of the issue is that, by subsidizing exports or placing taxes (tariffs) on incoming goods, a country can gain significant economic advantage over other countries. The tariffs that the United States and the European Union place on many of their agricultural products are sufficiently high that many poor countries cannot compete. Brazil, well aware of this situation, challenged the United States before the WTO, which ruled against the United States in a preliminary decision. Although the United States will appeal this decision, there is good reason to believe that it will need to reduce its tariffs to give poorer countries an opportunity to sell their products, such as cotton, within the United States. High tariffs are considered a human rights violation because they disadvantage poor countries and directly affect their ability to raise their standards of living.

* * *

As illustrated by the WTO example above, there is an overlap of second-generation human rights (social and economic rights) and third-generation human rights (solidarity rights). Some students of human rights merge the second and third generations, assuming that if the second generation is realized, then so will be the third. This is evident in the UN's 2000 *Human Development Report* (UNDP 2000), in which it is argued that human rights cannot be realized without human development, and that human development cannot be realized without human rights.

The final two articles of the UDHR affirm the universality of and responsibility for the three generations of rights described in Articles 1–28 (see Figure 4.4).

Figure 4.4 UDHR Articles 29–30

29. (1) Everyone has duties to the community in which alone the free and full development of his personality is possible.
(2) In the exercise of his rights and freedoms, everyone shall be subject only to such limitations as are determined by law solely for the purpose of securing the recognition and respect for the rights and freedoms of others and of meeting the just requirements of morality, public order and the general welfare in a democratic society.
(3) These rights and freedoms may in no case be exercised contrary to the purposes and principles of the United Nations.
30. Nothing in this Declaration may be interpreted as implying for any State, group or person any right to engage in any activity or to perform any act aimed at the destruction of any of the rights and freedoms set forth herein.

Human Rights: Universalism vs. Relativism

Clearly, human rights have not emerged onto the world's political agenda without controversy. At every step since the signing of the Universal Declaration of Human Rights, there have been delays and denouncements. There is no reason to believe that this controversy will cease anytime soon.

The UN Charter guarantees state sovereignty, or self-determination and nonintervention; it also states that all individuals, regardless of their citizenship and status, have human rights. These principles are often found to be contradictory. The idea that everyone possesses these rights as found in the UDHR is referred to as *universalism.* On the other hand, some countries and cultures follow traditions that are considered inconsistent with the UDHR, and they claim exception for their traditions. These governments say that they are the final authority in determining what is right for their citizens: they plead state sovereignty, which has been a global principle for three centuries. According to these countries, appropriate expectations for human rights are judged against, or relative to, local culture. In other words, certain customs that are thought by some to violate human rights, are considered legitimate, long-standing cultural or religious practices by others. This view is referred to as *relativism.* Two such customs are child marriage and female circumcision.

In South Asia, young girls are often betrothed by their families to marry at an early age, without consideration of the desires of the child. This is generally considered a violation of the child's rights, but it is often defended as a cultural (or relativist) tradition. For those who participate in this practice, the determining factor is tradition, not an abstract rule that is considered to apply universally.

Female circumcision is also practiced by millions of people throughout Africa and the Middle East, and has also been defended as a cultural practice that should not be subject to a universal human rights rule. Female circumcision (sometimes called female genital mutilation) involves a procedure that may include the complete removal of the clitoris and occasionally the removal of some of the inner and outer labia. It is estimated that this procedure has affected some 137 million women, mostly in Africa. Those who defend this practice culturally claim that it makes girls "marriageable" (because it ensures their virginity) and also diminishes their sex drives.

UN agencies disagree with this relativist position and have begun to take steps to eliminate female circumcision. Many African countries, such as Egypt, Senegal, Eritrea, and Sierra Leone, have either declared against the practice or passed laws in an effort to terminate it. But despite this seeming rush to eliminate female circumcision, the practice has found its way into Western countries also. It is estimated that some 66,000 women and girls in Britain have been at risk for the procedure, which often occurs outside the country. The British government has declared female circumcision a form of child abuse, but its perpetrators operate clandestinely and are difficult to stop ("£20k Reward" 2007).

As the debate proceeds over what rights are universal, most decisions will fall somewhere between the two extremes of universalism and relativism. At the same time, however, there is general agreement that acts such as genocide (indiscriminate killing of an entire people, such as Jews, Gypsies, Hutus, or Tutsis), torture, and summary executions are violations of human rights.

The United Nations and Human Rights Implementation

"For all its faults," says Darren O'Byrne, "the United Nations is probably the most important agency involved in the protection of human rights worldwide" (2003: 85). Tom Farer (2002) observes that the United Nations operates at four levels in supporting human rights. First, it formulates and defines international standards by approving conventions and making declarations. Second, it advances human rights by promoting knowledge and providing public support. At the third level, it supports human rights by protecting or implementing them. Although the task of directly enforcing human rights is primarily left to the states themselves, the UN does become involved in various means of implementation. During the 1990s and early 2000s, UN enforcement took on new meaning and controversy. The UN's efforts in the Persian Gulf, Somalia, Rwanda, the former Yugoslavia, and Afghanistan, often under pressure from the United States and its allies, are examples of this. They include boycotts of aggressor states, military action, military support for delivery of humanitarian aid, and protection of refugees. Some of these are controversial extensions of the UN mandate and will provide material for discussions about the security role of the UN in the future (see Chapter 5). Fourth, the UN has taken additional enforcement steps that some consider to be structural and economic aspects of human rights issues—that is, the third generation of rights, including economic development for poor countries as described previously. Though the UN has expended considerable resources on development, such efforts receive little public attention compared to more dramatic (such as military) actions. Many poor countries rely heavily on support and training provided by the UN.

Beyond the positive actions that the UN takes in supporting human rights, there is the significant criticism that human rights are often not enforced. Critics point out that some of the countries that have signed UN human rights conventions have made little progress toward instituting them. There are several reasons that countries may sign these conventions even if they have no apparent intention to enforce them. First, most countries want to appear to the world as though they treat their citizens justly.

Second, some countries, regardless of their human rights records, are reluctant to subject themselves to the jurisdiction of world bodies such as the UN. The United States, which generally has a good human rights record, often fits into this category. These countries take the stance that what happens within their own borders is their own business and not the concern of other political bodies (an example of the principle of state sovereignty).

Third, the UN is not an independent body or a world government; it is subject to the whims of its members and has no more power or resources than it is given by its members. For instance, many UN decisions on human rights violations are made in the Security Council, any of whose five powerful permanent members—Britain, China, France, Russia, and the United States—can veto an action (the ten nonpermanent members do not have such veto power). Thus, all five permanent members must approve any action taken, a difficult task.

The situation in Sudan is a case in point. It is estimated that 200,000–400,000 people have been killed, and over 2 million displaced, by what is considered an ethnic or tribal conflict. After decades of drought, desertification, and rapid population growth, this conflict erupted in 2003 in the Darfur region. The central government of Sudan has been implicated in this situation, which has been identified as a genocide by the United States. The UN Security Council has passed resolutions, and it approved a

UN peacekeeping mission in 2006, but the Sudanese central government, with the support of China, has been able to limit any serious impact on the situation. China's primary interest in Sudan is its oil, of which China receives approximately 70 percent.

This has all the characteristics of a major human rights dilemma. A country (Sudan) is implicit in the deaths of hundreds of thousands of its own citizens, but pleading state sovereignty or noninterference. The five permanent Security Council members are unwilling or unable to come to agreement on actions to take. One member of the Security Council (China) is benefiting from inaction. In the meantime, hundreds of thousands die and millions are displaced. This sounds too much like the situation in Rwanda in 1994, when some 800,000 were killed as the world stood by.

A final inhibiting factor is the availability of money for UN operations. Several countries are either unable or unwilling to pay their rightful shares. For instance, the United States is behind in its payments for three categories of UN operations—budget, peacekeeping, and international tribunals. This debt, over $1 billion, is the largest owed by any country. In this case the cause is not a lack of money, but rather opposition to UN actions and policies by some US politicians.

■ State Sovereignty, the United Nations, and the International Criminal Court

State sovereignty is a cornerstone of the contemporary international system. But like so many things in our globalizing world, it is being challenged and slowly altered. For instance, at the close of the Gulf War in April 1991, the UN Security Council passed Resolution 688, which permitted the establishment of temporary havens for refugees inside Iraq, without Iraq's permission. The rationale was that the Iraqi government's violent treatment of Kurds (a large ethnic group living in Iraq) threatened international peace and security. This clearly contradicts the traditional understanding of state sovereignty. A more recent example was the establishment of tribunals to try persons responsible for crimes against humanity in the former Yugoslavia and Rwanda.

Due to state sovereignty, some heads of states who have perpetuated massive human rights abuses have escaped responsibility. Those who have killed tens of thousands and even millions of their own citizens—such as Idi Amin in Uganda, Augusto Pinochet in Chile, Pol Pot in Cambodia, and Slobodan Milosevic in the former Yugoslavia—have rarely been held responsible for their crimes. They have not gone unnoticed, but because of the Cold War and the fact that indicting them would confront state sovereignty, they have received limited attention. Yet these kinds of violations have not always been overlooked.

Crimes that were labeled crimes against humanity were dealt with after World War II in both Germany and Japan. Those who perpetrated these crimes were tried before tribunals and found guilty. After that, however, there was a long lull in much human rights work, lasting from 1946 to 1976, a period that Geoffrey Robertson refers to as "thirty inglorious years" (2000: 37). During that time, the United States supported many governments that were responsible for massive human rights violations, but justified its support on the basis of fighting communism. The Soviet Union did the same thing with different justification—fighting Western imperialism and capitalism.

Thus these heads of state and those who supported them and carried out their grisly commands have generally been able to walk away from their crimes with little fear of accountability. Often, such heads of state order passage of legislation that absolves their actions, as well as the actions of their cohorts, allowing them to depart their posts with impunity.

In early 1993 and again in late 1994, two international tribunals were established by the UN Security Council to deal with crimes against humanity, one in the former Yugoslavia and one in Rwanda. In the former Yugoslavia some 200,000 people were killed in what was called "ethnic cleansing" (separating ethnic groups by killing or forced migration), and in Rwanda approximately 800,000 were massacred in tribal violence. The timing of these tribunals was especially important, since they were established before the end of the conflicts and thus constituted a form of early intervention. Also, these tribunals were established based on international law, which supersedes state sovereignty, at least in principle. Though their establishment was an important step in dealing with the impunity so pervasive until recent times, these were ad hoc tribunals—established for specific cases of crimes and for limited time periods.

In 1998 the International Criminal Court (ICC) was established via the Rome Statute. According to Court literature: "The International Criminal Court is the first ever permanent, treaty based, international criminal court established to promote the rule of law and ensure that the gravest international crimes do not go unpunished. The ICC will be complementary to national criminal jurisdictions" (ICC 2004). The ICC Statute entered into force on July 1, 2002, after sixty countries ratified it. The Court is an independent international organization, seated at The Hague in the Netherlands. By the end of 2004 nearly a hundred countries had ratified the ICC Statute. President Bill Clinton signed the Rome Statute on December 31, 2000, the last day before signature expiration, but the US Senate did not ratify it. Later, the George W. Bush administration "'nullified' the US signature by sending a letter to UN Secretary-General Kofi Annan on May 6, 2002, expressing its intention not to be bound by the treaty" (CGS 2004).

Since that time, the United States has made many efforts to keep the

ICC Statute on the periphery of UN operations, and has pressured countries to sign bilateral agreements meant to ensure immunity for US citizens. Why does the United States feel so threatened by the ICC? One argument is that it does not want its past or future actions to be subject to an international court. For instance, it is possible that the Court could charge high-ranking US officials with criminal behavior for crimes covered by the ICC Statute. However, since the ICC Statute, like other treaties, allows countries to register reservations upon signature, the United States could prevent such criminal charges against itself. Given this adamant opposition by the United States, it appears the world's only superpower largely objects in principle to opening itself up to legal criticism by a foreign court.

Human Rights Implementation Outside the United Nations

With all of its problems, one cannot speak of human rights without invoking the name of the United Nations. It is here that human rights issues emerge and major discussions take place. With all of its weaknesses, efforts by the UN must be seen within the total context of human rights efforts. Because of the centrality of human rights issues in the world (greatly due to the work of the UN), governments and individuals have founded other institutions to supplement this important work. These groups have taken different forms, and have seen varying degrees of approval and success.

Regional Human Rights Organizations

One non-UN force to enter the human rights arena comprises the regional human rights structures. The most advanced and effective of these is in Europe and operates under the Convention for the Protection of Human Rights and Fundamental Freedoms. It was established in 1950 and functions under the European Commission of Human Rights, whose mandate has been extended on several occasions. Under this system, it is possible for citizens of European countries to register complaints directly to the commission (many human rights agencies will receive complaints only from *governments*). According to human rights advocates, this sophisticated, well-funded, and very successful regional structure is a model that other regions of the world should emulate.

A similar but much less successful structure exists in the Americas. It includes the seven-member Inter-American Commission of Human Rights and the Inter-American Court of Human Rights. Its decisions have often been resisted or ignored by countries of the Americas on the principle that state sovereignty predominates. The commission and court are left with the power of publicity and moral influence, which have been quite limited. There are signs, however, that these institutions are now being taken more seriously, given their increasing decisions that pressure regional governments to take human rights into account.

Also notable at the regional level is the African Charter on Human and Peoples' Rights, which was approved in the 1980s. It is an interesting document in that, unlike other regional documents, it includes the rights of "peoples," that is, the third-generation, or solidarity, rights. Article 19 states: "All peoples shall be equal; they shall enjoy the same respect and shall have the same rights. Nothing shall justify the domination of a people by another." In January 2004 an African human rights court came into existence.

Nongovernmental Human Rights Organizations

A second and promising human rights development outside the United Nations has been seen in the activities of nongovernmental organizations (NGOs). These organizations have emerged to fill the gap left by the UN and its many agencies, and confront the reluctance or inability of sovereign states to enforce human rights. As their names indicate, these groups are active on a variety of global issues and their impact is growing. There are hundreds of human rights NGOs throughout the world, comprising millions of members.

Human rights NGOs take many forms and operate in many different ways. By operating externally, they are able to monitor the actions and policies of governments and exert pressure for change. They gather information, provide advocacy and expertise, lobby, educate, build solidarity, and provide services and access to the political system (Wiseberg 1992).

Amnesty International, Human Rights Watch, the International League for Human Rights, Cultural Survival, the International Commission of Jurists, the International Committee of the Red Cross, and Physicians for Human Rights are just a few examples of human rights NGOs. Provisions have been made for these organizations to enjoy official representation at the United Nations and at UN-sponsored conferences. Geoffrey Robertson, a human rights scholar who has been quite critical of the United Nations, argues that experts from human rights NGOs should be eligible for appointment to UN committees and commissions: "The best way forward is to bring non-government organizations (which do most of the real human rights fact-finding) into the appointments process, thereby providing some guarantee that members are true experts in human rights, rather than experts in defending governments accused of violating them" (2000: 47).

Human Rights and the "War on Terror"

From a book titled *Human Rights in Turmoil: Facing Threats, Consolidating Achievements* comes the following: "A number of alarming signs are clouding the sky. Actions, statements and initiatives questioning the validity of human rights, or even threatening their very existence, have become a regular part of current political realities, even in states traditionally dedicated to the rule of law" (Lagoutte, Sano, and Smith 2007: 1). This concern is repeated from many quarters. Irene Khan, secretary-general of Amnesty International, a nongovernmental human rights organization, said in her 2007 annual report that "the politics of fear is fuelling a downward spiral of human rights abuse in which no right is sacrosanct and no person safe. . . . The 'war on terror' and the war in Iraq, with their catalogue of human rights abuses, have created deep divisions that cast a shadow on international relations, making it more difficult to resolve conflicts and protect civilians" (Amnesty International 2007b).

UN High Commissioner for Human Rights Louise Arbour stated in September 2007 "the US war on terror is constantly being used by other countries as justification for torture and other violations of international human rights laws" ("UN High Commissioner" 2007). Another article states that there are three ways in which the "war on terror," as led by the United States and Britain, is resulting in challenges to human rights (Tujan, Gaughran, and Mollett 2004: 68). First is the introduction in many countries of antiterrorism legislation, which often permits arbitrary arrest and prolonged detention without trial—for example, the United States holding prisoners at Guantanamo in Cuba. Second is the tolerance of human rights abuses in combating terrorism such as torture. The outsourcing of torture, or "extraordinary rendition," could be included in this category—for example, the United States sending suspected terrorists to other countries for the purpose of detention and interrogation (cases of such outsourced torture have been documented—see FCNL 2007). The final challenge is the growth of arms sales and military aid to regimes with poor human rights records.

Women's Rights

A domain of human rights that has received some official attention from the United Nations and its members, but that has seen relatively little success in implementation, is women's rights. From the initial UDHR document, to the many other agreements and specific conventions directed toward women's rights, it still seems quite clear that "millions of women throughout the world live in conditions of abject deprivation of, and attacks against their fundamental human rights for no other reason than that they are women" (HRW 2004b). This statement by Human Rights Watch, an NGO, is reiterated by Amnesty International and many other NGO human rights groups throughout the world. Still, the UN is committed to this issue, as Chapter 10 indicates. The single most important UN document related to women's rights is the Convention on the Elimination of All Forms of

Discrimination Against Women. Over 90 percent of UN member states are party to this convention. President Jimmy Carter signed it in 1980, but much like other human rights conventions it has not been ratified by the US Senate.

The plight of women in our world relies heavily on the work of NGOs and their ability to bring abuses to the attention of the United Nations, sovereign states, and the world community.

Women and War

Rape often goes hand in hand with war; some would say it always does. Social norms are weakened and the enemy is demonized, thereby providing a climate in which rape and other forms of antisocial and illegal behavior often become the norm. It is estimated that 200,000–400,000 Bengali women were raped in 1971 by Pakistani soldiers. Many believe this represents the worst case of wartime rape, but reports on the genocide in Rwanda indicate that possibly half a million women were raped during a few months in 1994. During World War II, in what came to be called the "Rape of Nanking," 20,000–80,000 Chinese women were raped by Japanese soldiers. These events, however, are not things of the past. It is estimated that at least 20,000 women were raped in the mid-1990s during the wars in Bosnia-Herzegovina. More recent cases have occurred in the Democratic Republic of Congo and the Darfur region of Sudan (Mitchell 2005).

Women, Rape, and HIV/AIDS

Many women who are raped during war contract HIV/AIDS. This was especially true in the Democratic Republic of Congo and in Rwanda, but the spread of HIV/AIDS does not result only from wartime rape. The "virgin cure," a myth that is prevalent in many countries where the incidence of HIV/AIDS is high, also contributes to this abuse. This myth maintains that having intercourse with a virgin, often the younger the better, will cleanse a male of AIDS. Although this myth is typically associated with South Africa, it is also prevalent in Asia and the Caribbean. It also has a long history with other venereal diseases in many other parts of the world, including the West.

Trafficking in Women

Trafficking in women can refer to a variety of illegal and profitable recruitments into forced labor, such as prostitution and forced marriage. Human Rights Watch estimates that "anywhere from 700,000 to four million persons are trafficked annually worldwide, and that approximately 50,000 women and children are trafficked annually for sexual exploitation into the United States. Women are particularly vulnerable to this slavery-like practice, due largely to the persistent inequalities they face in status and opportunity worldwide. . . . In all cases, coercive tactics, including deception, fraud, intimidation, isolation, threat and use of physical force, or debt bondage, are used to control women" (HRW 2004a). Southern Asia and eastern Europe, where opportunities for women are limited and their rights often go unenforced, are consistently identified as sources of such trafficking. The destination of such women is typically the more affluent West and Japan. Many NGOs are working to reduce trafficking in women, as are the United Nations and states themselves. For example, the US House of Representatives has conducted trafficking-related hearings, and the UN's Economic and Social Council (ECOSOC) sponsors a trafficking project.

Women, Migration, and Violence

Much of domestic and international migration is becoming feminized, due to the skills that women possess, their lack of opportunity, and their determination to assist their families. A case in point is the migration of Mexican women to the north to work in *maquiladores* (assembly plants) near the southern border of the United States. In the 1990s, there were nearly a million workers, overwhelmingly women, in these factories. In the past decade, an especially tragic series of murders has taken place in Ciudad Juarez, Mexico, just south of El Paso, Texas. Hundreds of bodies of young women have been found in the area, and hundreds more women have disappeared. Mexican state and federal agencies have made little progress in identifying the perpetrators. A recent report by a Mexican federal special prosecutor states that the Mexican public officials involved in investigating these cases are often inept if not complicit (Betancourt 2004).

* * *

Other human rights abuses against women include "honor killings" (the murder of women who disgrace their family due to perceived sexual misconduct), mistreatment of women in state custody (several NGOs have been focusing on the US penal system in particular), and the legal status of women (particularly where they lack the right to inherit property). Such human rights abuses will continue until all societies in the world seek to address the plight of women through education. As one scholar writes: "The history of the drive for women's human rights indicates that only when women are literate, when they can articulate their view of life in publications and before audiences, when they can organize and demand equality, when girls are educated and socialized to think of themselves as citizens as well as wives and mothers, and when men take more responsibility for care of children and the home, can women be full and equal citizens able to enjoy human rights" (Fraser 2001: 58).

Conclusion

Human rights have become, and will hopefully remain, an integral part of the international political landscape. However, human rights advocates are very concerned about the trend that has emerged since September 11, 2001. Although the events of that day presented the possibility of greater international cooperation on human rights, the opposite seems to have happened. The rift between the United States and so many countries of the world, including some close allies, does not bode well for human rights. The current criticism of the United States on human rights issues is directly related to the George W. Bush administration and its "war on terror" policies. The situation in the Middle East, particularly the US invasion of Iraq and uncritical US support of Israel, is central to this trend. Still, it must be kept in mind that the United States was central in the establishment of the UDHR in the 1940s, and its record on human rights has generally been good, especially when the United States is compared to China, Iran and many other third world countries, and the former Soviet Union states. The United States, because of its economic and military power, its history of commitment to democracy and especially the UDHR's first generation of human rights, and more recently its position as the sole remaining superpower, should be held to a higher standard in the domain of human rights. Much remains to be done if greater respect for human rights is to be achieved throughout the world.

Discussion Questions

1. Which generation of human rights do you think is most important?
2. Why are UN-approved human rights often not enforced?
3. What should the international community do to protect the rights of women?
4. Which is more important: state sovereignty or universal human rights?
5. Why does the United States refuse to sign many of the UN conventions on human rights?

Suggested Readings

Agosin, Marjorie, ed. (2001) *Women, Gender, and Human Rights: A Global Perspective.* New Brunswick, NJ: Rutgers University Press.

Amnesty International (annual) *Amnesty International Report.* London.

——— (2004) *It's in Our Hands: Stop Violence Against Women.* New York.

——— (2007) "Amnesty International Press Release 2007" (October 1). Available at http://thereport.amnesty.org/eng/press%20"area/secure/press%20release.

Claude, Richard Pierre, and Burns H. Weston, eds. (1992) *Human Rights in the World Community.* Philadelphia: University of Pennsylvania Press.

Felice, William F. (1996) *Taking Suffering Seriously.* Albany: State University of New York Press.

Friends Committee on National Legislation (2005) "Extraordinary Rendition: Outsourcing Torture" (March 10). Available at http://www.fcnl.org/issues/item_print.php?item_id=1249&issue_id=70.

Gutman, Roy (1993) *A Witness to Genocide.* New York: Macmillan.

Hayner, Priscilla B. (2001) *Unspeakable Truths: Confronting State Terror and Atrocity.* New York: Routledge.

Human Rights Watch (annual) *Human Rights Watch World Report.* New York.

Lagoutte, Stephanie, Hans-Otto Sano, and Peter Scharff Smith, eds. (2007) *Human Rights in Turmoil: Facing Threats, Consolidating Achievements.* Leiden: Koninklijke Brill NV.

Langley, Winston E. (1996) *Encyclopedia of Human Rights Issues Since 1945.* Westport: Greenwood.

O'Byrne, Darren J. (2003) *Human Rights: An Introduction.* London: Pearson Education.

Pollis, Adamantia, and Peter Schwab, eds. (2000) *Human Rights: New Perspectives, New Realities.* Boulder: Lynne Rienner.

Robertson, Geoffrey (2000) *Crimes Against Humanity: The Struggle for Global Justice.* New York: Norton.

Tujan, Antonio, Audrey Gaughran, and Howard Mollett (2004) "Development and the 'Global War on Terror.'" *Race and Class* 46, no. 1.

"£20k Reward Aims to Stop Female Circumcision" (2007) *The Independent* (July 11). Available at http://news.independent.co.uk/uk/crime/article2753969.ece.

"UN High Commissioner for Human Rights Louise Arbour: The US War on Terror Is Constantly Being Used by Other Countries as Justification for Torture and Other Violations of International Human Rights" (2007) *Democracy Now* (September 7). Available at http://www.democracynow.org/article.pl?sid=07/09/07/1349246.

5

Global Security

Sean Kay

WHAT DOES SECURITY MEAN IN TWENTY-FIRST-CENTURY INTERnational relations? Traditionally, students of international security have focused mainly on the relationship between nation-states and power. The dynamics of security in the modern international system, however, have become "global," because the meaning of power has changed. Power still includes that held by nation-states, and great powers remain dominant. However, power also includes soft power, asymmetric power applied both by states and by nonstate actors, information power, and the power of nature. These variations on power create a dynamic relationship between individual, national, and regional security that can impact global security. Through the processes and networks of globalization, security may be increased or decreased. This chapter introduces these new dimensions of power and illustrates them with specific examples. It uses the case of terrorism to illustrate the relationship between globalization and security. The chapter concludes with an assessment of how education can serve as a basis for optimism about the future of global security.

Power and Security in a Global Century

Global security references the integrated relationship between globalization and its impact on individual, national, or regional security. Globalization is not a new phenomenon. What makes the twenty-first-century security situation unique is the expansion and acceleration of relationships across borders, notably through the wide range of transportation, communication, and technology that continuously transits the globe. To frame international security in a global context does not imply that the traditional state quest for "national security" is no longer relevant. Large and small states still maintain substantial military power that could be employed against their adversaries. States are also threatened by the vulnerabilities created through interdependence—but at the same time, states also are presented with new means to exert power in the international system (Keohane and Nye 2001). There are five specific ways through which power has evolved: hard power, soft power, asymmetric power, information power, and the power of nature.

Hard Power

The traditional focus of international security studies targets the relationship between states and "hard" power—those assets that determine the position of a state relative to its peers in the international system, including geography, population, and particularly military capabilities. States will assess their security needs based both on their absolute power, such as the size of their army, navy, or air force, but also on their position relative to other states, such as their ability to transfer economic gains into hard military capacity. Traditionally, security analysis has placed primary emphasis on the relationship between states and power—asserting that all countries seek to maximize their power to advance their primary national interests, especially in terms of survival. In this sense, power is the key means of achieving security, either by seeking to influence others or by adopting a defensive posture that provides for sufficient capability to deter any potential adversary from attacking. Generally, the international system is thought to be anarchic, in that there is no higher form of government than that provided by national governments. Thus states must incorporate worst-case assumptions as they advance along their path toward self-help. The world's long history of failed peace efforts and resultant major wars aptly proves this point. Yet, at the same time, the steadfast quest for peace within the international system also has been a prevalent trend among the nation-states.

If one wants to understand the nature of the international system, one must understand the distribution of power among the most influential states. In the modern era, this means the United States, Russia, China, and the combined member states of the European Union. Some regional powers, including India and Brazil, are also becoming increasingly important. Some states, like Japan, are major economic powers, but do not play a substantial military role. Among the world's major powers in 2007, the United States maintained an active-duty military of 1,473,966 personnel, with an additional 1,290,988 personnel in reserve. Russia's military force comprised 1,037,000 active-duty personnel plus 20 million in reserve. China's active-duty military numbered 2,225,000, with about 800,000 in reserve (IISS 2005).

Large numbers can be misleading if one considers quantity versus quality of military power. For example, China has the largest army in the world—but its ability to project power beyond its borders is highly limited. Russia faces similar constraints. Europe's collective defense spending is about two-thirds that of the United States, but it achieves only one-third the capability. The United States is without peer in defense spending—allocating more money than the next fifteen major military powers combined—and it is particularly capable because of its dramatic technological superiority. Yet its technological advantages can also create disadvantages, as they can make it much more difficult for the United States to undertake coalition warfare alongside countries whose equipment is not compatible and whose pilots are not as well trained. Nor do these advantages mean that the United States is immune to great power challenges. For example, China is well positioned to transfer its economic gains into military gains in coming decades, and Russia has substantial energy reserves. And if Europe can combine its forces, it too will be able to exert more leverage in complementing—or constraining—US power.

The world's great powers share a unique technical advantage over other states in the international system—nuclear weapons, and systems to effectively deliver them. During the Cold War, the United States and Russia learned to live with "mutual assured destruction"—nuclear conflict was deterred because both would be destroyed. Other countries have maintained a minimal deterrent—China, Britain, and France—but their nuclear status is still important to the international hierarchy. China has been modernizing its military and is likely to expand its nuclear arsenal in the coming years. With the nuclear "genie" out of the bottle, other states have attained or are actively seeking to attain nuclear capabilities as well, including India, Pakistan, North Korea, and Iran.

Globalization has created complex political and economic dynamics that can impact traditional security concerns among the major powers. China, for example, needs a strong US dollar to help sell its products in the United States, but the United States is simultaneously highly dependent on Chinese investment in the US economy. This allows the United States to finance its debt, and allows for a strong dollar that is thus available to purchase Chinese goods. On the other hand, many Americans feel vulnerable to the export of labor and manufacturing to China. Meanwhile, the Chinese communist leadership is vulnerable to external pressures for democratic reform and human rights protections. Trade can thus be a great positive force for peace, but can also create major vulnerabilities. As the United States and China increasingly compete over diminishing sources of energy, tensions could escalate—especially if either side feels the other is making economic or security gains at the other's expense. If China develops into a major economic power, as is anticipated by 2050, this would allow China to develop the capability to deploy military forces at substantial distances—at least throughout Asia. In a broader, global context, the complex nature of the US-China relationship has served as a roadblock to international intervention in Darfur, Sudan, in Africa—where strong Chinese business interests have constrained what the West might do there. Meanwhile, around the world, there are many flashpoints where great powers could find themselves in conflict—such as the Korean Peninsula, Taiwan, the Persian Gulf, and Central Asia. Moreover, the danger of nuclear proliferation, particularly in North Korea and Iran, remains high. Thus, while the general climate of relations among great powers is currently peaceful, the potential for major state conflict is still a serious security problem.

Soft Power

The globalization of international security increases the need to work within the realm of "soft" power—the general reputation of a country in the international system, combined with its ability to achieve international outcomes through persuasion rather than military force. A particularly important component of soft power is the capacity to work effectively within multilateral coalitions and inspire other countries to emulate and align willingly with a particular state (Nye 2002).

The United States has a successful tradition of combining hard and soft power. After World War II, the United States built a global security and trading system that was multilateral, with a central emphasis on the role of the United Nations and regional defense arrangements. The end of the Cold War created a dilemma for the United States. It now had much more freedom of movement to advance its hard military power unchecked by the Soviet Union. Yet exercising hard power could result in significant loss of soft power. These soft power sensitivities have played into recent US strategy as well. For example, through fall 2002, the United States succeeded in persuading the United Nations to redeploy weapons inspectors to Iraq. However, when the United States refused to wait for the United Nations to authorize war against Iraq, its soft power status dissolved. The toppling of Saddam Hussein proved relatively easy—but when the postwar peace- and nation-building operations experienced serious security problems, the United States was left providing 90 percent of the troops, suffering 90 percent of the casualties, and paying 90 percent of the costs. Meanwhile, around the world, public opinion surveys showed that the United States had suffered serious blows to its reputation—even among traditional allies like Britain and Canada.

After the end of the Cold War, through its historical commitment to both hard and soft power, the United States found itself in an unprecedented position of global primacy. This commitment was of great benefit to core regions of the world that benefited from the stabilizing presence of the United States, such as Europe, the Middle East, and Asia. While the United States has recently suffered in terms of soft power, it is important to note the scant evidence of hard military efforts by other states to undermine US power. Nonetheless, the globalization of security has created a new opportunity for states to engage in "soft power balancing," which includes a range of actions that do not directly challenge the hard military power of another state, but instead delay, complicate, and increase the costs to that state of using its hard military power (Pape 2005). Through the effective use of international rules, small states can restrain the actions of a much larger power, as happened in the run-up to the invasion of Iraq when small countries used NATO rules to block a request from Turkey for increased territorial defense. NATO members France, Germany, and Belgium argued that the best way to prevent an attack by Iraq on Turkey was to avoid a war altogether. If NATO were to commit to defend Turkey in advance, this would only increase the likelihood of a US invasion of Iraq. Such denials of territorial defense can impose significant costs on the exercise of military power. Indeed, despite offers of $30 billion in cash and credit, Turkey refused to allow the United States to invade Iraq from its territory in the north, significantly complicating the US mission. The costs of the US invasion and occupation of Iraq—over $600 billion at the start of 2008—will eventually take a serious toll on the US economy, since the war has been funded off-budget and without any tax increase. Eventually, these costs will have to be paid, and the resulting economic impact will provide an important measure of the long-term soft power status of the United States.

Asymmetric Power

Globalization accelerates and heightens both the opportunities and the dangers posed by asymmetric exercises of power. Asymmetric power refers to the disproportionate capability of weak states and nonstate actors to challenge powerful states in the international system—often by not "fighting fair," such as through the use of terrorism or other unconventional means.

Through channels of global trade, transportation, technology, and communication, weak states and nonstate actors can do serious damage to powerful states and in the process set the international security agenda. Of course, few power relationships are ever purely "symmetrical," meaning completely equal in terms of power. While states could in the past rely on broad military power to deter threats, the application of asymmetric tactics can circumvent conventional power capabilities by acting outside the assumed norms of expected behavior—as is the case with terrorism.

Asymmetric tactics in warfare can include relatively small military actions that have dramatic strategic consequences. In 1992, for example, the United States deployed a humanitarian mission to feed starving famine victims in Somalia. Local political dynamics eventually led to conflict between US troops and Somali warlords. Somali militias—irregular forces that did not wear official army uniforms–used asymmetric tactics to kill eighteen US soldiers in October 1993. These Somalis then stripped bare the bodies of these US soldiers, desecrated them, and paraded them on worldwide television. The US response was to withdraw from Somalia. The killing and barbarity were sufficient to send the United States, the world's most advanced military superpower, into retreat. The lesson: the way to defeat the United States is to inflict casualties and to incite shock and horror through the networks of global communication. Given the overwhelming military power of the United States, the possibility of any group or state attempting to fight it in any conventional sense seems very unlikely. Terrorism and insurgency, in particular, are two clear examples of how unconventional and barbaric tactics by weak actors can have a major impact on the behavior of powerful states.

Nation-states can also adopt asymmetric tactics to promote their security interests. China has studied a range of ways—including terrorism, drug-trafficking, environmental degradation, and computer viruses—in which its relatively weak military might engage the superior US armed forces in the event of any conflict between the two countries. Chinese strategists have concluded that in a war with the United States, "China needs a new strategy to right the balance of power" (Pomfret 1999). These planners see complexity in warfare as a tactical advantage against US military superiority, and argue that any war with the United States will be fought by China as "unrestricted war," which takes "non-military forms and military forms and creates a war on many fronts" (Pomfret 1999). In turn, China itself is facing such asymmetric tactics from Taiwan, as evident in what some Taiwanese refer to as a possible "scorpion strategy." Rather than waiting for a large-scale Chinese invasion, Taiwan could launch preemptive attacks against Chinese missile sites. Or Taiwan could choose not to engage China's military at all and instead launch its own attacks on vital targets deep inside the mainland. By presenting a credible threat to the mainland, Taiwan might deter a Chinese effort to retake the island by force (Hogg 2004).

Information Power

The networks of globalization have given popular movements, nongovernmental organizations, and individual actors the capacity to engage in international security debates and affect security outcomes—positively and negatively. This capacity is facilitated by the enhanced role of the media and high-technology communications as means of transmitting information, which can translate into agenda-setting power. With such dramatically increased access to information flows, the aphorism that "one person can change the world" is no longer just an ideal. For example, in 1997 the Nobel Peace Prize was awarded to Jody Williams—a woman from Vermont who organized activists across borders, put her fax machine to work, and through tireless efforts mobilized support for an international treaty to ban the production and use of landmines. Other activists have successfully worked to engage the media and highlight ethnic cleansing in the Balkans, humanitarian crises in Haiti, and genocide in Rwanda and Sudan in ways that are hard for states to ignore. On the other hand, the media can also insufficiently play its investigatory role, as appeared to be the case in advance of the 2003 US invasion of Iraq regarding dubious claims about the presence of weapons of mass destruction there.

A growing security concern is the use of traditional weapons in new ways that target vital information systems and civilian infrastructure. The explosion of a nuclear bomb in the atmosphere above the US Midwest, for example, would create an electromagnetic pulse with the capacity to significantly disrupt power grids as well as civilian telecommunications in exposed areas, affecting computer systems, banks, food delivery, traffic lights, and the like. Similarly, countries that are highly reliant on computer information technology are also vulnerable to cyber attack. According to the US Department of Defense, in 2001 it faced about 40,000 Internet attacks, and in 2000 about 715 of such attacks successfully achieved some degree of access to Department of Defense systems (Gansler 2002). Overreliance on technology can create strategic and tactical problems on the battlefield. While the United States had complete control of the information battlespace in the 2003 invasion of Iraq, it inaccurately judged the country's possession of weapons of mass destruction, and failed to anticipate the insurgency that grew out of the invasion. Finally, government control of information technology also raises questions about individual security and privacy. In the United States, for example, the National Geospatial-Intelligence Agency has assembled visual information on more than 130 urban areas (Sharder 2005). Additionally in the United States, there has been considerable controversy over whether the government should be allowed to wiretap phones as part of its counterterrorism efforts.

Information infrastructure has become an important measure of power in the twenty-first century. Developing countries like China and India are making significant gains in information support networks, which give them increased international leverage in terms of their soft power. India is attracting international investment due to relatively low-cost information support systems, and is expanding its infrastructure investment into the Middle East. China is expanding its telecommunications infrastructure into emerging economies in Asia, Latin America, and Africa. Europe, meanwhile, is building the first major satellite navigation network, Galileo, which will soon be competing with the GPS systems developed in the United States. Information power can shift rapidly in the modern international system. For example, in 2002, 85 percent of the undersea communications infrastructure was owned by US companies; by 2004, China had become the major owner, having bought control of several major US telecommunications corporations (Nonow 2004). Such trends are vital in the context of global security, as these infrastructure tools can be used to enhance national power, deny other states access to international power, and engage in new kinds of nontraditional asymmetric warfare.

The Power of Nature

As Earth's population grows, basic human necessities are placing extraordinary stresses on the relationship between humans and their environment. In some key areas of the world, the demographics within certain populations are presenting increasingly acute problems. In the Middle East, for example, almost 39 percent of the population is under the age of fifteen, and 50 percent is under the age of twenty-five. In the Arab world, 62 million people live on less than one $1 a day, and 145 million live on less than $2 a day (UNFPA 2004). Worldwide, on average, a child dies of malnutrition every five seconds. There are over 800 million people who are undernourished, of which 170 million are children. In addition, malnutrition weakens the body's immune systems, making societies vulnerable to famine and disease. In the underdeveloped world, this dilemma is made worse by the lack of adequate medical services. There is a significant disparity between the developed world, where people consume far more calories than they need, and the underdeveloped world, where a billion people live on a dollar a day and worry where their next meal will come from. The disparity of income and food between the developed world and the underdeveloped world is illustrated by the $17 billion spent in the United States and Europe each year on pet food, versus the $19 billion needed in annual investment to reach the UN's development goals for elimination of hunger and malnutrition (Assadourian 2004).

Because of globalization, the potential for a worldwide outbreak of disease is growing. Left unchecked, tuberculosis is predicted to kill 35 million people by 2024. Currently, 1 million people die every year of malaria—of which 700,000 are children (Mitchell 2004). Meanwhile, HIV/AIDS ravages sub-Saharan Africa, which comprises about 10 percent of the world's population, but two-thirds of all people infected with HIV. By 2003, it was estimated that 3 million people in sub-Saharan Africa had been infected with HIV, and that 2.2 million had died of AIDS—representing 75 percent of the 3 million total global AIDS deaths that year (UNAIDS 2004). By

2000, the United States had identified AIDS as a national security problem—noting that the spread of AIDS can weaken foreign governments, contribute to ethnic conflict, and undermine efforts to expand international trade. According to the US National Intelligence Council, "At least some of the hardest-hit countries, initially in sub-Saharan Africa and later in other regions, will face a demographic catastrophe" by 2020. This would produce a strong risk for "revolutionary wars, ethnic wars, genocides, and disruptive regime transitions" in the developing world (Gellman 2000). While the situation in Africa is a humanitarian catastrophe, even more challenging strategically are worst-case projections that, by 2025, AIDS deaths in Russia, China, and India could reach 155 million (Eberstadt 2002). This level of potential instability in three of the world's most important economies, all of which are nuclear powers, would have a major impact on global security.

Relative supplies of fresh water, and competition for access to and control over them, may also impact global security, as water has the potential to rival oil as the most important global commodity. The United Nations Environment Programme predicts that by 2015, 40 percent of the world's population will have insufficient access to safe drinking water (Vergano 2003). Already, 1 billion people do not have access to clean water, and 40 percent of the world's people do not have access to basic sanitation. According to a study by UNICEF (2005), some 4,000 children die every day from illnesses caused by lack of clean water. Already, agricultural demand accounts for 7 percent of all water use worldwide, while industrial demand accounts for 22 percent. People in rich countries use ten times more water than those in underdeveloped countries (Kirby 2003). A key challenge for states will be how to share water when confronted with a dwindling supply. Though there are many transboundary freshwater basins in the world, and though conflicts over them have historically been resolved through diplomacy, access to water in the future may become a military affair, particularly in arid areas already ridden by conflicts over other matters, like the Middle East and Central Asia.

Meanwhile, the entire planet, according to a strong consensus among scientists, is confronting a significant challenge from the consequences of global climate change. While some debate remains about the pace of global warming, the scientific community is in agreement that human behavior is a significant source of this warming. Unchecked global warming over an extended period could pose one of the most serious threats to international security that humanity has ever experienced. Already, sea ice and glaciers are melting, the average global sea level is rising, and ocean temperatures are increasing. As well, the planet will likely see an increase in average precipitation over the middle and high latitudes of the Northern Hemisphere, and over tropical land areas, in addition to an increasing frequency of extreme precipitation in other regions of the world (UCS 2004). The US Department of Defense, in studying the worst-case scenario for abrupt climate change, has concluded that the end result would be a "significant drop in the human carrying capacity of the Earth's environment." The authors of the Defense Department study conclude that "technological progress and market forces, which have long helped boost Earth's carrying capacity, can do little to offset the crisis—it is too widespread and unfolds too fast. [Eventually] an ancient pattern reemerges: the eruption of desperate, all-out wars over food, water, and energy supplies." War itself may "define human life" (Stipp 2004).

Global Security in Action: Terrorism

Transnational terrorism is an international security problem that combines all elements of global security. The attacks of September 11, 2001, were a major example of the complex relationship between power and security. Even the general response was confusing: a "war on terror" was declared. But if terrorism is a tactic, not a strategy, then how can this war ever be won? Because wars are won by defeating an enemy's strategic objectives and tactics, and because of the diverse nature of terrorism as a threat, achieving victory in a global war on terror would require application of a wide range of tools—including military tools but also intelligence, diplomatic, economic, and police tools. States are important in this context, as terrorists often employ asymmetric tactics to circumvent traditional power advantages. But also, some states have historically supported terrorism—through proxy networks (informally linked groups) that again use asymmetric tactics to achieve their objectives. Soft power is relevant because winning a war on terror involves more than military power. States might respond to terrorism in military ways that can undermine their soft power legitimacy, at the risk of creating more terrorists in the process. By definition, terrorism is asymmetric and also takes advantage of nonstate actors' access to information power. Energy, in the context of environmental security, is central to the Middle East, and may become a highly valuable target for terrorists.

Terrorism is a serious international problem, though the phenomenon itself is not new. Some of the more "traditional" terrorist groups, such as the Irish Republican Army and the Palestine Liberation Organization, have been persuaded through diplomacy to renounce terrorism. What is new is that terrorism has become globalized. Terrorist groups are increasingly able to use the international financial system to move and hide money, and to use the Internet to transmit messages via encoded communications (Williams 2003). Dispersed ethnic groups can provide local cover and recruitment for terrorist operatives. Moreover, leaders of terrorist groups often rail against the perceived negative effects of globalization (and especially US and European dominance in the globalized world) in order to recruit new members.

Religion plays an increasingly important, inspirational role for terrorists. In 1968, none of the international terrorist organizations identified in a study by the Rand Corporation were considered religiously motivated. By 1994, one-third of forty-nine international terrorist groups were classified as primarily religious (Hoffman 1997). Religion is especially important for recruitment, but not necessarily an end in and of itself. One comprehensive study shows that the main group responsible for the most suicide attacks globally is not a religious movement, but rather an ideological (Marxist-Leninist) movement—the Liberation Tigers of Tamil Eelam (LTTE) in Sri Lanka. Of 186 suicide terrorist attacks worldwide between 1980 and 2001, the LTTE accounted for 75 (Pape 2003). Nonetheless, the possibility that an extreme religious movement might utilize the networks of globalization to attain weapons of mass destruction, while improbable, is a very dangerous worst-case scenario.

Since September 11, 2001, the Al-Qaida terrorist network has shown resilience, evolving from an elaborate hierarchical organization based in Afghanistan, into a looser linkage of groups based in Egypt, Libya, Algeria, Saudi Arabia, Oman, Tunisia, Jordan, Iraq, Lebanon, Morocco, Somalia, and Eritrea (Frantz et al. 2004). Meanwhile, most of the money used to support Al-Qaida's operations has been raised through legal, global networks. Showing the nature of high-profile asymmetric tactics, Al-Qaida's 2001 attack against the United States, which killed 3,000 innocent civilians, was estimated by the 9/11 Commission to have cost $500,000 to carry out. The much lower-profile but still deadly Al-Qaida bombings in Morocco in May 2003, which killed forty-five, cost only about $4,000 to carry out. Meanwhile, the Internet has also become a vital tactical communications and propaganda tool for international terrorist movements. One observer of the fall 2001 US war in Afghanistan against Al-Qaida watched "every second al Qaeda member carrying a laptop computer along with a Kalashnikov" as they dispersed around the globe to flee the US military assault (Coll and Glassner 2005). The Internet has been used by Al-Qaida both to disseminate propaganda and to coordinate training and attacks. The central challenge of this and other modern technology is that, unlike a traditionally organized conventional army, there is no clear target to deter or retaliate against. Moreover, the Internet can also be used to spread false information that elevates the level of fear in a society. Producing strategic "chatter" for national intelligence agencies to collect can do psychological and economic damage to a country without the need to undertake any actual attack. When the United States elevates its terrorist warning alert level from "yellow" to "orange," it costs the US economy, on average, $1 billion a week. Meanwhile, mistakes in warfare can proliferate into massive strategic blunders—such as with the worldwide distribution of photos of US soldiers grossly mistreating Iraq prisoners at Abu Ghraib prison.

Terrorism is tragically effective given its disproportionate effects, often at the global level. Following the horrific attacks against the United States on September 11, 2001, terrorism was understandably elevated to the number-one international issue in world politics. However, fear often trumps facts and reality when discussing the issue of terrorism. This is unfortunate, because the actual risk that anyone will be involved in a terrorist incident is minimal to nonexistent. During the entire twentieth century, fewer than twenty terrorist attacks killed more than 100 people, and no single attack killed more than 400 people. Excepting the one-day horror of September 11, more Americans are killed by lighting strikes in any given grouping of years than by all forms of international terrorism combined (Mueller 2002). Since fall 2001, no Americans have been killed by foreign terrorist attacks on US soil.

Despite such relatively reassuring numbers, the globalization of terrorism has raised the stakes—there is the danger that terrorists might one day acquire weapons of mass destruction, either manufactured, such as nuclear weapons, or natural, such as deadly disease. Built-in safeguards in US and Russian nuclear weapons present significant complications to untrained use if stolen. However, the black market and the weak security systems adopted by new nuclear powers, such as Pakistan, could pose a serious risk of proliferation of components that could be assembled into a nuclear device. Indeed, there are at least 200 locations worldwide where terrorists could acquire a nuclear weapon or the fissile material to make one (Allison 2004). Short of a nuclear explosion, there is greater risk of the use of a conventional explosive with a radioactive material—in the form of a so-called dirty bomb, which would not kill many people, but could cause serious economic damage and general panic. The use of biological agents could pose an additional mass destruction scenario—though such efforts have to date proven limited, as in the fall 2001 anthrax attacks in the United States. Biological agents such as viruses are especially problematic, because many occur naturally and are hard to detect until someone has been infected. The 2001 anthrax attacks in the United States killed very few (five, out of eighteen infected), but resulted in the shutdown of parts of the US Congress and Supreme Court, preventive antibiotic treatment for 33,000 people, and $3 billion in damage to the US postal service.

A nightmare nuclear or biological scenario would be a massive blow to any targeted country—and would have global economic ripple effects. More likely would be a smaller detonation, with the device itself, or parts of a device, being smuggled into a country in a shipping container and exploded in a port or a major urban area. This threat would be difficult to detect in a country like the United States, with its daily traffic of about 300,000 trucks, 6,500 rail cars, and 140 ships delivering some 50,000 containers holding more than 500,000 items from around the world. Another possibility is an attack on a nuclear power plant—though this too would be difficult to achieve and also depend on terrorist target priorities. Indeed, on September 11, 2001, Al-Qaida could have flown its hijacked airplanes into any number of nuclear power stations in the northeastern United States, but chose not to. Even the more likely dirty-bomb scenario poses relatively minimal physical risk. Such devices are not easy to make. Key ingredients, such as cesium 137, cobalt 60, and radioisotopes, are hard to obtain—and anyone constructing such a device would likely become very ill with radiation poisoning before they could deliver it. Second, the effects of radiation dispersal are not necessarily catastrophic. Humans live every day with tolerable levels of radiation simply by traveling on jets or walking in the sun. Just four medical CAT scans create an increased cancer risk of 4 in 1,000, which is greater than the risk for most functional dirty-bomb scenarios. A terrorist would likely do far more human and economic damage by using some sort of air or ground assault on a petrochemical facility, or by stealing and exploding a truck full of chlorine gas in a populated area. Car bombs, assassinations, and suicide bombers remain the most likely threats, and can be further exploited using the means of modern technology and the networks of global security. Indeed, the bombs exploded in London in July 2005 were built from materials that anyone could obtain at local stores.

Ultimately, terrorism will be a reality of global security for the extended future. The challenge for societies will be to identify vulnerabilities without compromising basic values and freedom. Open democracies are especially vulnerable to major terrorist shocks. In the United States, for example, there are 600,000 bridges to protect, and 14,000 small airports that could be used by terrorists. There are 4 million miles of paved roadways, 95,000 miles of coastline, and 361 ports to be policed, in addition to an open 4,000-mile border with Canada. There are some 260,000 natural gas wells and 1.3 million miles of pipeline that could be attacked. In New York, 1.2 billion people ride the subways every year, and more than 77 million people use its three airports annually. Meanwhile, though the US federal government has hardened the defense of obvious, high-profile targets, the private sector remains highly vulnerable to attacks designed to generate mass fear—targets such as movie theaters, shopping malls, hotels, sporting events, concerts, and schools (Brzezinski 2004). Ultimately, there is no such thing as perfect security—but an understanding of the relative risks does help to reassure societies as they establish appropriate defensive measures. Indeed, immediately after September 11, the most important, and least expensive, counterterrorism measure was accomplished easily—locking the doors to the cockpits on passenger airlines.

Education for Global Security

There are times when the scope of challenges to global security can seem overwhelming. Indeed, the very thought that about 10 million children die every year from preventable and curable diseases—simply because of where they were born—challenges even the most hardened of sensibilities. Yet there is also good reason for optimism about general security trends in the twenty-first century. Hundreds of millions of lives have been bettered by the collapse of the Soviet Union, and a billion people in India and China are on the verge of benefiting from tremendous opportunities to engage in the world economy. While terrorism is a major international problem, there are other dangerous threats in the world. Some of them have been better managed—as demonstrated by the dramatic reduction in the risk of a thermonuclear exchange between the United States and the Soviet Union since the Cold War. Thus, in many ways, people today are more secure than ever before. However, this reality is often not felt—serious security challenges continue to generate fear and uncertainty about the future, eroding optimism and hope.

There is no magic bullet that can resolve the various and complex dilemmas inherent in global security—but the one asset that is essential to any such effort is education. Education can defeat ignorance, which breeds fear, and thus facilitates a realistic perspective about various threats, and creative problem solving to address them. Particularly when confronting a problem like international terrorism, defeating fear is one of the greatest strategic assets available. Ironically, Americans are much more afraid of terrorism today than they were before September 11, 2001, even though they are now actually safer—and as a result mainly of common sense (like locking airliner cockpit doors), not massive technological innovations. Moreover, when one gains a deeper understanding of global security concepts, one can better judge what really to worry about. Approximately 8,000 Africans die every day from AIDS, but this rarely receives sustained media attention. Terrorism, conversely, kills on average several hundred people a year. If one were to rank threats in terms of the capacity to affect the lives of tens of millions of people in the world, first-order priorities would be the risk of a war between India and Pakistan, conflict in Asia over Taiwan or the Korean Peninsula, transnational disease, and worst-case environmental scenarios. Terrorism does require very high-level priority, of course, given the deadly mix of religious fanaticism, weapons of mass destruction, and the proven willingness of terrorists to conduct extreme and barbaric attacks against civilian populations.

Because terrorism involving weapons of mass destruction is a significant concern, facts also help to inform judgments about how to effectively address the problem with policy. For example, the United States was spending more money every three days in Iraq by 2005 than it had in the previous three years combined to secure its own 361 seaports. These ports are the primary means through which a terrorist group might bring a nuclear weapon into the United States (Flynn 2004). By 2007, the United States had committed over $600 billion to Iraq, a relatively prosperous country sitting on one of the world's largest oil reserves—and a country that had no weapons of mass destruction, and no link to terrorism before the United States invaded. Ironically, the United States could have opted to solve the problem of insecure nuclear weapons in the former Soviet Union, and at a fraction of the Iraq cost—$30 billion spread out over ten years (Pincus 2001). Meanwhile, if the United States is to win a war on terror, it will also need to enhance its soft power appeal. For example, in 2004, the Global Campaign for Education advocated an action plan to ensure access to primary school for every child in the world by 2015, at a projected cost of about $80 billion—a comparatively small cost for helping to educate an entire generation of young people, but the United States declined the opportunity.

Countries that invest strategically in educating their own citizenry are likely to gain significant advantages in managing global security challenges. Physics, biology, geology, medicine, communications, language, math, and information technology are all areas where expertise and applied skills are essential for national security in the twenty-first century. This also means that almost anyone who wants to can make major contributions to their country or region, or even to the world, toward promoting peace and security. The world caught a clear glimpse of what can go wrong on September 11, 2001—and has witnessed many ongoing crises since. However, a close look shows that an educated individual can make a positive difference. The challenge to the next generation of leadership is to utilize those opportunities for good, and thus to ensure that the way the twenty-first century began actually reflected the last shot of a dying age of terror. While global security challenges can seem insurmountable, the opportunity to make a difference is there, whether for the student, soldier, diplomat, intelligence analyst, peace-corps volunteer, customs worker, firefighter, police officer, teacher, or political leader. The globalization of security allows opportunities for every individual to prove to those who would use the networks of global power for evil that they will be outmatched by someone advancing the search for peace.

Discussion Questions

1. How is international security traditionally understood? What do you think "security" means?

2. Is globalization a positive or negative development in terms of the quest for worldwide peace and security?

3. How should the United States harness its "hard" and "soft" capacities to best assert itself in contemporary global security issues?

4. How should the great powers of the world organize to best meet the challenges posed by asymmetric security threats?

5. With such unprecedented access to information, one person truly can change the world. How would you like to change the world, and how would you go about it?

6. To what extent should environmental problems like global warming be seen as security problems? How should states organize to best meet the challenges emerging from civilization's interaction with nature?

7. Do you think it is possible to "win" a war on terror? What tools should be used to best meet this threat?

Note

This chapter draws on and updates key material from the author's previous work. See Kay 2006.

Suggested Readings

Allison, Graham (2004) *Nuclear Terror: The Ultimate Preventable Catastrophe.* New York: Times Books.

Buzan, Barry (1991) *People, States, and Fear: An Agenda for Security Studies in the Post–Cold War Era.* 2nd ed. Boulder: Lynne Rienner.

Hoffman, Bruce (2006) *Inside Terrorism.* New York: Columbia University Press.

Homer-Dixon, Thomas (2006) *The Upside of Down: Catastrophe, Creativity, and the Renewal of Civilization.* New York: Island.

Kay, Sean (2007) *Global Security in the Twenty-First Century: The Quest for Power and the Search for Peace.* Lanham: Rowman and Littlefield.

Keohane, Robert O., and Joseph S. Nye (2001) *Power and Interdependence.* 3rd ed. New York: Longman.

Kupchan, Charles (2003) *The End of the American Era: US Foreign Policy and the Geopolitics of the Twenty-First Century.* New York: Vintage.

Mearsheimer, John J. (2003) *The Tragedy of Great Power Politics.* New York: Norton.

Nye, Joseph S. (2004) *Soft Power: The Means to Success in World Politics.* New York: Public Affairs.

Pape, Robert A. (2005) *Dying to Win: The Strategic Logic of Suicide Terrorism.* New York: Random House.

Power, Samantha (2003) *A Problem from Hell: America and the Age of Genocide* New York: Perennial.

Slaughter, Anne-Marie (2004) *A New World Order.* Princeton: Princeton University Press.

PART 2

The Global Economy

6

Free Trade vs. Protectionism: Values and Controversies

Bruce E. Moon

INTERNATIONAL TRADE IS OFTEN TREATED PURELY AS AN ECOnomic matter that can and should be divorced from politics. This is a mistake, because trade not only shapes our economy but also determines the kind of world in which we live. The far-reaching consequences of trade pose fundamental choices for all of us. Citizens must understand those consequences before judging the inherently controversial issues that arise over trade policy. More than that, we cannot even make sound consumer decisions without weighing carefully the consequences of our own behavior.

The Case for Trade

The individual motives that generate international trade are familiar. Consumers seek to buy foreign products that are better or cheaper than domestic ones in order to improve their material standard of living. Producers sell their products abroad to increase their profit and wealth.

Most policymakers believe that governments should also welcome trade because it provides benefits for the nation and the global economy as well as for the individual. Exports produce jobs for workers, profits for corporations, and revenues that can be used to purchase imports. Imports increase the welfare (well-being) of citizens because people can acquire more for their money as well as obtain products that are not available from domestic sources. The stronger economy that follows can fuel increasing power and prestige for the nation as a whole. Further, the resultant interdependence and shared prosperity among countries may strengthen global cooperation and maintain international peace.

Considerable historical evidence supports the view that trade improves productivity, consumption, and therefore material standard of living (Moon 1998). Trade successes have generated spurts of national growth, most notably in East Asia. The global economy has grown most rapidly during periods of trade expansion, especially after World War II, and has slowed when trade levels have fallen, especially during the Great Depression of the 1930s. Periods of international peace have also coincided with trade-induced growth, while war has followed declines in trade and prosperity.

However, more recent evidence casts doubt on whether trade has had such positive effects during the era of globalization, especially among poor countries. Globally, trade as a percentage of gross domestic product (GDP; the total goods and services produced in a given year) has more than doubled since 1960, and average incomes are now nearly 2.5 times larger. Yet the groups of countries that have lagged behind over that period began as the most active traders in the world. Africa's trade levels were more than twice the global average in 1960 and remain well above the average now, but for nearly half a century, per capita income has grown only about 0.5 percent per year, and it was lower in 2005 than in 1975. Among the heavily indebted poor countries, whose trade levels are also very high, average income today is lower than it was in 1965. A World Bank report indicating that the median per capita growth of developing countries was 0.0 percent between 1980 and 1998 also suggests that trade expansion may be considerably more beneficial for developed nations than for poorer ones (Easterly 1999). Another report establishes that inequality has grown both within and between societies, estimating that since 1980 the number of people living on an income of less than $2 a day has grown from 2.4 billion to 2.7 billion, a figure that represents more than 45 percent of humankind (Chen and Ravallion 2004).

Still, the private benefits of trade have led individual consumers and producers to embrace it with zeal for the past half century. As a result, trade has assumed a much greater role in almost all nations, with exports now constituting about a quarter of the economy in most countries and over half in many (World Bank 2002). Even in the United States, which is less reliant on trade than virtually any other economy in the world because of its size and diversity, the export sector is now about 10 percent of GDP.

Since World War II, most governments have encouraged and promoted this growth in trade levels, though they have also restrained and regulated it in a variety of ways. All but a handful of nations now rely so heavily on jobs in the export sector, and on foreign products to meet domestic needs, that discontinuing trade is no longer an option. To attempt it would require a vast restructuring that would entail huge economic losses and massive social change. Furthermore, according to the "liberal" trade theory accepted by most economists, governments have no compelling reason to interfere with the private markets that achieve such benefits. The reader is cautioned that the term *liberalism,* as used throughout this chapter, refers to liberal economic theory that opposes government interference with the market, and is not to be confused with the ambiguous way the term *liberal* is applied in US politics, where it often means the opposite.

From its roots in the work of Scottish political economist Adam Smith (1723–1790) and English economist David Ricardo (1772–1823), this liberal perspective has emphasized that international trade can benefit all nations simultaneously, without requiring governmental involvement (Smith 1910). According to Ricardo's theory of *comparative advantage* (1981), no nation need lose in order for another to win, because trade allows total global production to rise. The key to creating these gains from trade is the efficient allocation of resources, whereby each nation specializes in the production of goods in which it has a comparative advantage. For example, a nation with especially fertile farmland and a favorable climate can produce food much more cheaply than a country that lacks this comparative advantage. If it were to trade its excess food production to a nation with efficient manufacturing facilities for clothing production, both nations would be better off, because trade allows each to apply its resources to their most efficient use. No action by governments is required to bring about this trade, however, since profit-motivated investors will see to it that producers specialize in the goods in which they have a comparative advantage, and consumers will naturally purchase the best or cheapest products. Thus, liberal theory concludes that international trade conducted by private actors free of government control will maximize global welfare.

Challenges to the Liberal Faith in Trade

Though trade levels have grown massively in the two centuries since Adam Smith presented his theories on political economy, no government has followed the advice of liberal economic theorists to refrain from interfering with trade altogether. That is because governments also have been influenced by a dissenting body of thought known as *mercantilism,* which originated with the trade policy of European nations, especially England, from the sixteenth century to the middle of the nineteenth.

While mercantilists do not oppose trade, they *do* hold that governments must regulate it in order for trade to advance various aspects of the national interest. The aspirations of mercantilists go beyond the immediate consumption gains emphasized by liberals, to include long-term growth, national self-sufficiency, the vitality of key industries, and a powerful state in foreign policy. Because most states accept the mercantilist conviction that trade has negative as well as positive consequences, they try to manage it in a fashion that will minimize its most severe costs yet also capture the benefits claimed for it by liberal theory. It is a fine line to walk.

In particular, mercantilists observe that the rosy evaluation of trade advanced by Smith and Ricardo was predicated on their expectation that any given nation's imports would more or less balance its exports. However, when a nation's imports are greater than its exports—meaning that it buys more from other nations than it sells to them—mercantilists warn that this "trade deficit" carries with it potential dangers that may not be readily apparent. On its face, a trade deficit appears as the proverbial free lunch: if a nation's imports are greater than its exports, it follows that national consumption must exceed its production. One might ask how anyone could object to an arrangement that allows a nation to consume more than it produces. The answer lies in recognizing that such a situation has negative consequences in the present and, especially, dangerously adverse repercussions in the future.

For example, the United States has run a substantial trade deficit for three decades, with imports surpassing exports by more than $5 trillion over that period, including about $760 billion in 2006 alone. That trade deficit allowed US citizens to enjoy a standard of living more than $2,500 per person higher than would otherwise be possible. But mercantilists observe that these excess imports permit foreigners to obtain employment and profits from production that might otherwise benefit US citizens. For example, since the US trade deficit began to bloom in the 1970s, the massive sales of Japanese cars in the United States have transferred millions of jobs out of the US economy, accounting for high levels of unemployment in Detroit and low levels of unemployment in Tokyo. Corporate profits and government tax revenues also accrue abroad rather than at home.

However, the longer-term impact of trade deficits produces even greater anxiety. Simply put, trade deficits generate a form of indebtedness. Just as individuals cannot continue to spend more than they earn without eventually suffering detrimental consequences, the liabilities created by trade deficits threaten a nation's future. Unfortunately, the consequences of trade imbalances cannot be evaluated easily, because they trigger complex and unpredictable flows of money, including some that occur years after the trade deficit itself.

To understand this point, consider that the trade deficit of the United States means that more money flows out of the US economy, in the form of dollars to pay for imports, than flows back into the US economy, through payments for US goods purchased by foreigners. The consequences of the trade deficit depend in large part on what happens to those excess dollars, which would appear to be piling up abroad.

In fact, some of these dollars are, literally, piling up abroad. About $360 billion in US currency is held outside the United States. However, this cash held abroad is a mere drop in the bucket, less than 7 percent of the $5.4 trillion that has flowed out of the United States to pay for the excess of imports over exports since 1985. That year—the last time that Americans owned more assets abroad than foreigners owned of US assets—marked the transformation of the United States from a net creditor nation to a debtor nation. About 93 percent of that $5.4 trillion has already found its way back into the US economy as loans to Americans and purchases of US financial assets. For example, the US Treasury has borrowed $600 billion from foreign citizens and more than $1.5 trillion from foreign governments by selling them US Treasury bonds. Not only must this debt be repaid someday, but foreigners now receive more than $130 billion in interest payments annually from the US federal government. US businesses owe foreigners another $2.7 trillion as the result of the sale of corporate bonds. About $2.5 trillion in stocks—about 10 percent of outstanding US equities—are owned by foreigners. Such capital flows can offset a trade deficit temporarily and render it harmless in the short run, but they create future liabilities that only postpone the inevitable need to balance production and consumption. The United States is being sold to foreigners piece by piece to finance a trade deficit that continues to grow.

Economists disagree about whether these developments ought to raise alarm. After all, the willingness of foreigners to invest in the United States and to lend money to Americans surely is an indication of confidence in the strength of the US economy. More generally, as Chapter 7 shows, capital flows can be beneficial to the economy and its future. Indeed, foreign capital is an essential ingredient to development in many third world countries. Whether capital inflows produce effects that are, on balance, positive or negative, depends heavily on the source of the capital, the terms on which it is acquired, the uses to which it is put, and the unpredictable future behavior of foreign lenders and investors.

For example, should foreigners decide to use their $360 billion worth of holdings in US currency to purchase US goods, the result could be catastrophic: the increased demand for US products would bid up prices and unleash massive inflation. Alternatively, should they try to exchange those dollars for other currencies, the increased volume of dollars available in currency markets would constitute excess supply that could trigger a violent collapse of the external value of the dollar. If owners of US Treasury certificates—including Japan ($600 billion) and China ($400 billion)—sell their dollar-denominated holdings and invest in euro- or yen-denominated assets, US interest rates would rise and the dollar would plummet. In July 2007, Chinese officials threatened to use their dollar-denominated holdings, estimated at $900 billion in various forms, to achieve political leverage in negotiations with the United States over other matters.

Even if none of these scenarios occur suddenly—as they did in

Mexico, Argentina, Russia, Thailand, Malaysia, Indonesia, and Korea in recent years—over time the excess supply of dollars is bound to erode the value of the dollar more gradually. No one can predict the timing or severity of this decline, but it has been long under way already: the dollar was once equivalent to 360 Japanese yen, but traded at just over 100 yen in early 2005. The dollar has declined by nearly half just since 2000, from 1.18 euros to under 0.65 euros by 2008. As the purchasing power of the dollar continues to decline, the prices paid by Americans for foreign products, services, and investment assets increase. The net worth of Americans—their national wealth—declines.

Thus a trade deficit provides immediate benefits but also risks reducing the standard of living for future generations. Americans who have grown accustomed to consuming far more than they produce will be forced to consume far less. Because these consequences are uncertain, nations vary somewhat in their tolerance for trade deficits, but most try to minimize or avoid them altogether, as counseled by mercantilists.

Options in Trade Policy

To achieve their desired trade balance, nations often combine two mercantilist approaches. They may emphasize the expansion of exports through a strategy known as *industrial policy*. More commonly, they emphasize minimizing imports, a stance known generally as *protectionism* (Fallows 1993).

Protectionist policies include many forms of import restriction designed to limit the purchase of goods from abroad. All allow domestic import-competing industries to capture a larger share of the market and, in the process, to earn higher profits and to employ more workers at higher wages. The most traditional barriers are taxes on imports called *tariffs,* or import duties, but they are no longer the main form of protectionism in most countries.

In fact, after a long decline from their peak in the 1930s, tariff levels throughout the world are now generally very low. In the United States, the average tariff rate reached a modern high of 59 percent in 1932 under what has been called "a remarkably irresponsible tariff law," the Smoot-Hawley Act, which has been widely credited with triggering a spiral of restrictions by other nations that helped plunge the global economy into the Great Depression of the 1930s. The average rate in the United States was reduced to 25 percent after World War II and declined to about 2 percent after the Uruguay Round of trade negotiations (discussed in greater detail later in the chapter) concluded in 1994. Most other countries have followed suit—and some have reduced rates even further—so that average rates above 10 percent are now quite rare.

However, in place of tariffs, governments have responded to the pleas of industries threatened by foreign competition with a variety of nontariff barriers (NTBs), especially voluntary export restraints (VERs). In the most famous case of VERs, Japanese automakers "voluntarily" agreed to limit exports to the United States in 1981 (had Japan refused, a quota that would have been more damaging to Japanese automakers would have been imposed).

A favorable trade balance also can be sought through an industrial policy that promotes exports. The simplest technique is a *direct export subsidy,* in which the government pays a domestic firm for each good exported, so that it can compete with foreign firms that otherwise would have a cost advantage. Such a policy has at least three motivations. First, by increasing production in the chosen industry, it reduces the unemployment rate. Second, by enabling firms to gain a greater share of foreign markets, it gives them greater leverage to increase prices (and profits) in the future. Third, increasing exports will improve the balance of trade and avoid the problems of trade deficits.

Liberals are by no means indifferent to the dangers of trade deficits, but they argue that most mercantilist cures are worse than the disease. When mercantilist policies affect prices, they automatically create winners and losers and in the process engender political controversies. For example, to raise the revenue to pay for a subsidy, the domestic consumer has to pay higher taxes. As noted above, protectionism also harms the consumer by raising prices even while it benefits domestic firms that compete against imports.

The Multiple Consequences of Trade

As nations choose among policy options, they must acknowledge liberal theory's contention that free trade allows the market to efficiently allocate resources and thus to maximize global and national consumption. Nonetheless, governments almost universally restrict trade, at least to some degree. That is because governments seek many other outcomes from trade as well—full employment, long-term growth, economic stability, social harmony, power, security, and friendly foreign relations—yet discover that these desirable outcomes are frequently incompatible with one another. Because free trade may achieve some goals but undermine others, governments that fail to heed the advice of economic theory need not be judged ignorant or corrupt. Instead, they recognize a governmental responsibility to cope with all of trade's consequences, not only those addressed by liberal trade theory. For example, while trade affects the prices of individual products, global markets also influence which individuals and nations accumulate wealth and political power. Trade determines who will be employed and at what wage. It determines what natural resources will be used and at what environmental cost. It shapes opportunities and constraints in foreign policy.

Because trade affects such a broad range of social outcomes, conflict among alternative goals and values is inevitable. As a result, both individuals and governments must face dilemmas that involve the multiple consequences of trade, the multiple goals of national policy, and the multiple values that compete for dominance in shaping behavior (Moon 2000).

The Distributional Effects of Trade: Who Wins, Who Loses?

Many of these dilemmas stem from the sizable effect that international trade has on the distribution of income and wealth among individuals, groups, and nations. Simply put, some gain material benefits from trade while others lose. Thus, to choose one trade policy and reject others is simultaneously a choice of one income distribution over another. As a result, trade is inevitably politicized: each group pressures its government to adopt a trade policy from which it expects to benefit.

The most visible distributional effects occur because trade policy often protects or promotes one industry or sector of the economy at the expense of others. For example, in response to pleas from the US steel industry, President George W. Bush imposed a temporary 30 percent tariff on various types of imported steel in March 2002. Because the import tax effectively added 30 percent to the price of steel imports, the US steel industry could benefit from this protection against foreign competition by increasing its share of the market, by raising its own prices, or by some combination of the two. A larger market share and/or higher prices would certainly increase the profits of US steel firms, which would benefit steel executives and stockholders, and perhaps permit higher levels of employment and wage rates, which would benefit steel workers. Steel producers argued that the respite from foreign competition brought idled mills back on line and kept teetering plants from shutting down, resurrecting 16,000 steel jobs.

Distributional effects are often regional as well as sectoral. The entire economy of steel-producing areas would be boosted by the tariff, because steel companies would purchase more goods from their suppliers, executives and workers would purchase more products, and the multiplier effect would spread the gains in jobs and profits throughout the regional economies where the steel industry is concentrated—Pennsylvania, Ohio, and West Virginia. In fact, critics contended (and White House officials only halfheartedly denied) that the main purpose of the tariffs was to boost the president's reelection prospects in those key electoral states.

However, these gains represent only one side of the distributional effect, because there are losers as well as winners. For example, by making foreign-produced steel more expensive, the tariffs also harm domestic automakers, who must pay higher prices for the steel they use. Indeed, the representatives of auto-producing states like Michigan and Tennessee denounced the tariffs. The president's own economic advisers, led by Treasury Secretary Paul O'Neill, also opposed the tariffs, bolstered by liberal theory's contention that the total losses would outweigh the total benefits. A report by the International Trade Commission estimated the cost to industries that consume steel at more than $680 million per year. A study backed by steel-using companies concluded that higher steel prices had cost the country about 200,000 manufacturing jobs, many of which went to China, where Chinese steel remained cheap. This episode illustrates that most barriers to trade harm consumers because of higher prices, a point always emphasized by proponents of free trade.

Trade policy also benefits some classes at the expense of others, a point more often emphasized by those who favor greater governmental control. For example, the elimination of trade barriers between the United States and Mexico under the terms of the North American Free Trade Agreement (NAFTA) forces some US manufacturing workers into direct competition with Mexican workers, who earn a markedly lower wage. Since NAFTA guarantees that imports can enter the United States without tariffs, some US businesses move to Mexico, where production costs are lower, and US workers lose their jobs in the process. Facing the threat of such production shifts, many more US workers will accept a decline in wages, benefits, or working conditions. The losses from such wage competition will be greatest for unskilled workers in high-wage countries employed in industries that can move either their products or their production facilities most easily across national boundaries. But they also affect skilled workers in industries like steel and autos. Others, particularly more affluent professionals who face less direct competition from abroad (such as doctors, lawyers, and university professors), stand to gain from trade, because it lowers prices on the goods they consume. Of course, the greatest beneficiaries are the owners of businesses that profit from lower wage rates and expanded markets.

Proponents of free trade tend to de-emphasize these distributional effects and instead focus on the impact of trade on the economy as a whole. That is partly because liberal theory contends that free trade does not decrease employment but only shifts it from an inefficient sector to one in which a nation has a comparative advantage. For example, US workers losing their jobs to Mexican imports should eventually find employment in industries that export to Mexico. Proponents of free trade insist that it is far better to tolerate these "transition costs"—the short-term dislocations and distributional effects—than to protect an inefficient industry. Workers are not so sure, especially because "short-term" effects seem to last a lot longer to those who actually live through them, and because future prospects rarely compensate for present losses when security, stability, and peace of mind are factored in.

Because these distributional consequences have such obvious political implications, the state is also much more attentive to them than are economic theorists. That is one reason why all governments control trade to one degree or another. Of course, that does not mean that they do so wisely or fairly, in part because their decisions are shaped by patterns of representation among the constituencies whose material interests are affected by trade policy. In general, workers tend to be underrepresented, which is why trade policies so often encourage trade built on low wages that enrich business owners but constrain the opportunities for workers. Moreover, as the discussion of trade deficits has indicated, the economic activities shaped by trade policies tend to affect current generations very differently from future ones—and the latter are seldom represented at all.

The Values Dilemma

These distributional effects pose challenging trade-offs among competing values. For example, the effects of NAFTA were predicted to include somewhat lower prices for US consumers but also job loss or wage reduction for some unskilled US workers. The positions taken on this issue by most individuals, however, did not hinge on their own material interests; few could confidently foresee any personal impact of NAFTA, since the gains were estimated at well under 1 percent of GDP, and job losses were not expected to exceed a few hundred thousand in a labor force of more than a hundred million. However, the choice among competing *values* was plain: NAFTA meant gains in wealth but also greater inequality and insecurity for workers. Some citizens acceded to the judgment of liberal theory that the country as a whole would be better off with freer trade, while others identified with the plight of workers, who were more skeptical of liberal theory simply because, for them, the stakes were so much higher. After all, it is easy for a theorist to postulate that job losses in an import-competing industry would be matched by job gains in an exporting firm, but it is far harder for a worker who has devoted his life to one career to pack up and move to a strange town, hopeful that he *might* find a job that requires skills he may not possess in an unfamiliar industry. In the final analysis, NAFTA became a referendum on what kind of society people wished to live in. The decision was quintessentially American: one of greater wealth but also greater inequality and insecurity. Other nations, which assess the trade-off between values differently, might have chosen an alternative policy toward trade.

Of course, distributional effects gave rise to other value choices as well. Since the gains from NAFTA were expected to be greater for Mexico than for the United States, the conscientious citizen would also weigh whether it is better to help Mexican workers because they are poorer, or to protect US workers because they are US citizens. As Chapter 8 implies, such issues of inequality in poor societies can translate directly into questions of life or death. As a result, the importance of trade policy, which has such a powerful impact on the distribution of gains and losses, is heightened in poor, dependent nations where half of the economy is related to trade.

Perhaps the most challenging value trade-offs concern the trade policies that shift gains and losses from one time period to another. Such "intergenerational" effects arise from a variety of trade issues. For example, as discussed earlier, the US trade deficit, like any form of debt, represents an immediate increase in consumption but a postponement of its costs. The Japanese industrial policy of export promotion fosters a trade surplus, which produces the opposite effect in Japan. The subsidies the Japanese government pays to Japanese exporters require Japanese citizens to pay both higher prices and higher taxes. However, the sacrifices of Japan's current generation may benefit future ones if this subsidy eventually transforms an "infant industry" into a powerful enterprise that can repay the subsidies through cheaper prices or greater employment.

The values dilemma encompasses much more than just an alternative angle on distributional effects, however (Polanyi 1944). The debate over "competitiveness," which began with the efforts by US businesses to lower their production costs in order to compete with foreign firms, illustrates how trade considerations may imply a compromise of other societal values. Companies could lower their costs if the abolition of seniority systems or age and gender discrimination laws allowed them to terminate employees at will. But that would leave workers vulnerable to the whim of a boss. Labor costs would be reduced if the minimum wage and workplace safety regulations were cut, if collective bargaining and labor unions were outlawed, and if pensions, healthcare, paid vacations and holidays, sick leave, and workers' compensation for accidents were eliminated. But such actions entail a compromise with fundamental values about the kind of society in which people want to live. Government regulations that protect the environment, promote equality and social harmony, and achieve justice and security may add to production costs, but surely achieving economic interests is not worth abandoning all other values. Choosing between them is always difficult for a society, because reasonable people can differ in the priority they ascribe to alternative values. Still, agreements on such matters can usually be forged *within* societies, in part because values tend to be broadly, if not universally, shared.

Unfortunately, trade forces firms burdened by these value choices to compete with firms operating in countries that may not share them. This situation creates a dilemma for consumers, forcing them to balance economic interests against other values. For example, continuing to trade with nations that permit shabby treatment of workers—or even outright human rights abuses—poses a painful moral choice, not least because goods from such countries are often cheaper. As Chapter 4 documents, foreign governments have often declared their opposition to human rights abuses but have seldom supported their rhetoric with actions that effectively curtail the abuses. In fact, the maintenance of normative standards has fallen to consumers, who must unwittingly answer key questions daily: Should we purchase cheap foreign goods like clothing and textiles even though they may have been made with child labor—or even slave labor? Of course, we seldom know the conditions under which these products were produced—or even *where* they were produced—so we ask government to adopt policies to support principles we cannot personally defend with our own consumer behavior.

Where values are concerned, of course, we cannot expect everyone to agree with the choices we might make. As Chapter 11 describes, child labor remains a key source of comparative advantage for many countries in several industries prominent in international trade. We cannot expect them to give up easily a practice that is a major component of their domestic economy and that is more offensive to us than to them. Unfortunately, if trade competitors do not share our values, it may prove difficult to maintain these values ourselves—unless we restrict trade, accept trade deficits, or design state policies to alleviate the most dire consequences. After all, it is hard to see how US textile producers can compete with the sweatshops of Asia without creating sweatshops in New York. That point inevitably animates a complex debate over whether eliminating sweatshops would really benefit the poor, a dilemma of international trade that cannot be avoided merely by refusing to think about it.

Foreign Policy Considerations: Power and Peace

Some of the most challenging value choices concern the effect of trade on the foreign policy goals pursued by states, especially power, peace, and national autonomy. Policymakers have long been aware that trade has two deep, if contradictory, effects on national security. On the one hand, trade contributes to national prosperity, which increases national power and enhances security. On the other hand, it has the same effect on a nation's trade partner, which could become a political or even military rival. The resulting ambivalent attitude is torn between the vision of states cooperating for economic gain, and the recognition that they also use trade to compete for political power.

While a market perspective sees neighboring nations as potential customers, the state must also see them as potential enemies. As a result, the state not only must consider the absolute gains it receives from trade, but also must weigh those gains in relative terms, perhaps even avoiding trade that would be more advantageous to its potential enemies than to itself. For this reason, states are attentive to the distribution of the gains from trade and selective about their trade partners, frequently encouraging trade with some nations and banning it with others.

While understandable, such policies sometimes lead to open conflict. In fact, US president Franklin Roosevelt's secretary of state, Cordell Hull, contended that bitter trade rivalries were the chief cause of World War I and a substantial contributor to the outbreak of World War II. Both wars were precipitated by discriminatory trade policies in which different quotas or duties were imposed on the products of different nations. Hull, who believed that free multilateral trade would build bridges rather than create chasms between peoples and nations, thus championed the nondiscrimination principle and urged the creation of international institutions that would govern trade in accordance with it. Moreover, the Great Depression of the 1930s made it plain that international institutions are required to establish the rules of trade and create the international law that embodies them.

In response, the United States sponsored the Bretton Woods trade and monetary regime, centered on the General Agreement on Tariffs and Trade (GATT), the World Bank, and the International Monetary Fund (IMF). Since 1946, GATT, which evolved into the World Trade Organization (WTO) in 1995, has convened eight major negotiating sessions (referred to as "rounds") in which nations exchange reductions in trade barriers. This bargaining is necessary to overcome the inclination of most nations to retain their own trade barriers while hoping other countries will lower theirs. The World Bank supports the effort by lending money to nations that might otherwise seek trade-limiting solutions to their financial problems. The IMF facilitates trade by providing a stable monetary system that permits the easy exchange of national currencies and the adjustment of trade imbalances. The result has been a dramatic increase in global trade.

In between negotiating rounds, the GATT/WTO dispute resolution mechanism has provided a forum for diverting the inevitable skirmishes over trade into the legal arena rather than the military realm. Such was the case with the controversial steel tariffs introduced by President George W. Bush in March 2002. This episode demonstrated that if mercantilist policies are controversial in the nations that enact them, they are met with even greater hostility by the nations with which they trade.

Immediately after the steel tariffs were announced, the European Union (EU) lodged a complaint with the WTO. It pointed out that the White House based its tariffs on the "safeguard" provision of GATT, which allows governments to provide *temporary* protection to industries that are overwhelmed by *sudden* surges of imports. The principle is that such industries must be given time to adjust to new competition. However, steel imports actually crested in 1998, and had already fallen by about 25 percent over the following four years—even before the tariffs were imposed. The EU, contending that the tariffs were unnecessary and therefore illegal, sought compensation for the damage done to its economies. Not only did the tariffs prevent European steel producers from selling in the US market, but they also blocked entry into the United States by big producers in China and South Korea. These firms would naturally seek instead to dump their products on the European market, thus doing double damage to European producers. The EU thus imposed steel tariffs similar to the US ones, which spread the chaos and enlisted international support for the WTO case against the United States.

In addition to protecting its home market, the European Union initiated WTO proceedings to secure compensation in the form of reduced US tariffs on other EU products, as required by GATT. In the event that no agreement on compensation could be reached, the European Union instead asked the WTO to authorize retaliatory duties on selected US products. Since it saw the steel tariffs as motivated by President Bush's desire to impress voters in swing states like Pennsylvania and Ohio, the EU proposed that the retaliation be directed at US goods that would inflict political pain on the White House. The EU chose citrus fruit and clothing because they were produced in battleground states such as Florida and the Carolinas. In March 2003 the WTO ruled against the United States and in December President Bush removed the tariffs, though in defiance he declared that the WTO had nothing to do with the decision.

At the regional level, a similar belief in the efficacy of free trade as a guarantor of peace was an important motivation for the initiative that eventually led to the creation of the European Union. In both the EU and Bretton Woods, policymakers saw several ways that an open and institutionalized trading system could promote peace among nations. The institutions themselves could weaken the hold of nationalism and mediate conflict between nations. Trade-induced contact could break down nationalistic hostility among societies. Multilateralism (nondiscrimination) would tend to prevent grievances from developing among states. Interdependence could constrain armed conflict and foster stability, while the economic growth generated by trade could remove the desperation that leads nations to aggression.

European integration was launched in 1951 with the founding of the European Coal and Steel Community (ECSC), which internationalized an industry that was key not only for the economies of the six nations involved, but also for their war-making potential. With production facilities scattered among different countries, each became dependent on the others to provide both demand for the final product and part of the supply capacity. This arrangement fulfilled the liberal dream of an interdependence that would prevent war by making it economically suicidal. In fact, the ECSC was an innovative form of peace treaty, designed, in the words of Robert Schuman, to "make it plain that any war between France and Germany becomes, not merely unthinkable, but materially impossible" (Pomfret 1988: 75). The most recent step to encourage European trade was the creation of a new regional currency, the euro, to replace national currencies in 2001.

Institutional efforts to secure global peace require the exercise of power. According to hegemonic stability theory, one dominant nation—a hegemon—will usually have to subsidize the organizational costs and will frequently offer side benefits in exchange for cooperation, such as the massive infusion of foreign aid provided to Europe by the United States under the Marshall Plan in the late 1940s. Maintaining the capability to handle these leadership requirements entails substantial costs. For example, US expenditures for defense, which have been many times higher than those of nations with which it has competed since World War II, erode the competitiveness of US business by requiring higher tax levels; they constrain the funds available to spend on other items that could enhance competitiveness; and they divert a substantial share of US scientific and technological expertise into military innovation and away from commercial areas. The trade-off between competitiveness and defense may be judged differently by different individuals, but it can be ignored by none.

International Cooperation and National Autonomy

International institutions may be necessary to facilitate trade and to alleviate the conflict that inevitably surrounds it, but they can also *create* conflict. Institutions require hegemonic leadership, but many critics complain that the United States benefits so much from its capacity to dictate the rules under which institutions operate that they have become extensions of US imperialism. Institutions seek to maintain fair competition among firms in different countries—which is essential to the international trading system—but they must also do so without undermining the national sovereignty and autonomy that are central to the modern state system.

Trade disputes test the capacity of institutions to balance these imperatives, because one nation will often defend its policy as a rightful exercise of national sovereignty, while another may challenge it as an unfair barrier to trade. Since governments have many compelling motives for enacting policies that affect trade, clashes of values often appear as struggles over the rightful boundaries of sovereignty. Such disagreements can be settled by appeal to GATT or, more recently, to the WTO, but not even the WTO's chief sponsor, the United States, accepts the dominion of the WTO without serious reservations about its intrusion into affairs historically reserved for national governments.

Indeed, even though the US administration strongly supported the creation of the WTO to prevent trade violations by other nations, a surprising variety of US groups opposed its ratification because it might encroach on US sovereignty. Environmental groups such as Friends of the Earth, Greenpeace, and the Sierra Club were joined not only by consumer advocates like Ralph Nader, but also by conservatives such as Ross Perot, Pat Buchanan, and Jesse Helms, who feared that a WTO panel could rule that various US government policies constituted unfair trade practices, even though they were designed to pursue values utterly unrelated to trade. For example, EU automakers have challenged the US law that establishes standards for auto emissions and fuel economy. Buchanan said, "WTO means putting America's trade under foreign bureaucrats who will meet in secret to demand changes in United States laws. . . . WTO tramples all over American sovereignty and states' rights" (Dodge 1994: 1D). Because the WTO could not force a change in US law, GATT director-general Peter Sutherland called this position "errant nonsense" (Tumulty 1994), but the WTO could impose sanctions or authorize an offended nation to withdraw trade concessions as compensation for the injury.

The most dramatic example occurred in the 1994 case known as "GATTzilla versus Flipper," in which a GATT tribunal ruled in favor of a complaint brought by the EU on behalf of European processors who buy tuna from countries that use purse seine nets. The United States boycotts tuna caught in that manner because the procedure also kills large numbers of dolphins, but this value is not universally shared by other nations. In fact, the GATT tribunal ruled that the US law was an illegal barrier to trade because it discriminates against the fishing fleets of nations that use this technique. The United States saw this as an unwarranted intrusion into its domestic affairs and an affront to US values.

Soon thereafter the United States found itself on the other side of the clash between fair competition and national sovereignty when it appealed to the WTO to rule that the EU's prohibition of beef containing growth hormones violated the "national treatment" principle contained in GATT's Article 3. Since almost all cattle raised in the United States are fed growth hormones and very few European cattle are, the United States contended that the EU rule was simply disguised protectionism that unfairly discriminated against US products. The EU contended that such beef was a cancer risk and that as a sovereign power it had the right to establish whatever health regulations it chose to protect its citizens. The WTO ruled in favor of the United States, incurring the wrath of those who saw this as an example of national democratic processes being overruled by undemocratic global ones. Can it be long before Colombia challenges US drug laws as discriminating against marijuana while favoring Canadian whiskey?

Neither can regional agreements avoid this clash between fair competition in trade and national autonomy. The first trade dispute under NAFTA involved a challenge by the United States to regulations under Canada's

Fisheries Act, established to promote conservation of herring and salmon stocks in Canada's Pacific Coast waters. Soon thereafter the Canadian government challenged US Environmental Protection Agency regulations that require the phasing-out of asbestos, a carcinogen no longer permitted as a building material in the United States (Cavanaugh et al. 1992).

Similarly, critics of the EU worry that its leveling of the playing field for trade competition also threatens to level cultural and political differences among nations. Denmark, for example, found that free trade made it impossible to maintain a sales tax rate higher than neighboring Germany's, because Danish citizens could simply evade the tax by purchasing goods in Germany and bringing them across the border duty-free. Competitiveness pressures also make it difficult for a nation to adopt policies that impose costs on business when low trade barriers force firms to compete with those in other countries that do not bear such burdens. For example, French firms demand a level playing field in competing with Spanish firms whenever the French government mandates employee benefits, health and safety rules, or environmental regulations that are more costly than those in Spain. In fact, free trade tends to harmonize many national policies.

Some trade barriers are designed to protect unique aspects of the economic, social, and political life of nations, especially when trade affects cultural matters of symbolic importance. For example, France imposes limits on the percentage of television programming that can originate abroad, allegedly in defense of French language and custom. The obvious targets of these restrictions, US producers of movies and youth-oriented music, contend that the French are simply protecting their own inefficient entertainment industry. They argue that programming deserves the same legal protection abroad that the foreign television sets and DVD players that display these images receive in the United States. But if we restrict trade because we oppose child labor or rainforest destruction, how can we object when other countries ban the sale of US products because they violate *their* values—such as music and Hollywood films that celebrate sex, violence, and free expression of controversial ideas, or even blue jeans, McDonald's hamburgers, and other symbols of US cultural domination?

Conclusion: Choices for Nations and Individuals

Few would deny the contention of liberal theory that trade permits a higher level of aggregate consumption than would be possible if consumers were prevented from purchasing foreign products. It is hard to imagine modern life without the benefits of trade. Of course, it does not follow that trade must be utterly unrestricted, because the aggregate economic effect tells only part of the story. As mercantilists remind us, trade also carries with it important social and political implications. Trade shapes the distribution of income and wealth among individuals, affects the power of states and the relations among them, and constrains or enhances the ability of both individuals and nations to achieve goals built on other values. Thus, trade presents a dilemma for nations: no policy can avoid some of trade's negative consequences without also sacrificing some of its benefits. That is why most governments have sought to encompass elements of both liberalism and mercantilism in fashioning their trade policies. The same is true for individuals, because every day each individual must—explicitly or implicitly—assume a stance on the dilemmas identified in this chapter. In turn, trade forces individuals to consider some of the following discussion questions, questions that require normative judgments as well as a keen understanding of the empirical consequences of trade. We must always remember to ask not only what trade policy will best achieve our goals, but also what our goals should be.

Discussion Questions

1. Are your views closer to those of a liberal or a mercantilist?
2. Is it patriotic to purchase domestic products? Why or why not?
3. Does one owe a greater obligation to domestic workers and corporations than to foreign ones?
4. Should one purchase a product that is cheap even though it was made with slave labor or by workers deprived of human rights?
5. Should a country surrender some of its sovereignty in order to receive the benefits of joining the WTO?
6. Should one lobby the US government to restrict the sales of US forestry products abroad because these products compromise the environment?

Suggested Readings

Chen, Shaohua, and Martin Ravallion (2004) "How Have the World's Poorest Fared Since the Early 1980s?" Policy research working paper. Washington, DC: World Bank.

Easterly, William (1999) "The Lost Decades: Explaining Developing Countries' Stagnation, 1980–1998." Policy research working paper. Washington, DC: World Bank.

Fallows, James (1993) "How the World Works." *Atlantic Monthly* (December).

Moon, Bruce E. (1998) "Exports, Outward-Oriented Development, and Economic Growth." *Political Research Quarterly* (March).

——— (2000) *Dilemmas of International Trade.* 2nd ed. Boulder: Westview.

Polanyi, Karl (1944) *The Great Transformation.* New York: Farrar and Reinhart.
Ricardo, David (1981) *Works and Correspondence of David Ricardo: Principles of Political Economy and Taxation.* London: Cambridge University Press.
Smith, Adam (1910) *An Inquiry into the Nature and Causes of the Wealth of Nations.* London: Dutton.
World Bank (2007) *World Development Indicators 2007.* Washington, DC.

7

The Political Economy of Development

Mary Ellen Batiuk

FOR THE PEOPLE IN ANY COUNTRY TO PROSPER, IT IS CRITICAL that they have *capital,* in the form of goods, services, and money. Economic productivity and trade most often generate this capital, but some countries also receive loans and aid (official development assistance [ODA]) from other countries in order to increase the supply of capital. In addition, private corporations can invest directly into a country by building a company there (foreign direct investment [FDI]), or invest in the stock market in that country (foreign portfolio investment [FPI]). These flows of money into a country through trade, loans, aid, and investments are critical to the ability of any country to develop. Of course, capital also flows out of all countries in the form of payments for goods and services and to reduce debt. Government officials, business leaders, and citizens all want to see more capital flowing into their country than out, so as to foster growth and development.

But in today's world there is a great gap between those countries that have experienced significant development, the more-developed countries (MDCs), and those countries still trying to develop, the less-developed countries (LDCs). For example, in 2006, total FDI in the United States amounted to about $591.00 per capita (UNCTAD 2007a) (the total amount invested divided by the number of people in the country) (PRB 2007). By contrast, FDI in Kenya was about $1.44 per capita (UNCTAD 2007b). By this measure, people in the United States have a lot more financial capital to work with—capital that helps create the jobs that eventually support education, healthcare, and a higher standard of living. Annual per capita gross national income (GNI) in the United States in 2006 was $44,970 (World Bank 2008b), and in Kenya it was $250 (World Bank 2008a) (gross national product [GNP] was used by reporting agencies until 2000, when it was replaced by GNI, which equals GDP [the total of goods and services produced in a given year] plus any foreign income, minus any interest or debt payments). In Wilmington, Ohio, which has a population of 11,921 (US Census Bureau 2008a), the biggest employer (more than 7,000 people) is DHL, a company owned by the German company Deutsche Post World Net, which has invested (FDI) in the United States. Per capita income in Wilmington was $17,346 in 2006—close to the US average of $21,587 (US Census Bureau 2008a). International capital flows have a direct effect on the well-being of individuals living in every city and in every country in the world.

Generally speaking, the more financial capital flowing into a country, the better off the country will be. The single biggest foreign investor in the United States in 2006 was the European Union—of which Germany is a member. Today, most investment capital flows from MDC to MDC. In 2006, the developed economies accounted for 65.5 percent of all global FDI inflows. The whole of sub-Saharan Africa accounted for only 1.8 percent of all global FDI inflows (UNCTAD 2007a). Countries with too much capital flowing out and not enough coming in fail to develop over time.

Global Capital Flows Before World War II

This current gap in FDI between LDCs and MDCs has had a long history—one that dates back to before World War II and the system of colonialism that dominated global capital flows at that time.

Economic and Political Structure of Colonialism

Colonialism is a political, economic, and social system in which one group of people, the core or dominant group, controls the political and economic lives of another group of people, the peripheral or dominated group. Colonialism usually signifies direct political control as well as economic control of another group. Colonies usually provide the core countries with cheap labor, raw materials unavailable at home, cheap commodities, and markets for goods produced in the core country. Typically, vast revenues flow out of the periphery while relatively little investment flows in. Colonialism has been around for a long time. The Greeks and Romans possessed colonies in the ancient world, and the Chinese and Japanese also possessed colonies. During the period 1500–1900, however, much of the world's population lived in colonies controlled by Western Europeans—primarily English, Spanish, French, Belgian, German, Dutch, and Portuguese. The colonies were located primarily in North America, South America, Africa, and Asia.

From 1500 to 1800, mercantilist trade policies and the slave trade dominated trade between the core and periphery. From the point of view of the colonizer, the idea was to protect industry at home and extract as much profit from the colony as possible. Colonizers ignored traditional native claims to property and created huge plantations for sugar, coffee, tobacco, cotton, and rice. Along with these agricultural commodities, lumber, furs, gold, and silver flowed out of the periphery and into the core. It is estimated that 9–15 million Africans were forcibly transported to the New World colonies during this time as slaves—forming what has come to be known as the Triangular Trade (see Figure 7.1). The slave trade was also incredibly profitable. Slaves made huge agricultural plantations possible because

Figure 7.1 Capital Flows in the Triangular Trade, 1500–1800

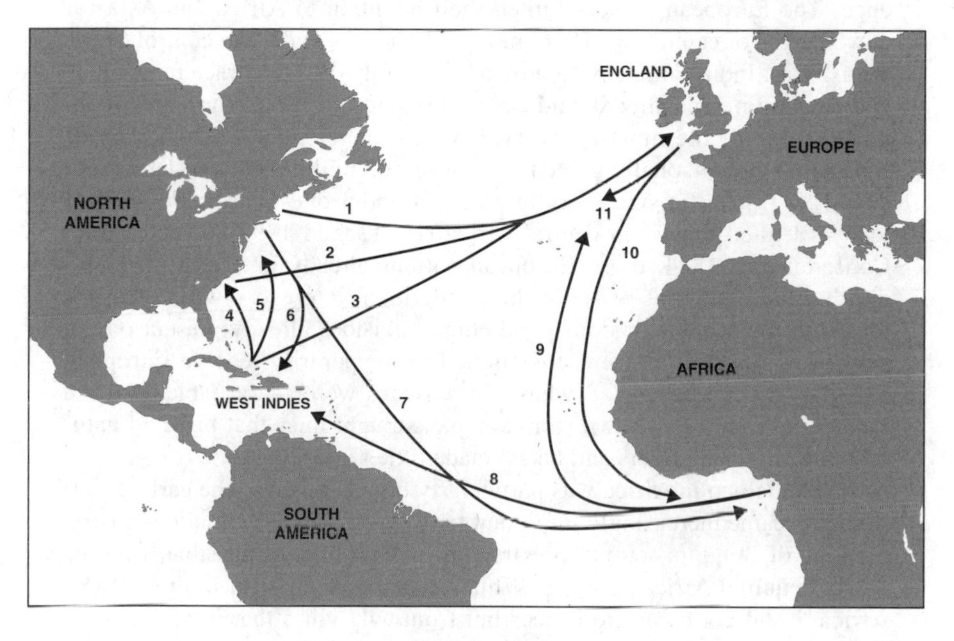

Trade Routes

1	whale oil, lumber, furs	7	rum, iron, gunpowder, cloth, tools
2	rice, silk, indigo, tobacco	8	slaves
3	sugar, molasses, wood	9	gold, ivory, spices, hardwoods
4	slaves	10	guns, cloth, iron, beer
5	slaves, sugar, molasses	11	manufactured goods, luxuries
6	fish, flour, livestock, lumber		

Source: National Archives (2003), available at http://www.nationalarchives.gov.uk/pathways/blackhistory/africa_caribbean/britain_trade.htm.

native populations often escaped when pressed into service. In addition, native populations had no natural defenses for diseases like smallpox and measles that the colonists brought with them.

Colonies in the Americas were especially lucrative for countries like Spain, Portugal, France, and England. For example, estimated revenue extracted from the Spanish colony of New Spain (today's Mexico), from 1795 to 1799, exceeded 400 million Spanish pesos. At that time, a Spanish peso was roughly equivalent to the US dollar. Total estimated revenue in the newly independent United States during the same period was about $43 million. Money flowing into Europe from these colonies financed wars and personal luxury as well as considerable industrial development—especially in England.

But by the beginning of the 1800s, England had outlawed slave trading, and populations in North and South America were achieving independence. The European powers turned their attention to Africa and Asia for new sources of capital. By 1856, most of India was under the control of the British East India Company, and by the end of the 1800s a "race for Africa" had begun. England already laid claim to vast territories in North and South Africa. One of the most aggressive African colonizers was the king of Belgium, who personally gained control of 568 million acres of land (seventy-five times the size of Belgium, and about one-quarter the size of today's United States) in central Africa (Hochschild 1999). At the Berlin Conference of 1884, those European nations already in Africa officially divided the continent to serve their individual interests—without much regard to the tribes, kingdoms, and ethnic divisions already present on the continent. Boundaries were constructed to temporarily resolve European disputes about who owned what. No Africans were at the table, and the boundaries that were drawn (for example, straight lines that bisected natural boundaries like rivers and lakes) made little sense.

Colonialism in Africa was particularly brutal. When in the early 1900s rubber became increasingly important to the emerging automobile industry, the king of Belgium set out to extract as much rubber as possible from his newly acquired African empire. While rubber was plentiful in this area of Africa, it did not come from trees but from wild vines that were scattered throughout the rainforest, so supervision of the laborers was not as straightforward as on the rubber plantations in places like Brazil. New methods of forcing workers to return to their villages at the end of grueling days of collecting rubber evolved into a program of terror carried out by Belgium's Force Publique, whose tactics included burning villages, murder, rape, flogging, holding women and children hostage, and severing the limbs of individuals who did not meet their quotas. Severed human hands were also taken to prove that bullets had not been wasted in murdering underachieving workers. Baskets of severed limbs and heads were collected and openly displayed by the Force Publique to warn workers about the importance of meeting ever-growing quotas. Unfortunately, these kinds of tactics typified the European approach to the colonization of Africa and Asia, because they worked. Whether the commodities were rubber, cotton, or diamonds, native people were forced to collect, grow, and mine them under horrific conditions.

Economic, Political, and Social Consequences of Colonialism

The consequences of colonialism were mixed for both the core and the periphery. For the peripheral territories (not yet countries), colonialism left a legacy of capital drain, exploitation, bureaucratic corruption, and ethnic and religious rivalries. Colonies were most often developed to provide one particular resource, or cash crop, to the dominant country. Instead of evolving diverse and resilient economic systems, colonial economies were actually quite fragile. If a population in the periphery produced primarily cotton, and global cotton markets collapsed, the people had little else to fall back on and suffered bitter consequences.

At the same time, peripheral peoples in the Americas, Africa, and Asia were introduced to European models of government, and European ideas about individualism, equality, and justice. English poet Rudyard Kipling had written in 1899 about "the white man's burden" of bringing civilization and Christianity to indigenous (native) populations. Whether they liked it or not, and often under extreme duress, indigenous peoples around the globe were dragged into the white man's world (Rodney 1983).

For core countries, colonialism offered raw materials for growing industrialization, cheap labor from subjected populations, and new markets for industrialized goods, including the weapons needed to keep subjugated populations at bay. However, this mounting need to feed industrial appetites would eventually contribute to the tensions that led to World War I and World War II.

On the one hand, Europeans drained their colonies of human capital, commodities, and money. Africa alone lost an estimated 12 million men to the slave trade. Commodity revenues that could have been used for the development of African peoples, such as from rubber, were lost forever in the race for profits. On the other hand, Europeans did develop considerable infrastructure in their colonies in the form of roads, shipyards, railways, and cities. After all, commodities must be collected, stored, and shipped, workers must be supervised, and life for the colonizers must be supported.

But how did a relatively small number of Europeans control such vast territories, markets, and capital flows? Core countries used both direct and indirect methods to rule the periphery. Direct rule created strong centralized bureaucratic administrations in the urban areas populated by Europeans.

These bureaucracies generally followed European norms of property, individual freedom, civil rights, and justice. Still, direct rule was founded on principles that were clearly racist. Indigenous peoples participated in urban colonial bureaucracies only at the lowest levels, with little hope of advancement. Indirect rule ceded power in the rural areas to local indigenous chieftains and warlords. In doing so, indirect rule created deep divisions among ethnic groups, tribes, and villages, as these strong and ruthless leaders were granted almost unlimited powers by colonial governments. In rural areas, systems of brutality and corruption became a way of life, supported and reinforced by distant urban administrators. Direct and indirect rule worked more or less side by side to enable a relatively small group to exert control over a much larger population. Terror, corruption, racism, and ethnic tensions turned out to be powerful tools that the core countries used to divide, conquer, and exploit their vast holdings in the periphery. All the while, capital kept flowing back to the core (Mamdani 1996).

Breakdown of Colonialism

Ironically, colonial revenues and rivalries ultimately fueled political tensions already simmering among Europeans themselves. Throughout the late 1800s, European leaders tried to keep a "balance of power" among themselves. But since each country wanted a military edge over its rivals, what really resulted was a huge arms buildup paid for in part by colonial revenues. Of all the countries in Western Europe, only Germany was without significant colonies in Africa and Asia. Lack of colonies meant lack of access to markets and raw materials. War appealed to some as an opportunity for colonial realignment and increased capital flows from abroad.

With each European government looking warily over its shoulder, secret alliances were forged to counter the Austro-Hungarian Empire, which controlled much of Eastern Europe, and the Ottoman Empire, which controlled the Middle East. When the heir to the Austro-Hungarian throne was assassinated in 1914, what might have been a minor event turned into World War I.

World War I (1914–1918) was the most expensive and deadly war in history up to that point. One reason the war was so deadly was because of the degree to which advanced industrial economies mobilized huge capital flows to finance it. Although US president Woodrow Wilson came to the 1919 peace talks in Paris promising to work for "a free, open-minded and absolutely impartial adjustment of all colonial claims," postwar colonial boundaries remained remarkably similar to those that had been established before the war. In addition, the Allies of 1919 carved up the defeated Austro-Hungarian and Ottoman Empires as recklessly as they had Africa in 1884. While fragile independent Eastern European countries emerged,

Britain and France gained control over much of the Middle East and began to exploit its oil wealth.

In the end, World War I settled nothing. Capital flows from the periphery to the core resumed their prewar patterns, and war wounds festered. Colonial tensions between Japan and China also became aggravated in Asia. A worldwide economic crisis crippled global economies in the 1930s, and in 1939 World War II began, less than twenty years after the "war to end all wars." World War II was even more devastating than its predecessor in both military and civilian deaths. The economies of Europe and Asia all but collapsed because of the extent and severity of the aerial bombing. Without European armies and capital to support colonial domination, populations in the periphery saw their chance to achieve sovereignty. Through rebellion and negotiation, most colonies would eventually seek and achieve independence over the next twenty-five years.

Global Capital Flows After World War II

Rebuilding the Industrialized World

World War II ended the formal colonial systems that had so strongly influenced global capital flows for 500 years. By the end of the war, it was clear to many leaders that new institutions needed to be created to help rebuild the economies of Europe and Asia and restore stable capital flows around the world. For the first time, the United States took a leadership role in these global negotiations, and in 1944 hosted the United Nations Monetary and Financial Conference among the Allied nations in Bretton Woods, New Hampshire. These meetings resulted in the Bretton Woods agreements.

The basic philosophical ideas behind the Bretton Woods agreements were that liberal economics, freely flowing capital, and open markets held the keys to a more secure and peaceful world. Instead of secret alliances and trading blocs that resulted in nations fighting over raw materials and markets for finished goods, tariffs and protectionism would be minimized and all nations would have equal access to markets. There were three basic tasks to be accomplished under the agreements: stabilizing all national currencies, creating institutions and mechanisms for nations to manage their currency valuations, and financing the reconstruction of the battered European economies. Most of the resultant new rules and institutions offered advantages for the US economy. At the end of World War II, after all, the United States had by far the strongest economy. Having built up considerable industrial capacity during the war, the United States clearly wanted more trade and needed access to markets that had been artificially limited by prewar colonial systems and trading blocs. Bretton Woods helped make that possible.

Trade, however, can only flourish when currencies are easily convertible and free from extreme fluctuations. In the Bretton Woods agreements, the first order of business was to stabilize currencies. Stable currencies could be achieved with fixed exchange rates. Fixed exchange rates facilitate trade by giving all nations the confidence that the currencies they hold today will continue to have a stable value tomorrow. If the British bought French cheese using British pounds, it would be important to the French that those pounds held their value over time—otherwise the French would lose money on the deal. Too many currency losses could eventually bring trade to a standstill. Connecting the value of a currency to an independent commodity like gold could solve the problem. In fact, a gold standard had existed in Europe before World War I, but was abandoned when countries started printing money to finance the war. After World War II, gold production was not sufficient to guarantee all existing currencies, so they were instead pegged to US dollars, which were redeemable nominally in gold, at $35 to one Troy ounce.

But currencies naturally tend to fluctuate whenever countries experience severe budget imbalances. So the second thing the Bretton Woods agreements created was an institution called the International Monetary Fund (IMF), to advise countries and loan them money to help pay their debts and balance their budgets. Before the IMF, when countries owed money to other countries, they were tempted to devalue their currencies (usually by printing more money). This left their creditors holding nearly worthless currency. Through financial subscriptions and quotas from member countries, the IMF amassed a pool of money to be used for loans to help cover these debts. Though countries then became indebted to the IMF, national currencies remained stable. To aid nations in paying off their debts, the IMF would also make recommendations for cost-cutting measures that would free up money in national budgets for debt repayment. The original ten country members of the IMF were known as the Group of Ten, and still meet periodically today (now as the Group of Twelve).

Finally, the Bretton Woods agreements created the International Bank for Reconstruction and Development (IBRD), to aid in the reindustrialization of Europe. The IBRD would eventually become part of a group of institutions known simply as the World Bank Group—or World Bank. Initially, however, the IBRD was underfunded, and greater flows of capital were needed to jump-start the economies of Europe. A plan developed by then–US secretary of state George Marshall, called the Marshall Plan, sent billions of extra dollars ($17 billion between 1948 and 1954) in grants to sixteen Western European countries. The decision to help rebuild even the economies of defeated Germany and Japan shows the degree to which US economic growth in this period depended on these rebounding markets. Without markets, economies simply cannot continue growing.

To keep the Bretton Woods negotiations an ongoing part of international financial stability, the Allied countries resolved to keep meeting to make adjustments to the Bretton Woods agreements as needed. This resolution became known as the General Agreement on Tariffs and Trade (GATT). Twenty-three countries signed the first GATT treaty in 1947, and eventually 125 countries participated in the 1986–1994 GATT round of negotiations.

For those who participated in them, the Bretton Woods agreements, coupled with the ambitious Marshall Plan and other development aid programs, did in fact help to rebuild war-torn economies and sustain US levels of production.

Cold War Politics and the Periphery

While Bretton Woods represented a big step forward in international financial cooperation and coordination, some major parts of the world remained outside of these capital flows. Neither Russia nor the European colonies in Africa and Asia became fully integrated into these new economic arrangements, largely because of the Cold War, which dominated international politics from the end of World War II (1945) through 1991. The Cold War, whose roots lie in World War I, had as much impact on post–World War II international capital flows as did the World Bank and International Monetary Fund in the latter half of the twentieth century.

In 1917, at the end of World War I, Vladimir Lenin successfully led a communist revolution in czarist Russia, which was eventually consolidated into a totalitarian socialist state by Josef Stalin. Pure socialism is an economic system in which the state owns and controls all the capital used in the production of goods and services. Almost all individuals are employed directly or indirectly by the state, and government assumes the major responsibility of feeding, clothing, housing, and caring for people. By contrast, pure capitalism is an economic system in which almost all capital is owned and controlled privately, and incentives for increasing social wealth are directly linked to the desire to increase one's personal fortune. Government primarily acts as a watchdog to prevent abuses, and food, shelter, and healthcare are private concerns. Both systems can and do exhibit excesses. In socialism, governments can assume huge, sometimes totalitarian powers over the individual and society. In capitalism, the private drive for wealth can override and sometimes endanger the social good.

At the end of World War II, the Bretton Woods agreements fashioned a global arena for capitalist free trade and open competition for markets that was in direct opposition to the Soviet economic system. As a result, the Soviet government moved decisively to further extend its economic and political power into larger and larger "spheres of influence." It seized the territories that its armies had liberated at the end of the war, moving to take economic and political control of Poland, Estonia, Latvia, Lithuania, East Germany, Romania, Hungary, Bulgaria, and Albania. Eventually the Soviets would come to control Yugoslavia and Czechoslovakia as well, and, as Winston Churchill observed, an "iron curtain" fell between Eastern and Western Europe. When communism also spread to China, which embarked on its own program of political domination and consolidation in Asia, the United States assumed a new global role as the defender of free trade and open capitalist markets worldwide. The official US policy from the 1950s through the 1980s would be to "contain" the spread of communism in general, and Soviet and Chinese power in particular.

In the last half of the twentieth century, the United States and the Soviet Union confronted each other on several levels. Both engaged in a race to outer space. Both sides built up tremendous arsenals of conventional and nuclear weapons and stationed military forces throughout Europe. Both created political and military alliances—NATO in the West and the Warsaw Pact in the East. But perhaps most important, both sides tried to gain influence around the world, as former European colonies in Latin America, Africa, the Middle East, and Asia sought independence. In the last half of the twentieth century, these two huge economic and political systems dominated capital flows, each trying to extend its influence into new markets (see Figure 7.2). Two rival economic spheres emerged—the West and the East.

New World or New World Order?

World War II had massively weakened the ability of the core European colonial powers to exercise political control over the periphery. One by one, English, French, Belgian, and Dutch colonies sought independence. Europeans fought back, with mixed success, using the same brutal methods they had employed during their colonial rule. But the United States and the Soviet Union were not casual observers to these conflicts, since neither wanted the new independent states to fall under the economic and political control of the other. From the 1950s through the 1970s, the Cold War bubbled up in a series of small "hot" wars all over the global periphery.

Both the United States, through its newly formed CIA, and the Soviet Union, through its similar agency the KGB, sought to control postcolonial development in the periphery. Throughout the periphery, they pitted old colonial chieftains and warlords against one another, setting the stage for decades of civil strife, destruction, and instability. Operatives for both East and West disrupted elections, assassinated or plotted to assassinate elected officials, and financed ongoing civil wars with tons of military firepower. In these battles, both East and West relied heavily on strong but often ruthless dictators to control postcolonial populations whose expectations for

Figure 7.2 Western vs. Eastern Spheres of Influence, 1959

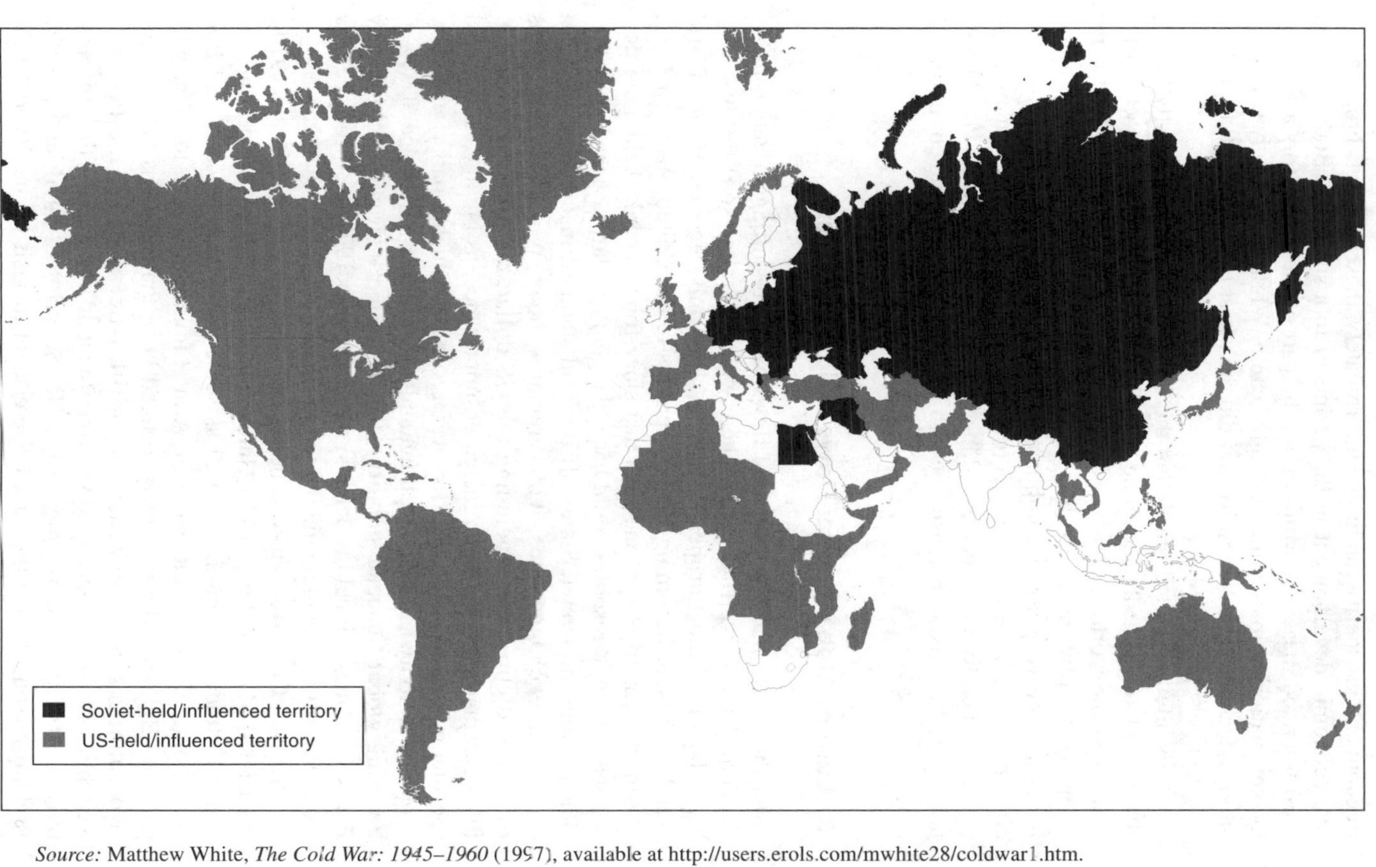

Source: Matthew White, *The Cold War: 1945–1960* (1997), available at http://users.erols.com/mwhite28/coldwar1.htm.

freedom and prosperity ran high. Unfortunately, these conflicts brought any real economic development in the periphery to a standstill. Both East and West tolerated massive corruption on the part of these dictators as long as the new nations remained in the "right" camp. While some newly sovereign nations declared their neutrality and did their best to maintain it, East/West forces largely prevented that.

So while the United States and the Soviet Union never directly fired a shot at each other, hence the name "Cold" War, they did engage each another all over the world for control of minds and hearts certainly, but also for the markets, raw materials, and cheap labor abundant in the periphery. Alongside the very real accomplishments of Bretton Woods, the Cold War created a new, increasingly dangerous global arrangement. Populations in the core generally lived secure and increasingly abundant lives. Populations in the periphery lived increasingly with civil war, disease, and corruption.

Current Global Capital Flows and Development

After the 1970s, patterns of capital flows began to shift again, into roughly the configuration we find today. There were several reasons for this shift. First, Japanese and European economies not only recovered from World War II, but were becoming major exporters to the United States. In 1971, President Richard Nixon reasoned that since foreign banks held more dollars than US gold reserves could back, the gold standard was no longer feasible. Taking the United States off the gold standard inaugurated a system of floating exchange rates, which remains today. When exchange rates float, world currencies, including the US dollar, can fluctuate in value. Financial markets take on an increased importance, as buying, selling, and speculating on the price of currencies becomes as much a part of everyday economic activity as buying, selling, and speculating on the price of wheat. Ensuring global financial stability—the job of the International Monetary Fund—becomes critical in such a system, and after 1971 the IMF became a more powerful force in regulating world finance.

Second, Cold War tensions began to ease a little. Again in the 1970s, the United States and the Soviet Union entered into a period of détente and began to trade art, culture, as well as some goods and services. The first arms limitation treaty, the Strategic Arms Limitation Treaty, together with the Anti–Ballistic Missile Treaty, offered hope that relations between the two countries could become less militaristic. In 1971, the Peoples' Republic of China replaced Taiwan (the Republic of China) as the representative of the Chinese people on the UN Security Council, and in 1972 Richard Nixon became the first US president to visit China. By 1989 the

Cold War would be officially over, and formerly socialist economies would gradually become more integrated into the world trading system.

Third, the struggling nations of the periphery began to attract some systematic attention from the rest of the world. After centuries of colonial brutality, two world wars, and decades of corruption and civil strife, it was becoming clearer that development in the periphery needed to become a part of a global agenda. In the 1970s, a consensus began to form around the belief that the World Bank and International Monetary Fund, with help from the United States, could replicate in the less-developed countries what had happened between 1950 to 1970 in Europe and Japan. By the 1980s, this approach became widely known as the Washington Consensus. Essentially, advocates of the Washington Consensus argued that opening markets in the developing world to unrestricted free trade, initiating programs for fiscal responsibility, and encouraging privatization and foreign investment would be the best approach to bringing the economic periphery into the twentieth century. The more-developed countries of the core felt that they could do for the LDCs of the periphery what Bretton Woods had done for Europe and Japan.

But there were problems with the Washington Consensus from the very beginning. Most important, diversified industrial economies had already been well established in the European countries and in Japan for decades, prior to receiving post–World War II financial aid from the World Bank. Their work forces were highly educated and highly skilled, and their legal systems protected individual private property rights and fought systematic corruption. Little of this existed in the LDCs, where agricultural, cash-crop economies remained fragile, work forces were primarily rural and uneducated, and legal systems had been corrupted repeatedly during centuries of colonialism and Cold War manipulation.

Nevertheless, GATT negotiations reflected the main principles of the Washington Consensus and moved forward with an aggressive program of trade liberalization in the LDCs. The World Bank, official development assistance, the International Monetary Fund, and foreign direct investment all played important and interlocking roles in this process.

The World Bank, Conditionality, and Comparative Advantage

The World Bank reasoned that developing exports in the less-developed countries required sizable loans for large-scale infrastructure and agricultural investments. While the LDCs were eager to obtain financial capital, the money itself was offered only if certain conditions were met—a process called *conditionality*.

The most basic condition of any loan is interest. Many LDCs were encouraged by the World Bank to borrow heavily during the 1970s, and hoped that inflation, which was high during this period, would eat away at the real interest they owed. For example, if a loan is taken at a nominal interest of 5 percent, and inflation is 6 percent, the real interest owed is actually a *negative* 1 percent—the loan is paid back with money that is worth less than the money borrowed. With inflation in the 1970s outpacing interest rates, borrowing heavily seemed to be a good bet. But the LDCs lost their gamble when inflation slowed in the 1980s, and real interest rates began to climb. A side effect of growing interest rates was that the economies in many of the more-developed countries entered a recession, and so those MDCs purchased fewer products from the LDCs, leaving the latter with less foreign currency to pay off their loans. Finally, in the 1970s the Organization of Petroleum Exporting Nations (OPEC) significantly raised the price of oil. The LDCs were forced to borrow even more money to meet their petroleum needs, as their debt kept mounting. Many private banks, flush with oil profits from the Middle East, loaned vast sums to the LDCs, and created even greater debt burdens.

A more important aspect of conditionality was the World Bank's power to decide precisely what projects would receive funding. In doing this, the World Bank employed the principle of *comparative advantage,* which maintains that if one country has a lower relative cost in producing a certain commodity, then it is to the advantage of all countries that that country specialize in producing that commodity. Thus the LDCs were encouraged to specialize in agricultural products like cotton and peanuts and use the export profits (foreign exchange) to buy whatever else they needed. This made sense because agriculture tends to be labor-intensive and LDCs generally have a surplus of labor. As another condition for the loans, the World Bank required that the LDCs remove all tariffs for imports, and all subsidies for exports—that is, develop free and completely open markets.

But these particular World Bank policies put the LDCs at a global disadvantage in at least five ways. First, land that had formerly been used for food production was placed into highly specialized and more mechanized crop production. In this process, many peasant farmers lost their livelihoods as small plots of land were consolidated into large agricultural complexes. These displaced peasants often migrated to overcrowded cities, which were unable to absorb this new influx of people into their industrial base, infrastructure, or education and healthcare systems. Without jobs or education, these individuals became a permanent underclass plagued by poverty, illiteracy, and disease.

Second, this intense industrialized agriculture demanded specialized seed, fertilizer, pesticides, herbicides, and machinery that had to be purchased from the MDCs, adding to already ballooning debt. Third, planting the same crop over and over often degraded fragile ecosystems. In many

LDCs, land began to lose fertility. Fourth, many LDCs became net importers of food with no effective means of national food distribution. Famine claimed more and more lives in the LDCs, not necessarily because there was no food, but because displaced peasants were too poor to buy relatively expensive imported food. Ironically, both food production and hunger grew globally during this time.

Fifth, the LDCs never fully realized enough profit from their agricultural exports, because although the World Bank required them to drop all tariffs and subsidies, many of the MDCs they wanted to trade with continued to heavily subsidize their domestic agriculture to protect their own farmers from cheap imports. For example, the United States continued to subsidize its cotton farmers, making US cotton some of the cheapest cotton in the world. Most LDCs, no matter how efficiently they produce any commodity, cannot compete with heavily subsidized commodities from MDCs. And LDCs cannot enact their own subsidies without alienating the World Bank. Without sufficient foreign exchange from trade to pay their bills, debt in the LDCs kept growing throughout the 1980s and 1990s.

To add to these problems, corrupt leaders in many LDCs robbed their own governments and people of needed capital. Political expediency during colonialism and the Cold War had not fostered strong democratic regimes in most of the LDCs. On the contrary, throughout the end of the twentieth century, ruthless dictators and warlords gained power and engaged in corruption as a part of standard economic practice in too many LDCs. These dictators siphoned off and redirected capital coming into foreign bank accounts in MDCs—a process called *capital flight.* As a result, in too many LDCs there was often little to show for years and years of continued loans and mounting debt.

Finally, poverty in some LDCs led to crime and social unrest, which came to dominate local government policies. Illegal crops like poppies in Afghanistan or coca in Latin America began to feed a worldwide drug trade. Some individuals in LDCs who felt their land had been taken from them unjustly openly rebelled. Some famine-stressed countries erupted into civil war. Thus, crime and civil strife have led some governments to purchase arms and weapons for controlling the violence. In Sudan, for example, land degradation has contributed to problems between residents of Darfur and the Sudanese government. Amnesty International has reported that the Sudanese government has been spending a large percentage of its oil revenues on arming militias against the people of Darfur (Amnesty International 2007b). At a time when education, healthcare, and jobs are desperately needed, many countries have used precious foreign exchange and even borrowed more heavily to arm themselves against rebels and criminals. All the while, the debt in the periphery has kept mounting.

Structural Adjustment and Foreign Direct Investment

By 1991, the LDCs had accumulated almost $1.4 trillion of total external debt to the World Bank, the International Monetary Fund, and other creditors, and were not experiencing sufficient economic growth to service their debt (meet interest payments), let alone retire it. In fact, debt was running at about 127 percent of total exports of goods and services. Some MDC economists began to fear that international economic crises might result if the LDCs defaulted on their debts (Hertz 2004). This is exactly what happened in Argentina in 2002, when the country defaulted on $93 billion. Foreign investment fled and the Argentine economy collapsed.

The International Monetary Fund, fearful that global economic crises might become epidemic if the less-developed countries unilaterally devalued their currencies, began an aggressive series of loans to the LDCs with even stricter conditions. These new conditions imposed by the IMF are often collectively called *structural adjustment programs.* Such programs have become very controversial in recent years, because through them the IMF has directly influenced the internal budgets of sovereign nations. Specifically, structural adjustment programs have required some national governments to make large budget cuts in education and healthcare programs. Some countries have been instructed by the IMF to cut their education and social service budgets by as much as 40 percent, to enable them to balance their budgets and make greater payments on their debts. At the very least, these kinds of cuts create severe hardships for the poorest of the poor, and at worst can exacerbate the types of social unrest many LDCs experience.

In addition, the IMF has pushed countries toward privatization, which essentially means countries sell the development rights to their own natural resources, like oil, to private (often foreign) corporations. Often, these private corporations are multinationals located in the MDCs. Social services, like education and healthcare, are also privatized. Privatization is closely linked with the larger global movement toward increasing trade liberalization. Both envision a world where private corporations have unlimited access to internal domestic economies. These policies of trade liberalization have increased the amount of foreign direct investment in LDCs, particularly by multinational corporations. But while these investments do create some internal jobs, the majority of the profits from them often leave the LDCs, migrating instead to the MDCs, where the multinational corporations originate.

Oil production in Africa and Latin America, for example, has become dominated by large multinational oil corporations like Exxon-Mobil (the United States), Royal Dutch Shell (the United Kingdom, the Netherlands), British Petroleum (the United Kingdom), and Chevron-Texaco (the United States). Relatively few LDCs (Saudi Arabia, Venezuela, Iran, Mexico) operate their own large oil corporations. Shell-Nigeria, for example, accounts for slightly more than 50 percent of the total oil production there, with Chevron-Texaco and Exxon-Mobil accounting for almost all other oil production. With trade liberalization policies as part of structural adjustment programs, LDCs are encouraged to grant huge concessions to multinational corporations, in the process often ceding not only profits, but environmental and social controls as well.

Recent structural adjustment programs have also called for the privatization of public services like healthcare, education, and even public utilities like water. Without financial support from the government, the cost of these services can rise beyond the reach of individuals with even moderate incomes, jeopardizing their well-being.

Until recently, trade liberalization was required by the World Bank and IMF as part of the conditionality for receiving needed capital through loans. The LDCs could avoid these conditions by opting out of the loan process altogether. But in 1995, during the last round of GATT talks, the World Trade Organization was created. The purpose of the WTO is to negotiate all global trade disputes with the goal of moving the entire global economy toward greater trade liberalization. Since the WTO now has the jurisdiction and the clout to force countries to continue liberalizing their trade policies, the Washington Consensus has in effect become the international law of trade.

Foreign Aid

Foreign aid, or official development assistance, has not really solved the financial problems of the LDCs. This is due to two major reasons. First, there is simply not enough of it to really help. Second, official development assistance is too often used as a tool of foreign or economic policy on the part of the donor nation.

Ideally, LDCs need large infusions of untied capital to break the cycle of borrowing and debt, and to develop enough social and industrial infrastructure to compete fully in the global economy. As early as 1970, the UN General Assembly noted "the role which can be fulfilled only by official development assistance" and urged each economically advanced country to increase its ODA to reach a net amount of 0.7 percent of its gross national product (GNP) by the middle of the decade. This same goal was repeated in the UN's Millennium Development Goals in 2000. But to date, the twenty-two largest MDC economies (collectively known as the Development Assistance Committee [DAC]) are averaging only about 0.33 percent of gross national income (GNI); this percentage is projected to rise to 0.36 by 2010, as shown in Figure 7.3. Among the largest contributors in terms of percentage of GNI are Sweden (1.86 percent in 2006), Luxembourg and Norway (0.89 percent), and the Netherlands and Denmark (0.80 percent).

Figure 7.3 DAC Members' Net ODA, 1990–2005, and DAC Secretariat Simulations of Net ODA, 2006–2010

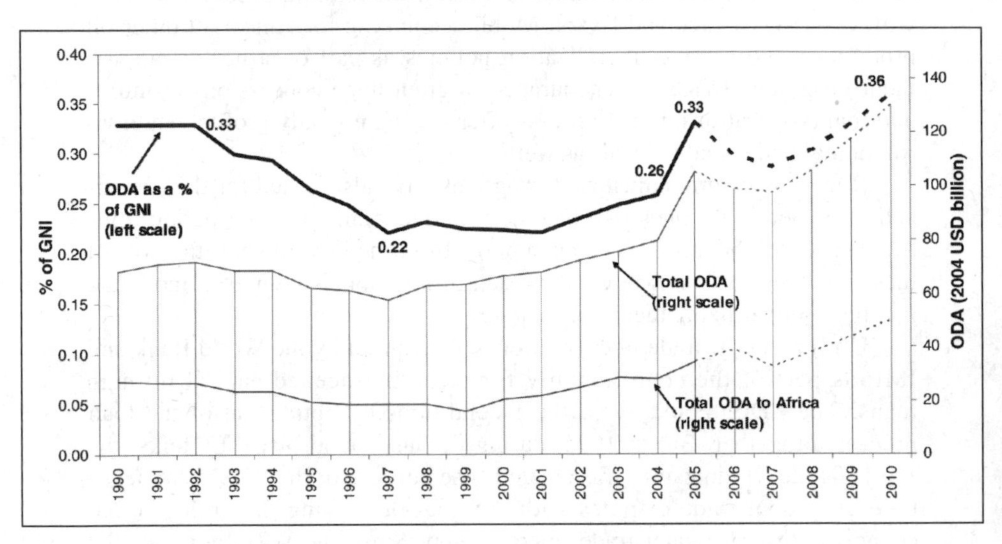

Source: Organization for Economic Cooperation and Development (2007), available at http://www.oecd.org/dataoecd/57/30/35320618.pdf.

The United States, while providing the largest amount of ODA in terms of dollars ($22 billion in 2006), is among the smallest contributors in terms of percentage of GNI (0.17 percent). Only Greece (0.16 percent in 2006) contributes less. The average ODA effort of all the DAC countries, as of 2006, is 0.46 percent of GNI.

Moreover, official development assistance is often tied to projects that directly benefit the donor nation at least as much as the intended LDC. This type of aid is sometimes called "phantom aid" and is often tied to debt relief, projects, and technical assistance in which capital flows in a circular fashion. A hypothetical example of this would be an MDC supporting a project in an LDC to build a dam for generating electricity. The money (given as ODA) is used to purchase supplies, technical assistance, and expertise from the MDC. While the LDC may or may not ultimately benefit from the dam (depending on the level of expertise already existing in that country), the multinational corporations located in the MDC do benefit directly through purchases of labor, materials, and technical assistance by the LDC. Even humanitarian food aid (which is not included in most ODA totals) often benefits the donor MDC more than the LDC. In 2007, one of the world's largest international relief agencies, CARE, declined $45 million in food aid from the United States because it believed that the aid would hurt the poor in LDCs more than it would help them (Dugger 2007). Cheap food from MDCs, often subsidized by their own governments, can drive local farmers in LDCs out of business—exacerbating problems of food production in very poor countries.

In the end, official development assistance has not relieved conditions of growing debt in many LDCs. Today, total external debt is about 108 percent of GNP in sub-Saharan Africa. This means that these countries owe more in debt than the total value of what they produce in any given year. Jubilee USA reports that Kenya has a total external debt of $7 billion. About 22 percent of Kenya's annual budget goes toward paying its debt service (interest plus principal). This amount is equivalent to Kenya's budgets for health, roads, water, agriculture, transport, and finance combined. Debt is 46 percent of GNP in North Africa and in the Middle East, 33 percent of GNP in East/South Asia and in the Pacific, and 34 percent of GNP in the Americas. Indonesia, the nation hit hard by the tsunami of 2004, currently has a debt of $132 billion and an annual debt service of $1.9 billion (Jubilee USA 2007).

One source of aid to LDCs that does not show up on any government balance sheet are the remittances that are annually sent home from emigrants living in MDCs. The Inter-American Development Bank, in association with the International Fund for Agricultural Development, recently reported that in 2006 the world's 150 million migrant workers sent more than $300 billion home to their families (more than 1.5 billion transactions of $100–300 each), almost triple the official estimate of the $104 billion in aid officially given to the LDCs (IFAD 2007).

Future Prospects

This new global structure of debt between more- and less-developed countries creates capital flows that unfortunately mirror centuries of economic activity between the core and the periphery. The dictates of the International Monetary Fund and the World Trade Organization recall the maneuvers of colonial overseers and Cold War politicians. Capital flowing from LDCs to multinational corporations and banking interests situated in MDCs raises the specter of a new imperialism.

Proposed solutions come from many quarters. Former IMF economist George Stiglitz (2002) contends that the World Bank and IMF need to rethink their "one size fits all" methods for liberalizing trade in LDCs. He notes that all developing economies need a certain amount of economic diversification and protectionism to become stable and strong. Former World Bank economist William Easterly (2006) argues that smaller, homegrown projects that strengthen local ongoing market activities have a greater chance of success in the small emerging economies of LDCs than do huge projects initiated from afar by MDCs. Former World Bank economist Noreena Hertz (2004) maintains that local institutions for economic regulation and protection of human rights need to be fostered in LDCs as a counterweight to unfettered liberalization of trade. Finally, former World Bank economist Paul Collier (2007) suggests that the countries composing the Group of Eight—Canada, France, Germany, Italy, Japan, Russia, the United Kingdom, and the United States, which together represent about 65 percent of the world economy—need to take the lead in meeting the current problem of global debt and poverty. Collier points out that, along with aid, LDCs need help creating security, appropriate laws and charters, and trade policies that emphasize export diversification and unreciprocated trade concessions to help strengthen weak economies.

Some groups, like Jubilee 2000, have argued for massive debt cancellation and enlisted celebrities like U2's Bono to present their message to the World Bank, the IMF, and MDC leaders. Partly in response to Jubilee and pressures for debt cancellation from other nongovernmental organization, the World Bank and IMF have inaugurated specific programs for the highly indebted poor countries (HIPCs). These programs have identified thirty-eight countries, thirty-two of which are in sub-Saharan Africa, for targeted aid and debt relief.

In 2002, US president George W. Bush created the Millennium Challenge Accounts, and in 2004 created the Millennium Challenge Corporation to administer the resultant 50 percent increase projected in US development assistance. Countries are selected through a set of sixteen indicators that are designed to measure their record on ruling justly, investing in people, and promoting economic freedom. Economist Jeffrey Sachs (2005) argues, however, that the World Bank and particularly the United States are not making a big enough commitment. The $5 billion generated by the 50 percent in US development assistance represents only 0.05 percent of US gross national product. Sachs strongly believes that with a contribution of 0.70 percent of gross national product, the United States could lead the fight to completely eliminate poverty in the LDCs. At the end of 2007, the *New York Times* reported that the Bush administration had spent only $155 million of the $4.8 billion it had approved.

Among the LDCs, regional development banks like the African Development Bank and the Asian Development Bank promise to focus money on the poorest countries with loans that are not tied so tightly to conditions of trade liberalization and comparative advantage. While these banks are not as well funded as the World Bank and IMF, they do bring a more focused perspective to global lending.

Whether or not suggested reforms and new institutions can succeed depends on the willingness of the more-developed countries to look beyond immediate profits and invest in the health and stability of the less-developed countries.

Discussion Questions

1. What are the lasting effects of colonialism on the flow of capital in today's world?
2. How did the politics of the Cold War impact less-developed countries?
3. Are current World Bank and IMF policies helping or hurting less-developed countries?
4. Do more-developed countries have a responsibility to assist less-developed countries? Why or why not?
5. Should less-developed countries be relieved of their current debts? What would be the pros and cons of such an approach?

Note

Special thanks to Joe Lehnert for his contributions to this chapter.

Suggested Readings

Collier, Paul (2007) *The Bottom Billion: Why the Poorest Countries Are Failing and What Can Be Done About It.* New York: Oxford University Press.
Easterly, William (2006) *The White Man's Burden: Why the West's Efforts to Aid the Rest Have Done So Much Ill and So Little Good.* New York: Penguin.
George, Susan (1988) *A Fate Worse Than Debt.* New York: Grove.
Hertz, Noreena (2004) *The Debt Threat: How Debt Is Destroying the Developing World . . . and Threatening Us All.* New York: HarperCollins.
Klein, Naomi (2007) *The Shock Doctrine: The Rise of Disaster Capitalism.* New York: Metropolitan.
Mamdani, Mahmood (1996) *Citizen and Subject: Contemporary Africa and the Legacy of Late Colonialism.* Princeton: Princeton University Press.
Rodney, Walter (1983) *How Europe Underdeveloped Africa.* Washington, DC: Howard University Press.
Sachs, Jeffrey (2005) *The End of Poverty: Economic Possibilities for Our Time.* New York: Penguin.
Stiglitz, Joseph E. (2002) *Globalization and Its Discontents.* New York: Norton.

8

Poverty in a Global Economy

Don Reeves and Jashinta D'Costa

- Poverty is a mother watching her kids get sick too often, or too hungry to pay attention in school.
- Poverty is to live crowded under a piece of plastic in Calcutta, huddled in a cardboard house during a rainstorm in São Paulo, or homeless in Washington, DC.
- Poverty is watching your child die for lack of a vaccination that would cost a few pennies, or never having seen a doctor.
- Poverty is a job application you can't read, a poor teacher in a rundown school, or no school at all.
- Poverty is hawking cigarettes one at a time on jeepneys in Manila, or being locked for long hours inside a garment factory near Dhaka or in Los Angeles, or working long hours as needed in someone else's field.
- Poverty is having a life expectancy below the national or world average.
- Poverty is to feel powerless—without dignity or hope.

Dimensions of Poverty

Poverty has many dimensions. Religious ascetics may choose to be poor as part of their spiritual discipline. Persons with great wealth may ignore the needs of those around them or may miss the richness and beauty of nature or great art and remain poor in spirit. But this chapter is about poverty as the involuntary lack of sufficient resources to provide or exchange for basic necessities—food, shelter, healthcare, clothing, education, and opportunities to work and to develop the human spirit.

Globally, poor people disproportionately live in Africa. The largest number live in Asia. A significant number are in Latin American and Caribbean countries. Three-quarters of the people in several sub-Saharan Africa countries and Haiti are poor and, depending on the threshold chosen, at least one in five and perhaps as many as one-half of all people in developing countries together are poor. The situation worldwide is shown in Figure 8.1.

But no place on the globe is immune to poverty. The United States, some European countries, and Australia also have large blocs of poor people. With few exceptions, the incidence of poverty is higher in rural than in urban areas, but is shifting toward the latter. Nearly everywhere, women and girls suffer from poverty more than do men and boys; infants, young children, and elderly people are particularly vulnerable. Cultural and discriminatory causes of poverty are immense, especially among minorities and indigenous people. The difficulties in changing long habits and practices should not be underestimated.

In this chapter, we look first at ways in which poverty is measured. Then we look at approaches to reducing poverty in the context of a global

Figure 8.1 Number and Percentage of Poor People in Developing Countries, 1981–2004

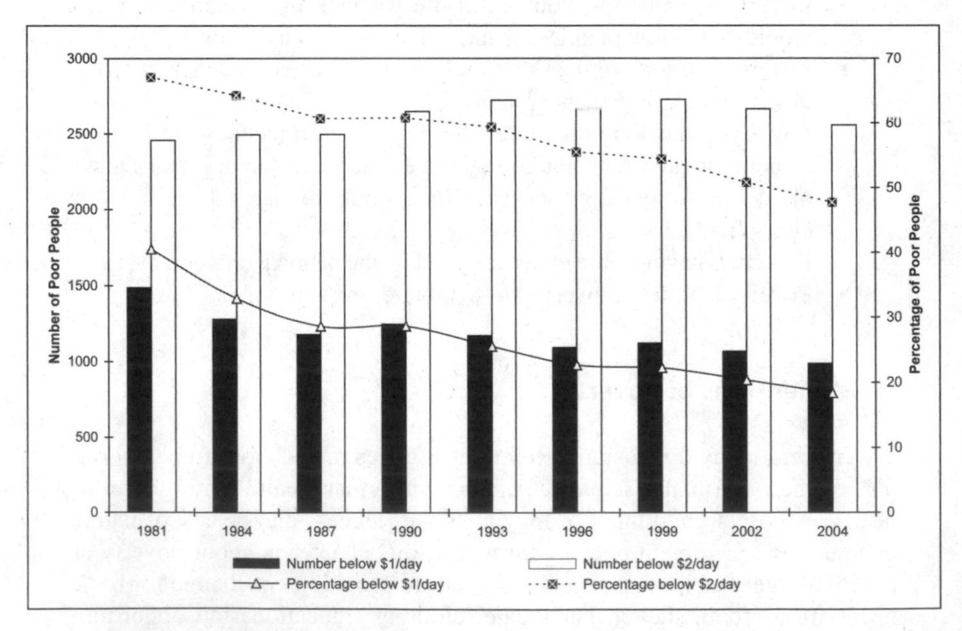

Source: World Bank, *World Development Indicators 2007* (Washington, DC, 2007).

economy, especially the relationship between economic growth and inequality. Finally, we examine a series of policy choices that developing country societies might consider as they attempt to reduce poverty.

Measuring Poverty and Inequality

Poverty is not the same in the United States or Poland or Zimbabwe. It will often be described differently by supporters or critics of a particular regime. Poverty does not lend itself to an exact or universal definition. Deciding who is poor depends on who is measuring, where, and why.

Poverty Thresholds

Poverty is usually measured by income or consumption. The World Bank estimates poverty using two thresholds. Worldwide, in 2004 about 986 million people lived on incomes equivalent to less than $1 per day. Also in 2004, about 2.6 billion lived on less than $2 per day (World Bank 2007). All the first group and the majority of the latter chronically lack some or all basic necessities. The rest live so close to the edge that any emergency—illness, work layoff, drought—pushes them from just getting by into desperation.

In the United States, poverty is defined as three times the value of a thrifty food plan devised by the US Department of Agriculture, adjusted for family size—$20,444 for a family of four in 2006. Some critics say the threshold is higher than necessary, partly because certain government program benefits—such as Medicare, housing subsidies, and school meals—are not counted. But poor people themselves feel hard-pressed. The thrifty food plan was worked out in the early 1960s to address short-term emergencies. Although it is adjusted annually for changes in food prices, other costs, particularly housing, have grown faster than food costs since the plan's base year (1955); so the threshold has represented a gradually declining standard of living.

The selection of the poverty threshold often makes a dramatic difference in the observed poverty rate. The World Bank, based on its cutoff of $1 per person per day, estimates that 128 million (9.9 percent) of China's total 1.3 billion people were poor in 2004. The Chinese government, using a lower cutoff point, claimed that less than 50 million of its people were poor (World Bank 2007).

Gross Domestic Product and Gross National Product

Two other widely used income measures are per capita gross domestic product (GDP) and gross national product (GNP). GDP is the value of all goods and services produced within an economy; GNP equals GDP plus or minus transfers in and out of the economy, such as profit paid to foreign investors or money sent home by citizens working abroad. (The World Bank now uses gross national income [GNI] instead of GNP, since transfers from abroad are not "produced" within a country's boundaries. In this chapter we use the more familiar measurement, GNP.)

Among the world's economies, large and small, the World Bank counts fifty-three low-income economies with an annual per capita GNI of $905 or less (as of 2004). At the other end are sixty high-income economies with per capita GNI of $11,116 or more. In between are ninety-six middle-income economies with per capita GNI between $906 and $11,115 (World Bank 2007).

GNP (or GDP) provides a quick measure of the capacity of an economy overall to meet people's needs. It also represents the pool from which savings and public expenditures can be drawn. But GNP is seriously flawed as a measure of poverty or well-being, because it gives no information about the quality of the production or the distribution of income within the country.

First, GNP and GDP fail to distinguish among types of economic activity. Manufacturing cigarettes, making bombs, and running prisons are scored as contributing to GNP or GDP the same as making autos, teaching school, building homes, or conducting scientific research. Second, many goods and services generate costs that are not reflected in their prices—polluted air from manufacturing or illness from overconsumption, for example. Third, many nurturing and creative activities—parenting, homemaking, gardening, and home food preparation—are not included, because they are not bought and sold. At best, GNP and GDP figures include only estimates for food or other goods consumed by producers, unpaid family labor, and a wide range of other economic activities lumped together as the informal sector. Illegal or criminal activities, such as drug-dealing or prostitution, are generally not included in estimates, but nonetheless contribute to some people's livelihood.

Purchasing Power Parity

GNP and GDP figures for various countries are usually compared on a currency exchange basis. In 2005, the per capita GNP in Bangladesh, at 32,000 taka, could be exchanged for $470. But 32,000 taka would buy more in Bangladesh than $470 would buy in the United States, primarily because wages there are much lower. Thus the World Bank and the United Nations Development Programme (UNDP) have adopted a measure—purchasing power parity (PPP)—that estimates the number of dollars required to purchase comparable goods in different countries. Bangladeshi PPP is estimated at $2,090, rather than $470 (World Bank 2007).

PPP estimates make country-to-country comparisons more accurate and realistic, and somewhat narrow the apparent gap between wealthy and poor countries. Even so, vast disparities remain. PPPs of $32,220 to $41,950 per capita in Canada, Switzerland, and the United States are forty to sixty times those in Tanzania and Ethiopia, at $730 and $1,000 (World Bank 2007).

Inequality

Estimates of poverty and well-being based on estimated GDP are at best crude measures. GNP, GDP, and PPP are all measured as country averages. But because poverty is experienced at the household and individual levels, the distribution of national incomes is crucial.

Detailed and accurate information is necessary for targeting antipoverty efforts and particularly for assessing the consequences of policy decisions in a timely fashion. But census data as comprehensive as those for the United States are a distant dream for most poor countries. Many of them do not keep such basic records as birth registrations, and may have only a guess as to the number of their citizens, let alone details about the conditions under which their citizens live. Representative household surveys are the only viable tool for most countries for the foreseeable future.

Household surveys, to be useful—especially for comparison purposes—need to be carefully designed, accurately interpreted, and usable for measuring comparable factors in different times, places, and circumstances. Private agencies, many governments, and even some international agencies are tempted to shape or interpret surveys to put themselves in the best light. Users of survey results need to be keenly aware of who conducted each survey and for what purposes.

Global inequality. Globally, we accept gross income inequality. The most used measure of inequality compares the income of the richest one-fifth, or quintile, of each population, with that of the lowest quintile. Measured in currency exchange value, the wealthiest one-fifth of the world's people, about 1.3 billion, control more than 85 percent of global income. The remaining 80 percent of people, about 5.3 billion, share less than 15 percent of the world's income. The poorest one-fifth receive barely 1 percent. The ratio between the average incomes of the top fifth and the bottom fifth of humankind is 80 to 1 (see Figure 8.2).

Using purchasing power parity as a measure, the poverty gap narrows, but it is still extreme. The first-ever global survey in 2001, based on household surveys and using PPP, estimated that during the 1990s the poorest 20 percent could purchase, on average, about one-twelfth as much as the top 20 percent, and the ratio increased during this period.

Figure 8.2 Distribution of Population and Global Income (percentages)

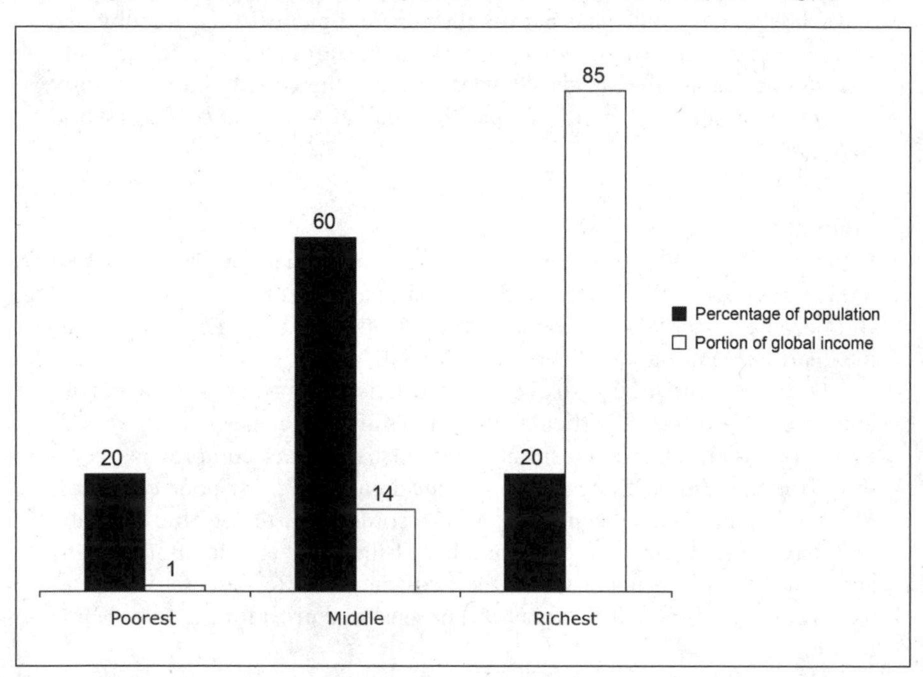

Source: World Commission on the Social Dimensions of Globalization, *A Fair Globalization: Creating Opportunities for All* (Geneva: ILO, 2004).

A more recent survey of 454 household studies, covering nearly all the developing countries since 1975, concludes that there was no clear overall trend in comparisons based on purchasing power from 1981 to 2004. Review again the trend lines in Figure 8.1. The number of people living on less than $1 per day (PPP) dropped moderately over this period, and their proportion dropped from 40.6 percent to 18.4 percent. But the number living on less than $2 per day actually rose, although the proportion dropped because of increasing population.

Two further caveats are necessary. First, nearly all gains have occurred in China and in India. Without China and India, the number of extremely poor has risen, especially in Africa but elsewhere as well. Second, the people below the threshold of $1 per day are on average much worse off, again especially in Africa but even in China and India.

Meanwhile, the incomes of a very small proportion of nearly every nation's population have skyrocketed. Recently, World Bank economist Branko Milanovic estimated that the ratio between the income of the top 5 percent and the bottom 5 percent of people in the world was 165 to 1 in 1993, up from 78 to 1 in 1988 (Milanovic 2006; UNDP 2003).

Inequality in the United States. Using US census data, in 1980 the richest one-fifth of US households received 43.7 percent of total US income, while the poorest one-fifth received 4.3 percent—a ratio of about 10 to 1. By 2005, the ratio had widened to nearly 15 to 1, with the top one-fifth receiving 50.4 percent of income, and the bottom one-fifth only 3.4 percent. Tax changes made between 2000 and 2004 are exacerbating this trend. As in the rest of the world, the poverty gap in the United States is widening rapidly between the very top and bottom. The poorest are growing even poorer. The richest are growing much richer.

The World Bank uses per capita income instead of household income to compare income shares. It found the ratio between the rich and poor quintiles' incomes in the United States to be 8.5 to 1 in 2000, the highest among industrial nations. Among other wealthy nations, the ratio ranged from 3.8 and 4.3 to 1 in Finland and Denmark, up to 7.0 and 7.2 to 1 in Australia and the United Kingdom (compare with the ratios in Table 8.1).

Inequality within developing countries. Among the low-income countries for which estimates are available, the ratios between the top and bottom quintiles range from 5 to 1 for Bangladesh and 5.6 to 1 for India, up to 21.8 to 1 for Brazil and 25.1 to 1 for Colombia (World Bank 2007).

Differences in income distribution make a big difference to poor people. Measured in PPP, Brazil's per capita GDP is more than double India's, but the poorest 20 percent of the population in India have one-quarter more purchasing power. Egypt's GDP is less than half Chile's, but Egypt's poor peo-

Table 8.1 Poverty Impact of Income Distribution, Selected Countries, 2005

	GNI PPP$/Capita	Income PPP$/Capita		Ratio, Highest 20% to Lowest 20%
		Lowest 20%	Highest 20%	
South Korea	21,850	8,631	40,969	4.7
Chile	11,470	2,179	34,410	15.8
Malaysia	10,320	2,270	28,019	12.3
Thailand	8,440	2,659	20,678	7.8
Brazil	8,230	1,152	25,143	21.8
Colombia	7,420	928	23,262	25.1
China	6,600	1,419	17,127	12.1
Philippines	5,300	1,431	13,409	9.4
Egypt	4,440	1,909	9,679	5.1
Indonesia	3,720	1,562	8,054	5.2
India	3,460	1,401	7,837	5.6
Bangladesh	2,090	899	4,462	5.0
Nigeria	1,040	260	2,558	9.8

Source: World Bank, *2007 World Development Indicators* (Washington, DC, 2007).

ple have nearly as much purchasing power. Thailand's per capita GDP is 15 percent more than Colombia's, but Thai poor have nearly three times the purchasing power. In China, the fastest-growing economy in the world, the income share of the richest quintile has grown while everyone else's has shrunk. Purchasing power for the poor has barely budged (see Table 8.1).

Direct Measures of Well-Being

Other indicators measure well-being even more directly than income or poverty rates—for example, infant or under-five mortality rates, life expectancy, educational achievement, and food intake.

Hunger. The Food and Agriculture Organization (FAO), part of the United Nations, estimates shortfalls in food consumption. In its 2006 report on the state of food insecurity in the world (FAO 2006a), the FAO estimated that the proportion of all people in developing countries who are hungry declined slightly over a period of two decades—from about 900 million in 1969–1971 to about 820 million in 1990–1992—but has stalled since then. Because population increased rapidly over the period, however, the proportion of hungry people in developing countries declined from about 35 percent to about 17 percent.

The most dramatic gains in reducing hunger over the period were in East and Southeast Asia, most notably China, where the percentage of hungry people dropped from 41 to 11 percent, and the number of hungry people dropped by two-thirds, from 476 million to 142 million (see Figure 8.3). Less dramatic gains by both measures were recorded in the Middle East and North Africa. In South Asia, Latin America, and the Caribbean, the proportion declined, but the absolute number increased slightly over the period. In Africa, the percentage of hungry people increased, while the number soared from 103 to 204 million (FAO 2006a).

In 2006, 89 percent of US households (103 million) were food-secure, meaning that they had access at all times to enough food for active, healthy life for all household members. Nearly 11 percent of US households (12.6 million) were food-insecure at some time during the preceding year. Of these, about one-third had very low food security, meaning that food intake of one or more adults was reduced or otherwise disrupted because the household lacked money or other resources to obtain adequate food (USDA 2007).

Human Development Index. The United Nations Development Programme has developed the Human Development Index (HDI), which gives equal weight to three factors: life expectancy at birth, educational attainment (based on the adult literacy rate and mean years of schooling), and per capita purchasing power.

Figure 8.3 Number of Undernourished People by Region, 2001–2003 (millions)

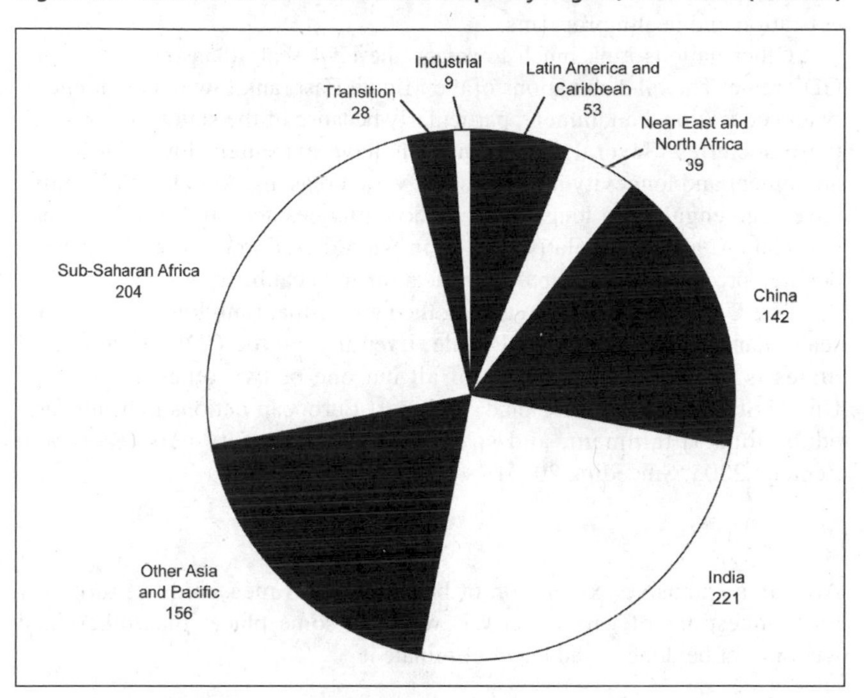

Source: Food and Agriculture Organization, *The State of Food Insecurity in the World, 2006* (Rome, 2006).

People's lives can be improved if even limited resources are focused on nutritional programs, public health, and basic education. Both China and Sri Lanka, for example, have invested relatively heavily in education and healthcare since independence. They rank alongside many industrial countries in terms of life expectancy and educational attainment.

Most of the former communist countries invested heavily in education and healthcare. Some former colonies continue to build on the educational systems established during the colonial era: Vietnam, Laos, and Madagascar (colonized by France); Tanzania, Uganda, and Burma (colonized by Britain); and the Philippines (colonized by Spain and the United States). Several Latin American countries have emphasized education more recently: Chile, Cuba, Costa Rica, and Uruguay. Each of these countries ranks higher on the HDI scale than other nations with comparable per capita GDPs.

But sustaining such improvements requires steady or improving economic performance. Many of these same nations have suffered recent economic downturns or are in the midst of drastic political and economic change. In the short term, at least, they are hard-pressed to maintain their education and health programs.

Other nations rank much lower on the HDI scale than on a per capita GDP scale. The oil-rich nations of the Middle East rank low in both longevity and educational attainment, particularly because of the status of women in these societies. Several African nations have extremely low educational attainment and longevity indicators, for varied reasons. Angola and Namibia have been engulfed in long independence struggles and civil war. Botswana and Gabon, although relatively rich in natural resource income, have not devoted proportional resources to education and healthcare services.

The United States is also among the nations that rank lower on the HDI scale than on a per capita GDP scale. Even though the GDP of the United States is higher than the GDPs of all but one or two other nations, the United States lags behind Canada and most European nations in healthcare, educational attainment, and strength of social safety nets (Bane and Zenteno 2005; Smeeding 2005).

* * *

After this extensive exploration of how poverty is measured, we turn now to the questions of why poverty is worse in some places than others and what might be done to reduce or eliminate it.

Economic Growth and Poverty Reduction in a Global Economy

Individual, community, and national efforts to reduce poverty must be set in the context of the threefold revolution during the past generation that has transformed national markets into a truly global economy:

- the evolution of a single worldwide system of producing and exchanging money, goods, and services;
- the shift from a resource-based economy to a knowledge-based economy; and
- the acceptance of market-based economics as conventional wisdom by most political leaders throughout the world.

The first two aspects of the global revolution are inextricably linked. New information, communication, and transportation technologies have dramatically changed the way many older businesses are managed (on-time delivery of manufacturing components and inventory, digital voice transcription, international call centers, and automatic business-to-business transmission of orders and billing, for example). These same technologies have also spawned whole new industries (distance learning and online marketplaces). Their immense capacity to process information has enabled new scientific advances (mapping the human genome, medical and space research, cross-species bioengineering). The knowledge factor outweighs the resource factor in an increasing number of endeavors.

At their peak in 1970, about 40 percent of the world lived in countries with centrally planned economies. By the turn of the twenty-first century, with a couple of minor exceptions, these countries had all introduced market-oriented reforms. Some, such as China, are introducing such reforms gradually. Others, such as the Soviet Union, held on until their economies collapsed. Despite rhetoric, no nation is attempting a true free market economy. Debates about political economies are usually about deciding which functions can be left to markets and which cannot, or shaping the context or rules under which markets function. A central question is how market-oriented political economies might contribute more to reducing poverty, especially in the poorest countries.

The Global Work Force: Need for 2 Billion Jobs

The route out of poverty for most people is through new economic opportunities—jobs or business ventures. As the world's population grows by one-third, from about 6.6 billion to over 8.0 billion in the next generation, the global labor force will grow from 3.1 billion to more than 4 billion workers (Kapsos 2007). The International Labour Organization (ILO) has estimated that more than a billion workers are unemployed or underemployed. More than half of the new workers have already been born, and the number of unemployed is still growing. The pressing need, therefore, is to create 2 billion new economic opportunities during the next decades. Most of the new jobs or businesses will be needed in developing countries, where more than 95 percent of the increase in population and labor force is taking place.

Virtually all of the opportunities for added income will need to be nonfarm jobs. Governments in developing countries, or developing markets, may increase incentives for food production, but farmers are likely to adopt technologies that increase their productivity and reduce farm employment even faster. More and more farmers, or their children, will seek nonfarm employment. Whether such nonfarm employment is urban or rural will depend on policy choices. Improvements in education, healthcare, and public infrastructure can provide some public service jobs. But most new income-earning opportunities, if they come to pass, will be in the private sector.

Every job or new opportunity, whether public or private, for employee or self-employment, requires savings and investment—in human resources and in physical capital (buildings and tools). The rates of savings and their allocation are crucial factors in determining whether enough decent income-earning opportunities can be created; these factors are determined in large measure by public policies.

Globalization and Poverty

Globalization, discussed more fully in Chapters 6–7, has had mixed effects on employment, poverty, and income distribution. Hundreds of thousands of new jobs have been created in poor countries as standardized manufacturing and, more recently, information technology jobs have been moved, especially to Asia. By the standards of the developed nations and human rights advocates, many of these jobs, especially in manufacturing, are not very desirable. But for most of the workers, they are an improvement over any alternative.

Low-wage jobs, by themselves, will not be enough to build thriving economies. Will poor nations be able to emulate Korea and Taiwan from a generation ago, with increasingly sophisticated products and services for themselves as well as for export?

One effect of globalization in the United States has been the "disappearing middle." Growing incomes and reduced inequality in the United States during the 1950s, 1960s, and 1970s reflected a growing middle class—most prominently, well-paid union workers in manufacturing. With labor-reducing technologies and increasing imports of standardized manufactures, more than half of US manufacturing jobs have disappeared since 1980. Most workers have found new jobs. About half of the new jobs are "tech-up" jobs, with comparable or higher wages; the other half are "dumb-down" jobs, some still in manufacturing, but primarily in the service sector, with lower wages and benefits, partly the result of competition with immigrants. This trend has contributed to the widening income gap described above, and to the growing number of working poor in the United States.

Some observers have noted a parallel phenomenon at the nation-state level. The poorest nations are actually gaining ground from a very low base with low-wage manufacturing. A few other nations are becoming competitive in more sophisticated goods and services, as have Korea and Japan. But a number of middle-income countries, especially those without highly skilled workers or valuable natural resources, seem to be stagnating. Refer once again to Figure 8.1.

Economic Growth and Poverty Reduction

Economic growth is often held up as the primary goal for economic development and as the means to increased employment opportunities. Some analysts, bankers, and political leaders *equate* development with economic growth. Most of these people expect poverty and other social problems to shrink as economies grow.

Economic growth is a necessary, but not sufficient, condition for reducing poverty. The distribution of the added income is also critical. Poverty has fallen rapidly in some fast-growing economies (Korea, Indonesia, China), while not changing much or rising in others (Brazil, South Africa, Oman). What lessons can be learned from the varied experiences?

Because poverty is experienced in households and by individuals, detailed and accurate information at those levels is critical. Both recent and earlier World Bank studies (Chen and Ravallion 2004; Ravallion 2001; Ravallion and Chen 1997), covering nearly a hundred developing countries, find the following consistent relationships:

- Poverty rates have consistently fallen as average incomes have risen, and have risen as average incomes have fallen.
- Poverty rates have not declined anywhere in the absence of economic growth.
- In developing countries, inequality increased about as often as it decreased as average incomes rose.
- In transition economies (former communist states in Eastern Europe and Central Asia), inequality consistently increased at the same time as average incomes fell, at least through a several-year period of adjustment.

Other studies of eight East and Southeast Asian countries show that it is possible to have both economic growth and decreasing inequality if the right policies are in place. In South Korea, for example, where per capita income has grown rapidly, the richest fifth of the population has about five times as much income as the poorest fifth. The ratio has narrowed over the past three decades; poor people have shared in the rapid growth. That is, their incomes have increased more than average on a percentage basis, although the absolute increase in income has been greater for wealthier persons.

In sharp contrast, Brazil's per capita GNP was twice Korea's in 1970. Since then, its economy has grown about half as fast; by 2002, Korea's per capita GNP was more than twice Brazil's, as measured by PPP. Meanwhile, the income ratio between Brazil's poorest and richest fifth has widened even further, to more than 20 to 1. Poor Brazilians have not benefited from growth and remain mired in deep poverty (see Table 8.1).

The Asian countries have reduced economic inequality, or at least have not increased it, by giving poor people the incentive and the means to improve their own earning power. Examples are land reform and support for small farmers in Korea and Taiwan; high school education, especially for women, in Singapore; and manufacturing for export, which has raised the demand for unskilled factory workers, plus a massive affirmative action program for the poorer ethnic groups, in Malaysia.

A Virtuous Circle

Declining inequality and economic growth support each other in three ways, in an ascending, or "virtuous," circle (see Figure 8.4):

- As poor families' incomes increase, they invest more in human capital—more education and better healthcare for their own (usually fewer) children.
- Improved health and better education, which usually accompany decreased inequality, increase the productivity of poorer workers and their communities and nations. This in turn further increases their income.
- Greater equality contributes to political stability, which is essential for continued economic progress.

Relative equality in distribution of national incomes also increases the likelihood that economic growth can be sustained. Widespread participation in political as well as economic activity reduces the likelihood of enacting bad policies and permits their earlier correction (Birdsall, Pinckney, and Sabot 1996).

Figure 8.4 A Virtuous Circle

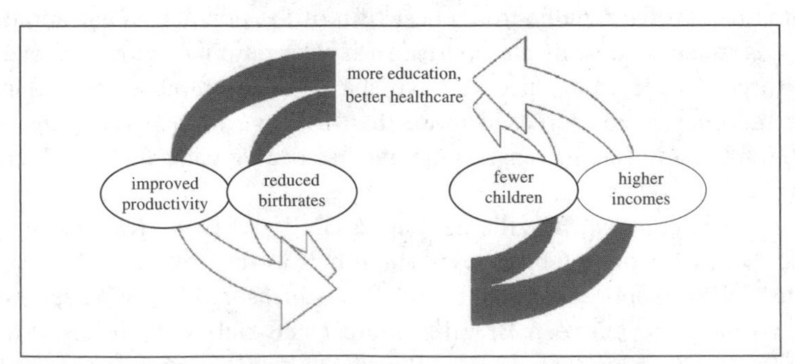

Source: Adapted from Bread for the World (BFW) Institute, *Hunger 1995: Causes of Hunger* (Washington, DC, 1994). Used with permission of the BFW Institute.

Sustainable Development

More than relatively equally distributed economic growth is needed to achieve long-term poverty reduction. The growth and distribution must be sustainable. The concept of *sustainable development* emerged in the 1970s to help ensure that development would not ruin the environment, which would, sooner or later, slow or reverse development. The UN's World Commission on Environment and Development (WCED) has defined sustainable development as "development which meets the needs of the present without compromising the ability of future generations to meet their own needs" (WCED 1987). More recently, and as used here, the concept has evolved to incorporate other aspects of development in addition to economic growth. Advocates of international justice, environmental protection, peace, sustainable population growth, democracy, and human rights have increasingly come to see that their goals are interlinked:

- There is no way to save the rainforests of Brazil without addressing poor Brazilians' need for land.
- There is no way to reduce rapid population growth in developing countries without improving living standards, especially for girls (see Chapters 9 and 10).
- There are no durable solutions to poverty and hunger in the United States without social peace, broader democratic participation, and a shift to economic patterns that will be environmentally sustainable.

Bread for the World is a nongovernmental organization focused on policy decisions that can reduce hunger and poverty. It defines sustainable development in terms of four interconnected objectives: providing economic opportunity for poor people, meeting basic human needs, ensuring environmental protection, and enabling democratic participation (BFW 1995). These concepts, and those from other chapters in this book, are reflected in the policy suggestions of the next section.

Antipoverty Policies in a Global Economy

Public policies aimed at reducing poverty fall into two broad categories: creating appropriate, effective guidelines for markets; and collecting and allocating public resources, especially for investment in human resources and in social safety nets.

The reality of achieving effective antipoverty policies is of course much more difficult than the assertion. Just as some actors in the marketplace can take advantage of their economic power, they and other powerful political actors can sway policies to their own self-interest, whether at the local, national, or international level. Meanwhile, poor people, whose well-being is the strongest evidence of whether policies are effective, often lack political access or clout. In addition, they and their allies are often unclear or divided on issues of national and international economic policy.

But as the global, knowledge-based, market economy reaches into the far corners of our planet, people of goodwill have only one option: to help create and implement policies that will direct a sizable portion of this economy toward creating income-earning opportunities for poor people. The most important areas for policies to help reduce poverty include: investing in people, sustaining agriculture and food production, creating a framework for sustainable development, and targeting international financing (BFW 1997; World Bank 2001).

Investing in People

Without question, if nations are to avoid the low-skills/low-wage trap, their people must be healthy and well educated. Investments in human capital (people) are foundational.

Healthcare and nutrition. Investments in basic healthcare and improved nutrition yield huge dividends. Healthy children learn better. Healthy adults work better. Improved healthcare begins with greater attention to basic public health measures: nutrition education, clean water and adequate sanitation, vaccination against infectious diseases, prevention and treatment of AIDS, distribution of iodine and Vitamin A capsules, and simple techniques of home healthcare. Delivery of these services can be relatively inexpensive, especially in developing countries, where village women with minimal training can be employed. These basic services should have priority over urban hospitals and specialized medical training.

In some instances, public health training can be delivered in conjunction with supplemental feeding programs such as the Special Supplemental Food Program for Women, Infants, and Children in the United States, or the Integrated Child Development Services in India.

The United Nations Children's Fund (UNICEF) provides another instance of the payoff from direct interventions that include education of mothers. In 1980, UNICEF estimated that 40,000 children died each day from preventable causes, about half of which were related to hunger. By 2003, the estimate had been reduced to about 27,400 per day, thanks in part to UNICEF's fourfold program of growth monitoring, oral rehydration therapy, breast-feeding, and immunization.

Education. Investments in basic education complement those in healthcare and improved nutrition, and yield huge payoffs in both developing and industrialized nations. Better education for youth, especially girls, leads to improved health awareness and health practices for their families on a life-long basis. Cognitive and other skills improve productivity, enable better management of resources, and permit access to new technologies. They also enhance participation in democracy.

One study of ninety-eight countries showed GDP gains up to 20 percent resulting from increases in elementary school enrollment, and up to 40 percent resulting from increases in secondary enrollment. In allocating educational resources, the highest payoff is for elementary education, because it reaches the most children (Fiske 1993).

In the United States, dramatic improvement has followed investments in Head Start, which provides preschool education and meals for low-income children, and the Job Corps, which provides remedial and vocational training for disadvantaged youth (see Figure 8.4).

Sustaining Agriculture and Food Production

Access to land. Widespread landownership by small farmers usually contributes directly to food security and improved environmental practices, as well as to increased incomes. The more successful land reform programs, as in South Korea and Taiwan, have provided at least minimum compensation to existing landlords.

Equitable prices for farm produce. Thriving agriculture is basic to successful development in most of the poorest nations. Much of new savings must be accumulated within agriculture, since savings constitute such a large share of the economy and are important for increasing agricultural productivity and for helping to finance rural, nonfarm businesses. Also, as their incomes rise, farmers expand their purchases of consumer goods, providing an important source of nonfarm employment.

In many developing countries, state-run marketing boards have taxed agriculture by setting farm prices very low and retaining for the government a large share of the value from farm exports. Meanwhile, the United States and the European Union have supported their own farmers in ways that generate surplus crops. They also subsidize exports of these crops, driving down prices around the world. Developing-country farmers, who usually are not subsidized, cannot match the low prices. Agriculture falters and with it the whole process of development. Both rich-country export subsidies and developing-country discrimination against agriculture should be phased out as quickly as possible.

In a scenario long anticipated, but only beginning to play out, competition between using cropland for food and using cropland as a source of renewable energy may raise prices for farmers everywhere. It will likely have negative impacts on food-importing nations, especially in urban or coastal areas. But higher food prices may also enable poor-country farmers to compete better for the markets in their own nations.

Creating a Framework for Sustainable Development

Access to credit. Equitable access to credit for small farmers and small businesses is probably the highest priority for the allocation of domestic savings or outside investment. Training in resource and business management is often part of successful credit programs.

Most informal economic activity results from the efforts of small entrepreneurs who cannot find a place in the formal economy. If they have access to good roads, markets, and credit, small farmers and small business people can create their own new income-earning opportunities in market economies.

Adequate physical infrastructure. Creating and maintaining an adequate physical infrastructure is essential to a viable, expanding economy. Important for rural areas are farm-to-market roads and food storage, oriented to both domestic production and, if appropriate, exports. For all areas, safe water, sanitation, electricity, and communication networks are needed.

Stable legal and institutional framework. Sustainable development requires a stable legal framework. This includes assured property titles, enforceable contracts, equitable access to courts and administrative bodies, and access to information networks.

Stable currency and fiscal policies. Neither domestic nor international investors, including small farmers and microentrepreneurs, are likely to invest in countries in which the political or economic environment is unsettled. High inflation or continuing trade deficits, which often go together, discourage needed investments and may even drive out domestic savings.

Effective, progressive tax systems. Effective, progressive tax structures are key to sustainable financing for investments in human resources and infrastructure. Taxes based on ability to pay are also key to stabilizing or reducing wide disparities in income distribution in both rich and poor countries. Such tax systems are usually difficult to enact where wealth and political power are controlled by a small minority.

Incentives for job-creating investments. The Asian countries that have grown so rapidly have all placed emphasis on labor-intensive exports—some in joint ventures with overseas partners, some with investments solicited from abroad, but many with subsidies from within their own economies. This is a distinct departure from their more general commitment to follow market signals.

A primary target for job-intensive investments will be processing operations for primary products—whether for domestic consumption or for crops and minerals now being exported. The success of these efforts will depend in considerable measure on further development of trade among poor nations and, especially, whether rich nations are willing to reduce higher tariffs on manufactured goods.

Development Assistance

The poorest countries will never catch up, or even escape extreme poverty, if they depend solely on their own resources. Their economic base, often primarily agricultural, is too tiny to permit the savings necessary for investing in human and other capital. Under the right conditions, private businesses will invest in plants, equipment, some kinds of infrastructure, and specialized training. But they cannot be expected to fund basic human development and much of the necessary social infrastructure.

For economic and security reasons, as well as humanitarian motives, wealthy nations must greatly increase the resources they provide through development assistance.

At the UN Millennium Summit in 2000, for the first time ever, all 198 attending nations committed themselves to specific Millennium Development Goals to be achieved by 2015. The first commitment is to cut poverty and hunger in half. Other goals include universal education for children, reducing child mortality, and promoting gender equality (see Table 8.2).

At the UN Financing for Development Conference in Monterrey, Mexico, in 2002, nations endorsed a program to reach these goals and made initial pledges of resources. By the standard of military spending, for example, the resources are not huge—on the order of two to three times the $55 billion directed to development assistance at that time. In 2005, the UN Millennium Project estimated that staying on track to meet the Millennium Development Goals by 2015 would require development assistance from donor countries totaling $121 billion for 2006, rising to $143 billion in 2010 and to $189 billion in 2015. In contrast, $88 billion had been pledged for 2006. If the United States and Japan were to raise their assistance to the same level of GNI as for most European countries, the gap would be nearly closed (UN Millennium Project, 2005).

Conclusion

Extreme poverty—the lack of resources or income to command basic necessities—is the condition of about one-sixth of the world's population,

Table 8.2 UN Millennium Development Goals

General	Specific
Eradicate extreme poverty and hunger	Reduce by half the proportion of people living on less than a dollar a day and who suffer from hunger
Achieve universal primary education	Ensure that all boys and girls complete a full course of primary schooling
Promote gender equality	Eliminate gender disparity in primary and secondary education
Reduce child mortality	Reduce by two-thirds the mortality rate among children under age five
Improve maternal health	Reduce by three-quarters the maternal mortality rate
Combat HIV/AIDS, malaria, and other major diseases	Halt and begin to reverse the spread of major diseases
Ensure environmental sustainability	Integrate principles of sustainable development; reverse loss of environmental resources; halve the proportion of people who lack access to safe drinking water
Develop a global partnership for development	Develop international trading and finance systems that meet needs of developing nations: nondiscriminatory trade and access for exports; debt relief; more development assistance; more productive work for youth; access to affordable essential drugs; access to new technologies, especially information and communications

Source: United Nations Development Programme, "Millennium Development Goals" (2000), available at http://www.undp.org/mdg/millennium%development%20goals.pdf.

or one-fifth of the people in developing countries. The absolute number of poor people has remained about steady in recent years, while their proportion has declined slightly, with considerable regional variation.

Countries need economic growth to overcome poverty, but other conditions are also critical. Relatively egalitarian distribution of national income among and within households matters greatly. Gains must be sustainable. Decisionmaking must be broadly shared.

Creating 2 billion good jobs or business opportunities is the biggest single challenge for this generation. The economic and policy tools to generate relatively equitable growth have been successfully demonstrated in recent years, particularly in East and Southeast Asia.

Meanwhile, some of the worst effects of poverty have been offset, and should continue to be, by public and private interventions. For example, infant mortality and overall hunger have declined, and literacy and longevity have increased in many instances, even in the face of continued poverty.

Adapting these tools and programs to particular circumstances, especially in Africa, is of utmost concern to everyone. In an increasingly global economy, the well-being and security of each person or community or nation is inescapably linked to that of every other.

Though most of the world has adopted the bold Millennium

Development Goals, reaching these goals will require a large measure of the one ingredient that seems in shortest supply: the political will to do so.

Discussion Questions

1. Are you more inclined to measure poverty in terms of absolute income, income distribution, or the capacity to reach more specific goals such as the Millennium Development Goals? If the latter, what would be your list of goals?
2. What would you consider a reasonable goal for the ratio between the top and bottom income groups within an economy? Within a business firm? What policies would be necessary to move toward these goals?
3. How important a problem is the "disappearing middle," both in the United States as well as globally?
4. Why is the United States falling behind Canada and many European countries in educational attainment and healthcare?
5. Should government policies encourage the redistribution of income? If so, to what extent?
6. Are you as optimistic as the authors of this chapter that poverty can be overcome? Do you concur that the well-being of everyone is inescapably linked, and that the principal missing ingredient in overcoming poverty is political will?
7. To which antipoverty policies would you give highest priority?

Note

This chapter is adapted from an earlier work by the authors in *Hunger 1995: Causes of Hunger* (BFW 1994) and is used with permission of Bread for the World Institute.

Suggested Readings

Bane, Mary Jo, and Rene Zenteno (2005) "Poverty and Place in North America." Luxembourg Income Study Working Paper no. 418. Available at http://www.lisproject.org/publications/newsletter/2005nov.pdf.

Bread for the World (annual) *Annual Report on the State of World Hunger.* Washington, DC.

Food and Agriculture Organization (2006) "The State of Food Insecurity in the World, 2006." Available at http://www.fao.org/docrep.

Friends Committee on National Legislation (monthly) *Washington Newsletter.* Available at http://www.fcnl.org.

Milanovic, Branko (2006) "Global Income Inequality: What It Is and Why It

Matters." DESA Working Paper no. 26. Washington, DC: World Bank.
Newman, Katherine S., and Victor Tan Chen (2007) *The Missing Class: Portraits of the Near Poor in America.* Boston: Beacon.
Sachs, Jeffrey (2005) *The End of Poverty: Economic Possibilities for Our Time.* New York: Penguin.
Sen, Amartya (1999) *Development as Freedom.* New York: Anchor.
Sklar, Holly, ed. (2004) *Putting Dignity and Rights at the Heart of the Global Economy.* Philadelphia: American Friends Service Committee, Working Party on Global Economics.
Smeeding, Timothy (2005) "Poor People in Rich Nations: The United States in Comparative Perspective." Luxembourg Income Study Working Paper no. 419. Available at http://www.lisproject.org/php/wp/wp.php#wp.
United Nations Development Programme (annual) *Human Development Report.* Available at http://www.undp.org/hdr.
United Nations Millennium Project (2005) "Investing in Development: A Practical Plan to Achieve the Millennium Development Goals." Available at http://www.unmillenniumproject.org/reports/index_overview.htm.
US Census Bureau websites: poverty statistics, http://www.census.gov/hhes/poverty; income statistics: http://www.census.gov/income;histpov.
World Bank (annual) *World Development Report.* Washington, DC.
——— (biennial) *World Development Indicators.* Washington, DC.

PART 3

Development

9

Population and Migration

Ellen Percy Kraly
and Fiona Mulligan

COMING TO GRIPS WITH THE IMPLICATIONS OF CURRENT POPUlation trends is an extremely important dimension of global studies. The process is neither easy nor comforting, because a significant population increase is an inevitable characteristic of the global landscape in the first fifty years of the twenty-first century. It is critical that students interested in global issues should appreciate both the causes of population growth and the consequences of population change for society and the environment. Such an appreciation will serve in developing appropriate and effective responses to population-related problems emerging globally, regionally, and locally.

This chapter addresses the interconnections among population change, environmental issues, and social, economic, and political change in both developing and developed regions of the world. Because population growth has momentum that cannot be quickly altered, it is important to begin by considering fundamental principles of population or demographic analysis, and to place recent global and regional population trends in historical perspective. Divergent philosophical and scientific perspectives on the relationships among population, society, and environment have been applied to past and current patterns of population growth, and also influence visions of the future. Debates on the implications of current growth have also influenced discussions about routes for population policy.

Principles and Trends

Demographic Concepts and Analysis

Demography is the study of population change and characteristics. A population can change in size and composition as a result of the interplay of three demographic processes: fertility, mortality, and migration. These components of change constitute the following equation for population change (P) between two points in time:

P = (+) births (–) deaths (+) in-migration (–) out-migration

On the global level, the world's population grows as the result of the relative balance between births and deaths, often called natural increase. The US population is currently increasing at about 1 percent per year; natural increase accounts for about two-thirds, and net international migration constitutes about one-third of this relatively low level of population growth.

Many people seeking routes to sustainable development advocate a cessation of population growth, often referred to as zero population growth. When viewed from a short-term perspective, zero population growth means simply balancing the components of the population equation to yield zero change in population size during a period of time. In a long-term perspective, however, population scientists usually consider zero population growth by examining a particular form of a zero-growth population: the stationary population, one in which constant patterns of childbearing interact with constant mortality and migration to yield zero population change. In such a case, fertility is considered "replacement" fertility, because one generation of parents is just replacing itself in the next generation. In low-mortality countries, replacement-level fertility can be measured by the total fertility rate, and is approximately 2.1 births per woman to achieve a stationary population over the long term.

It takes a relatively long time, perhaps three generations after replacement fertility has been achieved, for a population to cease growing on a yearly basis. Large groups of persons of childbearing age, reflecting earlier eras of high fertility, result in large numbers of births even with replacement-level fertility. Hence an excess of births over deaths occurs until these "age structure" effects work themselves out of the population. This is known as the "momentum" of population growth. K. B. Newbold (2002) estimates that by 2025 world population levels will grow to 7.5 billion, and projects a growth of up to 10.7 billion by 2050, an extraordinary increase due in part to the population momentum in the developing world.

Age structure is an important social demographic characteristic of a population. Both the very young and the very old in a population must be supported by persons in the working-age groups. The proportions of persons in different age groups in a population are depicted in a *population pyramid.* Figure 9.1 shows population pyramids for two countries, Tanzania and Spain. The pyramid for Tanzania reveals a youthful population, with nearly one-third under ten years of age. This reflects high fertility. In Spain, a low-fertility country, only 9 percent of the population is under ten years of age. At the other end of the age spectrum, less than 5 percent of the Tanzanian population is over the age of sixty years, compared to over one-fifth, 21 percent, of the Spanish population. The immigration of male workers into Spain is also evident in the large proportion of males aged thirty to thirty-four. These two age pyramids illustrate the history of past levels in fertility and migration as well as the different demands on society to support the young and the old.

Historical and Contemporary Trends in Population Growth

The world's population was estimated to be 6.7 billion at the beginning of 2007, and to be increasing at approximately 1.2 percent per year (UNPD 2007). These data represent a cross-sectional perspective on population characteristics—a snapshot that fails to capture the varying pace of population change worldwide and regionally. Over most of human history, populations have increased insignificantly or at very low annual rates of growth, with local populations being checked by disease, war, and unstable food supplies. Between the sixteenth and eighteenth centuries, population growth appeared to become more sustained as a result of changes in the social and economic environment: improved sanitation, more consistent food distribution, improved personal hygiene and clothing, political stability, and the like.

The world's population probably did not reach its first billion until just past 1800. But accelerating population growth during the nineteenth century dramatically reduced the length of time by which the next billion was added. According to the United Nations (UNDESIPA 1995; UNPD 2007), world population reached

- 1 billion in 1804
- 2 billion in 1927 (123 years later)
- 3 billion in 1960 (33 years later)
- 4 billion in 1974 (14 years later)
- 5 billion in 1987 (13 years later)
- 6 billion in 1999 (12 years later)
- 6.7 billion in 2007 (8 years later)

Rapid population growth occurred on a global scale during the second half of the twentieth century. Population data for years since 1950 are shown in Figure 9.2. Between 1950 and 2007, the world's total population

Figure 9.1 Ages of Males and Females as Percentage of Population, Spain and Tanzania, 2005

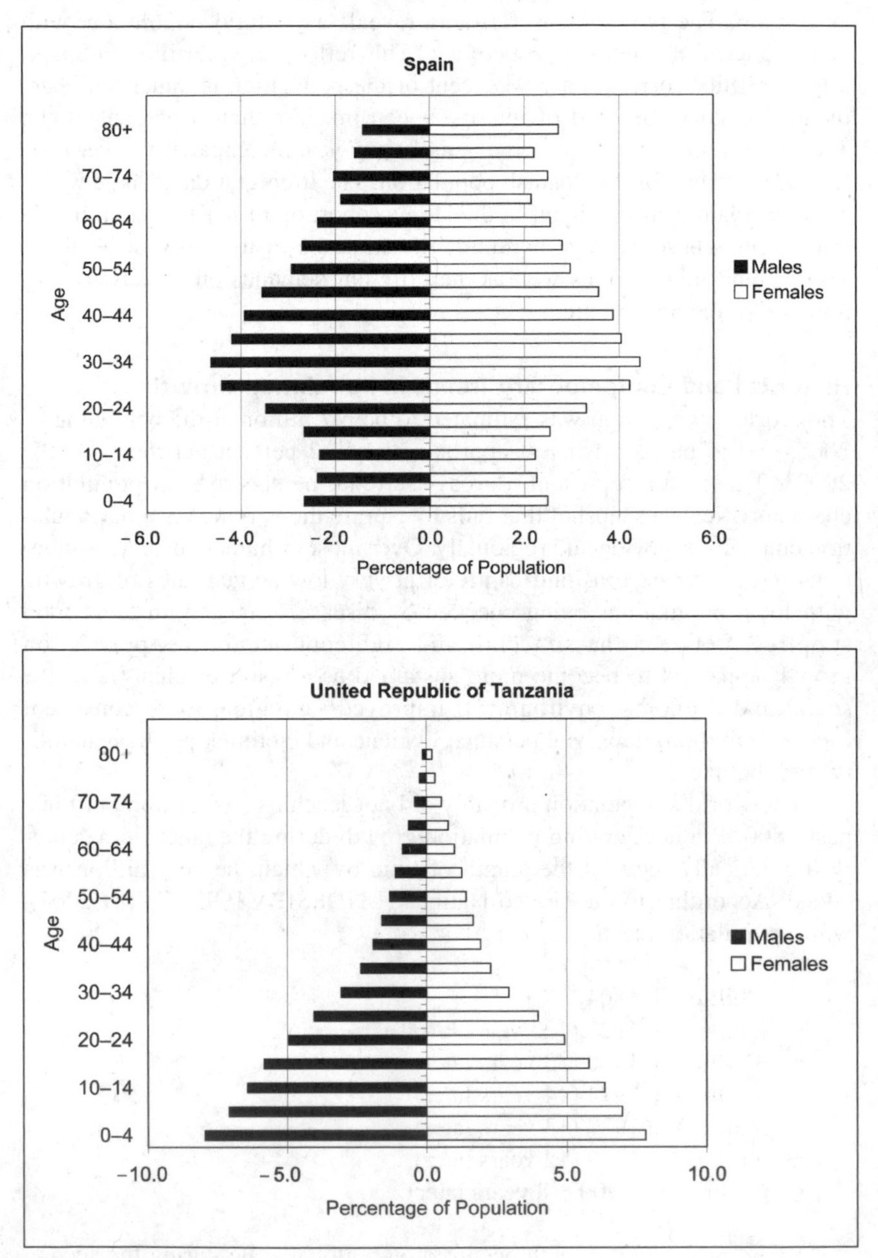

Source: United Nations Population Division, *World Population Prospects: The 2004 Revision,* vol. 2, *Sex and Age Distribution of the World Population* (New York, 2005).

more than doubled, increasing from 2.6 billion to 6.7 billion. The difference in height of the bars in the figure reveals the momentum of population growth that results in continued additions to the world's population, albeit in decreasing numbers: between 1985 and 1990, approximately 85 million persons were added to the world's population each year; in the late 1990s, the annual increase was estimated at 77 million (UNPD 2003).

It is important to note, however, that despite these large additions to the world's population, the rate of population growth is *decreasing*. The average annual rate of global population growth reached an all-time high, of about 2.2 percent, between 1962 and 1964. Since that time, the pace of growth of the world's population has decreased to the current rate of approximately 0.76 percent per year.

Patterns of population growth differ significantly between more- and less-developed regions of the world. Table 9.1 provides greater geographic detail and summarizes population size and distribution for major regions of the world for selected years since 1950. Dramatic shifts in the geography of world population have occurred during the past five decades and, as dis-

Figure 9.2 World Population for Development Categories, 1950–2007

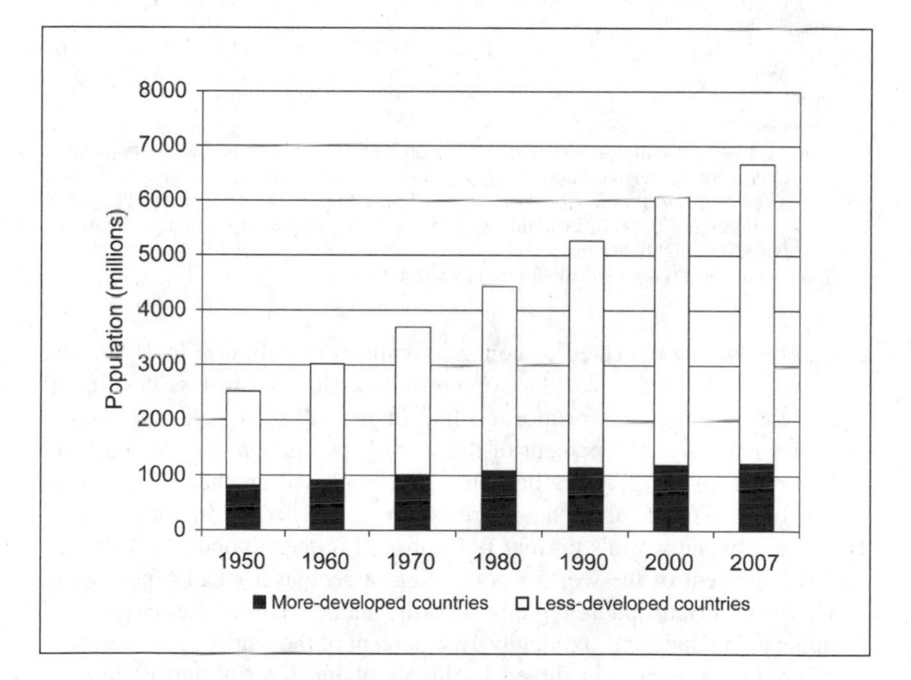

Source: United Nations Development Programme, *Human Development Report 2007* (New York: Oxford University Press, 2007).

Table 9.1 World Population by Geographic Region and for More- and Less-Developed Countries, 1950–2007

	1950	1960	1970	1980	1990	2000	2007
			Population (millions)				
World[a]	2,535	3,032	3,699	4,451	5,295	6,124	6,671
More-developed countries	814	916	1,008	1,083	1,149	1,194	1,223
Less-developed countries	1,722	2,116	2,690	3,368	4,146	4,930	5,448
Africa	224	282	364	480	637	821	965
Asia	1,411	1,704	2,139	2,636	3,181	3,705	4,030
Latin America and the Caribbean	168	220	288	364	444	523	731
Europe	548	605	657	693	721	729	572
North America	172	204	232	256	284	316	339
Oceania	13	16	20	23	27	31	34
			Percentage				
World	100.0	100.0	100.0	100.0	100.0	100.0	100.0
More-developed countries	32.1	30.2	27.3	24.3	21.7	19.5	18.3
Less-developed countries	67.9	69.8	72.7	75.7	78.3	80.5	81.7
Africa	8.8	9.3	9.8	10.8	12.0	13.4	14.5
Asia	55.7	56.2	57.8	59.2	60.1	60.5	60.4
Latin America and the Caribbean	6.6	7.3	7.8	8.2	8.4	8.5	11.0
Europe	21.6	20.0	17.8	15.6	13.6	11.9	8.6
North America	6.8	6.7	6.3	5.8	5.4	5.2	5.1
Oceania	0.5	0.5	0.5	0.5	0.5	0.5	0.5

Sources: United Nations Department of Economic and Social Affairs, "World Population Prospects: The 2006 Revision Population Database" (2007), available at http://esa/un.org/unpp; United Nations Population Division, "World Population Prospects: The 2006 Revision—Highlights" (2007), available at http://www.un.org/esa/population/publications/wpp2006/wpp2006_highlights_rev.pdf.

Note: a. Numbers for individual regions have been rounded.

cussed below, are expected to continue well into the future. In 1950, just over two-thirds of the world's population was located in less-developed countries; by 2007, this proportion had increased to 81.7 percent. Asian countries make up 60.4 percent of the world's population; nearly one-fifth, 19.2 percent, of the global village live in China; and another 17.5 percent live in India. Africa's share has increased from 8.8 percent in 1950 to over 14 percent of the world's current population. European populations constitute 11.0 percent of the world's population, a decline from 21.6 percent in 1950. Western Hemisphere regions—North America, Latin America, and the Caribbean—include approximately 16.1 percent of the world's population.

Population growth is fueled by levels of fertility, mortality, and net migration. The rapid population growth that occurred in the post–World

War II era reflected significant declines in mortality that resulted in large part from public health advances and the transfer of medical technology from more- to less-developed countries.

The total fertility rate measures the average number of births per woman of childbearing age, and is a strong indicator of overall population growth. In the period 2005–2010, based on a United Nations Population Division analysis of available national measures, the total fertility rate for the world as a whole is estimated at 2.55 births per woman, representing a significant decline from 4.47 in 1970–1975. Fertility in more-developed countries has been below replacement for some time, and is estimated at 1.60 births per woman. In less-developed countries, the rate has dropped from 5.41 in 1970–1975 to 2.75 in 2005–2010. Much of this decline is weighted by the aggressive fertility control campaign in China and by significant declines in fertility throughout Southeast Asia and in Latin America. Total fertility in India has also declined, from 5.23 in 1970–1975 to 2.81 for 2005–2010 (UNDP 2007). Fertility in least-developed countries, including many in sub-Saharan Africa, has declined in the past decades, yet still remains high, at 4.63 births per woman, well above replacement. Figure 9.3 provides a cartographic view of recent levels of fertility for countries of the world.

Perspectives on Population Growth

Reflections on the relationship between population and society can be found in the early history of many cultures. In early Greece, Plato wrote about the need for balance between the size of the city and its resource base; in China, Confucianism emphasized the social and economic advantages of large families. Concern about the implications of population growth for social progress became a focus of social theory in the nineteenth century, and continues in contemporary debates on the global effects of current levels of population growth.

One of the most influential thinkers in the realm of population growth was Thomas Malthus, who published an essay on the principle of population in 1798 (Malthus 1826). Malthus was a reactionary against the mercantilist philosophy that dominated eighteenth-century Europe and emphasized the value of large and increasing populations for economic growth and prosperity. Malthus instead argued that the inescapable human desire to reproduce would lead to starvation, poverty, and human misery if not halted by the "positive checks" of famine, war, and epidemics. Unlike today's neo-Malthusians, Malthus, a clergyman, was opposed to contraceptives as a means to limit family size and instead endorsed delayed marriage and abstinence.

Malthus's theories have permeated the population debate throughout

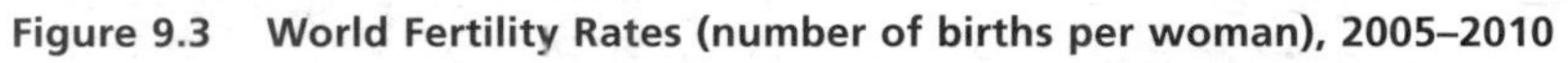

Figure 9.3 World Fertility Rates (number of births per woman), 2005–2010

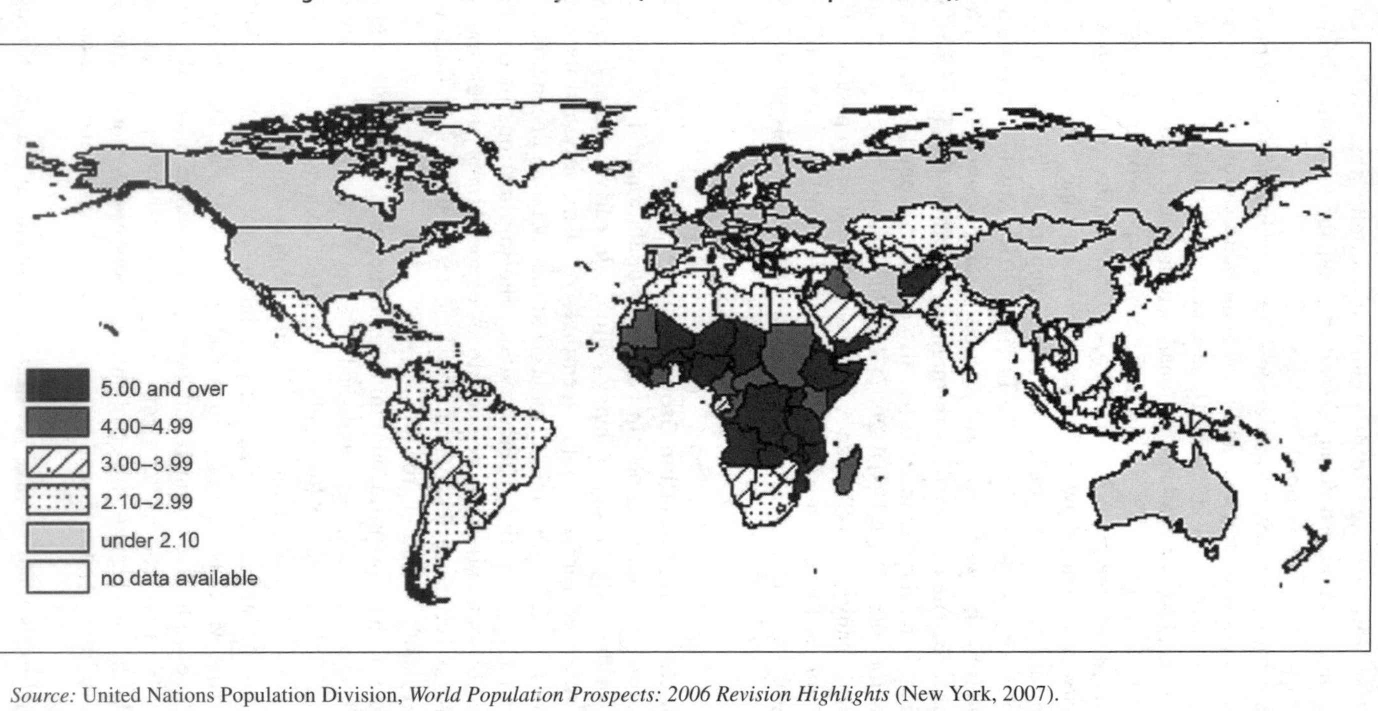

Source: United Nations Population Division, *World Population Prospects: 2006 Revision Highlights* (New York, 2007).

the decades, and the two lines of thought that have traditionally dominated the arena are the *neo-Malthusian* and the *Cornucopian*. The neo-Malthusian perspective has remained much the same, while the Cornucopian perspective emphasizes the role of technological innovations and market forces, which through economics will manage the use of natural resources. This would allow for population growth to actually solve global problems through increased economic productivity and capacity for economic progress.

A third and increasingly referenced perspective on population growth requires a focus on the *structural dimensions* of social change, including processes such as population displacement, health and disease, food security, and environmental issues as outcomes of broader social and economic structural processes and institutions. Population growth, in particular high fertility, thus becomes more of a consequence than a cause of slow economic development and restricted social and economic opportunities. In addition, there is increasing attention to the relationships among government, leadership and authority structures, and population processes on both the local and the global scale. Moreover, in spite of, or perhaps because of, processes of globalization addressed by many chapters in this volume, nation-states continue to exercise authority in areas of international migration, trade and employment, political enfranchisement, and human rights (Bailey 2005).

Although these perspectives might initially appear incongruent, there are a number of common themes that connect all models of thought, mostly regarding ways in which to address the issue of population expansion. Thus the reduction of poverty, the improvement of life choices and status of women, the increasing sustainability of food production, the improvement in water quality, and other aspects of social welfare all become strategies for reducing population growth. Similarly, there is an emerging recognition that slowing and ultimately ceasing world and regional population growth will eventually improve standards of living, stabilize food supplies, and halt environmental degradation (UNDP 2001; National Research Council 1986).

Coupled with these perspectives and theories of population growth are models of historically based expectations about the trajectory of population growth. The *demographic transition* model relies heavily upon the differences between developed and developing nations; it was initially advanced to describe population growth patterns in Europe and North America in the nineteenth and early twentieth centuries. It predicts that as societies undergo industrialization and urbanization, death rates will fall, followed by a lag of declining fertility, during which population growth continues to occur until norms and values shift from large-family to small-family ideals.

The demographic transition model has been widely criticized and revised, particularly in recognition of the fact that it depends on the experiences of non-Western societies mirroring or converging with those of Europe and North America. Revisions allow for a clearer understanding of the ways in which family size is influenced by cultural beliefs and gender, particularly educational opportunities for girls and young women. It is important to discern that within particular cultures, family size may have cultural as well as economic significance, and that there may not be one demographic transition model that is universally applicable.

Expectations About Future Population Growth

Theories of population change guide analyses of future population growth, usually in the form of population projections. Most demographers are quick to state that population projections are not predictions but rather represent a calculation of future population size based on a set of assumptions or variants. Shown in Figure 9.4 are population estimates and projec-

Figure 9.4 Projected World Population, 1950–2050

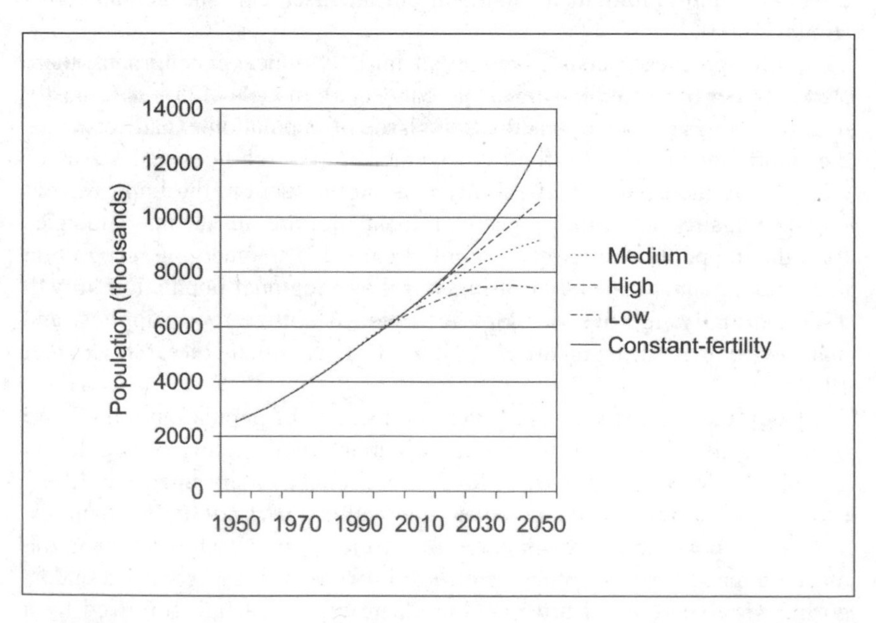

Sources: United Nations Development Programme, *Human Development Report 2003*, 3rd edition (New York: Oxford University Press, 2003); United Nations Department of Economic and Social Affairs, *World Population Prospects: The 2006 Revision* (2006) and *World Urbanization Prospects: The 2005 Revision* (2006), available at http://esa.un.org/unpp.

tions prepared by the United Nations Population Division in its most recent projection series for 1950–2050 (UNPD 2003). The world totals reflect the sum of projections conducted separately for 228 countries; the "fan" of population figures represents the four projection variants for the projection period 2000–2050. These variants reflect different assumptions about the pace and pattern of fertility change. The projections prepared by the UN in 2002 represent revised assumptions about both the pace and level of fertility decline, and the severity and geographic extent of the HIV/AIDS epidemic.

The medium variant projects the fertility of individual countries to follow the pattern of fertility decline of currently low-fertility countries. The ultimate target is a total fertility rate of 1.85, which is below a level of fertility that ensures population replacement (2.11 births per woman) at existing levels of mortality. In some cases, fertility of very low-fertility countries will actually have to increase to hit this target by the end of the projection period. In the high variant, fertility follows a path of 0.5 children *above* that projected for each country in the medium variant; in the low variant, fertility is projected at a level of 0.5 children *below* that in the medium variant. The constant fertility variant represents the status quo by assuming that fertility levels for 2000–2005 are maintained throughout the projection period. This variant provides a useful point of comparison. In all four projections, mortality levels decline, although at a slower pace in those countries where mortality is already low. In those countries in which the HIV/AIDS epidemic is severe, a slower pace of mortality decline is assumed. The projection series as a whole assumes convergence among countries and thus generally embodies a demographic transition model, although greater variation among countries in pattern and pace of change in demographic rates is present in these most recent UN projections, and significantly, fertility settles at *below* replacement.

The high-fertility variant results in growth in world population from 6.7 to 10.8 billion between 2007 and 2050, an increase of nearly 4 billion. The low-fertility variant projects an increase to 7.8 billion by 2050, an increase of 20 percent over the 2007 population. Thus, even with a declining trajectory of fertility to well below replacement, the world's population will continue to grow as a result of the momentum of population growth. The medium variant results in an increase to 9.2 billion by 2050, an increase of 46 percent over the current population.

The projections also reveal the future impact of the HIV/AIDS epidemic. In the 2006 revised projections, sixty-two countries are identified as being most severely affected by the disease, an increase over the fifty-three countries considered in the projections prepared five years earlier (UNPD 2007). Of the countries identified as most severely impacted by the disease, forty are in sub-Saharan Africa, five are in Asia, four are in Europe, one is in Oceania, and the remaining twelve are in North and South America. Among highly affected countries in southern Africa, the demography impact has been felt in overall decreases in life expectancy at birth: declining from an overall level of sixty-two years in 1990–1995 to approximately forty-nine years in 2005–2010. Because of high levels of fertility in many of these countries, overall population growth will continue to be positive. But in regions where HIV/AIDS is exceedingly acute, even high levels of fertility will not counter the effects of the epidemic. In southern Africa, the rate of population growth has decreased significantly in large part because of the epidemic, from 2.5 percent annually in 1990–1995 to 0.6 percent in 2005–2010 (UNDP 2007).

Shifts in the geographic distribution of the world's population are evident in all of the projections, as seen in Table 9.2. The results for the medium variant help make the point. The population in African countries, prima-

Table 9.2 Projections of World Population by Geographic Region and for More- and Less-Developed Countries, 2000 vs. 2050

	2007	Low, 2050	Medium, 2050	High, 2050	Constant Fertility, 2050
			Population (thousands)		
World[a]	6,671	7,792	9,191	10,756	11,858
More-developed countries	1,223	1,065	1,245	1,451	1,218
Less-developed countries	5,448	6,727	7,946	9,306	10,639
Africa	965	1,718	1,998	2,302	3,251
Asia	4,030	4,444	5,266	6,189	6,525
Latin America and the Caribbean	572	641	769	914	939
Europe	731	566	664	777	626
North America	339	382	445	517	460
Oceania	34	42	49	56	57
			Percentage		
World	100.0	100.0	100.0	100.0	100.0
More-developed countries	18.3	13.7	13.5	13.5	10.3
Less-developed countries	81.7	86.3	86.5	86.5	89.7
Africa	14.5	22.0	21.7	21.4	27.4
Asia	60.4	57.0	57.3	57.5	55.0
Latin America and the Caribbean	8.6	8.2	8.4	8.5	7.9
Europe	11.0	7.3	7.2	7.2	5.3
North America	5.1	4.9	4.8	4.8	3.9
Oceania	0.5	0.5	0.5	0.5	0.5

Source: United Nations Population Division, "World Population Prospects: The 2006 Revision—Highlights" (2007), available at http://www.un.org/esa/population/publications/wpp2006/wpp2006_highlights_rev.pdf.

Note: a. Numbers for individual regions have been rounded.

rily in sub-Saharan regions, will increase from 14.5 percent in 2007 to nearly 22 percent of the world's population by 2050, and the population of Europe is expected to decline from 11 percent to about 7 percent of the world's population. Countries in North America and Latin America and the Caribbean (combined) will continue to hold about 13 percent of the world's population. Asian countries will continue to hold the largest share of the world's population, although decreasing from over 60 percent in 2007 to approximately 57 percent by 2050. It is important to consider two Asian countries in particular, China and India, whose population policies are considered later in this chapter. The population of China is projected to increase from 1.3 billion in 2007 to 1.4 billion by 2050; India's population is projected to grow from 1.2 billion in 2007 to 1.5 billion by 2050, thus overtaking China as the largest national population in the world. Thus we can expect significant shifts in the world geography of population based on these projections.

Social and Environmental Dimensions of Geographic Mobility

As well as influencing population change, the movement of persons within and among countries is both a cause and a consequence of social, economic, political, and environmental factors. Geographic mobility is the general concept covering all types of human population movements. Migration is generally considered to refer to moves that are permanent or longer-term; internal migration within a country is distinguished from international population movements; international migration into a country is immigration, and international migration out of a country is emigration. Reasons for moving, such as labor migration, refugee migration, and seasonal migration, are often included in migration concepts. Internal and international migration are processes that are increasingly linked through the geographic and social dimensions of global economic development.

Internal Migration and Urbanization

A corollary of the demographic transition model is growth in the size of cities as well as increasing proportions of populations living in cities and metropolitan areas—that is, urbanization. The UN estimates that 49 percent of the world's population was living in urban areas in 2007; the difference between more- and less-developed countries, 74 percent and 44 percent respectively, is dramatic (UNPD 2008). Thus, while some of the largest metropolitan areas in the world are found in developing countries—for example, Mexico City and Bombay, each with 19.0 million in 2007; São Paulo, with 18.8 million; Delhi, with 15.0 million; and Shanghai, with 15.0

million—with the exception of Brazilians, over half the populations in these societies currently live in rural areas.

Demographers, however, are anticipating a shift in this rural-urban balance in the very near future. Cities will grow as a result of population growth and rural-to-urban migration throughout regions of the world. By 2050, for example, the level of urbanization in more-developed countries is expected to increase to 86 percent, and in less-developed countries even more steeply to 67 percent, representing a dramatic shift in patterns of residence and economic activity (UNPD 2008).

The causes of urban growth have varied among regions and during different historical periods. In Western societies, urbanization has been fueled in large part by technological change in both agricultural and industrial sectors, resulting in both a push from rural communities and a pull to emerging industrial centers (Harper 1995). In developing societies, rural-to-urban migration has been driven by many factors, including increasing population density (caused by high fertility rates) in rural areas, environmental degradation from practices such as overgrazing, and the hope for gainful employment in urban areas. In many developed countries, the pull of employment and higher wages is so great that levels of unemployment in Western cities are very high. Evidence of underemployment is shown in the large numbers of persons, including many children, attempting to earn livelihoods in what has been called by some the informal economy—for example, street vendors, curbside entertainers, and newspaper boys and girls.

International Population Movements

One of the most visible manifestations of globalization is the increasing scale of international population movements throughout all regions of the world. According to scholars Stephen Castles and Mark Miller:

> Millions of people are seeking work, a new home or simply a safe place to live outside their countries of birth. For many less-developed countries, emigration is one aspect of the social crisis which accompanies integration into the world market and modernization. . . . The movements take many forms: people migrate as manual workers, highly-qualified specialists, entrepreneurs, refugees or as family members of previous migrants. . . . Migrations can change demographic, economic, and social structures, and bring a new cultural diversity, which often brings into question national identity. . . . [Overall], migration ranks as one of the most important factors in global change. (2003: 3)

The consequences of international population movements for both the sending and the receiving nations and communities will have significant implications for emerging global issues.

Countries in both Western and Eastern Europe have been faced with large numbers of persons seeking political asylum from both European regions as well as from geographically distant sources, including East Africa and Southeast Asia. Significant labor migration flows have emerged between South and Southeast Asia and oil-producing regions of the Middle East, and throughout the Asia Pacific region. The United States has grappled with issues concerning large numbers of undocumented migrants drawing from Mexico and many other source countries. Refugee migration and population displacement are spatial characteristics of political, economic, and environmental change in many regions of Africa.

Castles and Miller (2003: 7–9) identify five "general tendencies" of contemporary international population movements that they expect to continue well into the twenty-first century. First, international population movements will involve an increasingly large number of countries, both as sending and receiving regions, hence the *globalization of migration.* Second, the volume of international migration can be expected to *increase* in volume. Third, international migration will continue to become *differentiated* by including a wider variety of migrants—for example, seasonal migrants as well as migrants seeking permanent resettlement. Fourth, as women throughout the world become increasingly involved in the global work force, international migration will become more *feminized.* Fifth, international migration is likely to become a more significant *political issue,* both on the international stage as well as in the politics of individual nations.

The significance of the scale of refugee migration and displaced populations in global population issues cannot be overstated. The international definition of a refugee is a person who, "owing to a well-founded fear of being persecuted for reasons of race, religion, nationality, membership of particular social group(s) or political opinion is outside the country of his nationality and is unable to or owing to such fear is unwilling to avail himself of the protection of that country; or who, not having a nationality and being outside the country of his former habitual residence . . . is unable or unwilling to return to it" (UNHCR 1995: 256). Refugees seek safety from war and oppression, but can also be a source of political and economic instability in border regions and countries of asylum. For instance, Afghanis seeking refuge following the US military response to the September 11, 2001, terrorist attacks faced resistance in neighboring Pakistan.

At the end of 2006, the Office of the UN High Commissioner for Refugees identified more than 32.9 million persons throughout the world who were of concern to the organization. Of this extraordinary number, 9.9 million are recognized as refugees living in asylum in other countries; the remainder are persons who are internally displaced within their own countries for complex political, economic, and environmental reasons, and persons outside their home countries living in refugee-like situations. Table 9.3 shows refugees and other persons of concern by broad category and geographic region. Many refugee settlements or camps have existed for many years. Some refugees have been repatriated to their homelands—for example, Guatemalans who had sought refuge in Mexico, and Muslims who had fled Myanmar (formerly Burma); others, including many Vietnamese during the 1970s and 1980s, have been permanently resettled in other countries such as Canada, Australia, and the United States. The majority of the world's refugees are women and children, whose voices are often not heard in discussions about programs to aid and resettle refugees (UNHCR 2006).

Population Policies

A population policy may be defined as a deliberately constructed or modified institutional arrangement or specific program through which a government influences demographic change, directly or indirectly (Demeny and McNicoll 2003). According to the UN's 2003 survey of government policies, seventy-five less-developed countries have policies or programs to reduce population growth; in contrast, eighteen more-developed countries have goals to increase growth (UNPD 2008). Many more countries have indirect population policies that, while not targeting population growth, have clear implications for mortality, fertility, or migration. The United States, for example, has not yet adopted a formal statement of goals concerning national population growth, but does have a long-standing policy for the permanent resettlement of immigrants and refugees (see discussion below), which in turn results in net additions to the population through international migration.

International Efforts to Reduce Population Growth

International population conferences bring government delegations and representatives of nongovernmental organizations (NGOs) together to discuss goals concerning population, and to develop strategies for achieving those goals. The most recent conference was held in Cairo in 1994. At each of these gatherings, there has been a general recognition, though not universal agreement, that (1) rapid population growth fueled by high fertility poses a challenge to economic development in less-developed countries; (2) mortality should be reduced regardless of the effect on population growth; and (3) international migration is an appropriate arena of national policy and control (Weeks 2005). Over the past three decades, however, important shifts in thinking about population growth have occurred, affecting population policies and programs within countries.

Table 9.3 Refugees and Other Persons of Concern to the UNHCR, by Geographic Region of Origin, 2006

	Total	Refugees	Asylum Seekers	Returned Refugees	Internally Displaced Persons	Internally Displaced Returned	Other Persons of Concern[a]
				Population			
World	32,861,500	9,877,700	740,100	733,700	12,794,300	1,864,200	6,851,500
Africa	9,752,600	2,607,600	244,100	312,200	5,373,000	1,043,900	171,800
Asia	14,910,900	4,537,800	90,100	408,900	3,879,100	811,600	5,183,400
Latin America and the Caribbean	3,542,500	40,600	16,200	100	3,000,000	0	485,600
Europe	34.236,700	1,612,400	240,200	12,500	542,200	8,700	1,010,700
North America	1,143,100	995,300	147,800	0	0	0	0
Oceania	85,700	84,000	1,700	0	0	0	0
				Percentage			
World	100.0	100.0	100.0	100.0	100.0	100.0	100.0
Africa	29.7	26.4	33.0	42.6	42.0	56.0	2.5
Asia	45.4	45.9	12.2	55.7	30.3	43.5	75.7
Latin America and the Caribbean	10.8	0.4	2.2	0.0	23.4	0.0	7.1
Europe	10.4	16.3	32.5	1.7	4.2	0.5	14.8
North America	3.5	10.1	20.0	0.0	0.0	0.0	0.0
Oceania	0.3	0.9	0.2	0.0	0.0	0.0	0.0

Source: United Nations High Commissioner for Refugees, Population Data Unit, "2006 UNHCR Population Statistics Provisional" (2007), available at http://www.unhcr.ch.

Note: a. Includes stateless persons.

The 1974 World Population Conference in Bucharest produced the first formal expression of a world population policy. The "World Population Plan of Action," however, embodied a wide range of perspectives on the ways to reduce population growth within developing societies. Some countries, notably the United States, advocated fertility control, specifically family planning programs, to reach population growth targets. Other countries, primarily in the developing world, emphasized the role of development in leading to fertility decline (hence, "development as the best contraceptive"). The 1984 International Population Conference in Mexico City found the United States reducing its support for family planning (which was linked in turn to the Ronald Reagan administration's views on abortion) and identifying population growth as having little hindrance on economic and social development. Many developing countries by this time, however, had instituted family planning programs in an effort to slow the retarding effects of rapid population growth on improving standards of living and educational levels and reducing mortality (Weeks 2005).

The 1994 International Conference on Population and Development recognized the global dimensions of population change. In its action program, the conference identified the connections among population processes, economic and social development, human rights and opportunities, and the environment, thus shifting attention away from targets concerning population growth to goals concerning sustained development, reduction of poverty, and environmental balance. The central role of women in the goals and programs to achieve sustainable development is underscored in the final conference report: "The key to this new approach is empowering women and providing them with more choices through expanded access to education and health services, skill development and employment, and through their full involvement in policy- and decision-making processes at all levels" (UNDESIPA 1995: 1).

Over many decades, the US government has implemented programs abroad concerning population, development, and human welfare. As mentioned earlier, US support for international family planning programs has wavered depending on the particular president's political lens. Under President Ronald Reagan, support for fertility control programs was barred from those countries that permitted women access to legal abortion. This restriction on US foreign aid was lifted in the early days of the Bill Clinton administration, only to be reinstated under President George W. Bush.

On the international scale, the emphasis on connecting population issues to the status of women has represented a significant forward step in embedding population analysis into broader discussions of the quality of human life and the balance between society and environment at local, national, and global levels. Given the demographic trends and patterns discussed above, three additional population issues are likely to become major policy concerns in the near future: population aging, excess deaths due to HIV/AIDS, and rapid population growth in already-large metropolitan areas.

National Population Policies

Most countries that have formal population policies seek to reduce population growth by reducing fertility. Beginning in the 1950s, providing contraception through family planning programs was initiated in many developing countries—often with significant contributions from developed countries, NGOs and foundations, and, in subsequent decades, international organizations such as the UN Fund for Population Activities (UNFPA). Increasingly, policies aimed at fertility reduction have encompassed broader perspectives on population dynamics, incorporating goals to increase the status of women through better health, enhanced educational and employment opportunities, greater access to credit, and so on.

The record of family planning programs has been variable throughout the developing world. India's national family planning program, maintained since 1952, has met with fertility trends that vary significantly throughout regions within the country and between rural and urban areas. In 1973, Mexico instituted a national policy to reduce population growth, with a focus on the reduction of fertility through maternal and child health programs, family planning services, sex education, and population information programs.

China's fertility control policy began in 1971 as a set of policy goals *(wan xi shao)* concerning later marriage *(wan),* longer intervals between births *(xi),* and fewer children *(shao),* following which the one-child policy was implemented in 1979. During the 1970s the Chinese birthrate declined significantly, reflecting the provision of contraception in combination with social and economic incentives to delay and reduce fertility. The birth control program also coincided with a general decline in fertility, which had been evident since the early 1960s. The Chinese fertility policy has been criticized harshly, however, for being coercive and for leading to selective abortion, abandonment, and infanticide of female infants. The current level of fertility in China is estimated to be 1.7 births per woman.

The demographic implications of rapid fertility decline in China can be anticipated using population projections. Despite below-replacement fertility, the Chinese population is projected to increase from 1.329 billion in 2007 to 1.409 in 2050 (medium-fertility variant). This nearly 10 percent increase in population size will be accompanied by significant shifts in age composition: in 2005, 11 percent of the Chinese population was over sixty years of age; by 2050, this proportion is expected to triple, to 31 percent (by contrast, the proportion of the US population over sixty years old in 2005 was 16 percent). This dramatic aging of Chinese society in the near future will require changes in Chinese social and economic arrangements to support and care for an increasingly elderly population (UNPD 2007).

International Migration and Refugee Policies

The UN's Universal Declaration of Human Rights (UDHR) recognizes the basic right of people to leave their homelands. The converse of this right to emigrate, however, is not recognized—that is, nation-states have the sovereign right to control the entry of nonnationals into their territory. Nearly all countries have clear policies concerning international migration and travel. The very few countries that continue to allow international migration for permanent resettlement include the United States, Canada, Australia, and New Zealand, often considered the traditional immigrant-receiving nations. A much larger range of countries provide humanitarian assistance to refugees in the form of permanent resettlement, political asylum, refugee camps, and financial and other resources for international organizations that seek to respond to refugee situations.

The demand for international migration, temporary labor migration, refugee resettlement, and temporary asylum can be expected to continue to grow in the next decade, with those who seek to move originating overwhelmingly from developing regions. Emerging and persistent patterns of undocumented migration throughout both the developed and the developing world are symptoms of the motivation of people to seek better opportunities through international migration. Matched with this demand are national doors that are gradually closing to international migrants. In the traditional receiving countries, concerns over the social, economic, political, and security effects of immigration have moved high up the political agenda at both the national and the state or provincial levels, often, but not always, alongside efforts to tighten migration controls, reduce immigration levels, and constrain access of immigrants to national and local social programs.

Worldwide, the United States accepts the largest number of immigrants for permanent resettlement. Between 2000 and 2006, immigration for permanent residence to the United States averaged approximately 1 million persons per year. In 2006, Mexico, India, China, the Philippines, and the Dominican Republic provided the largest numbers of immigrants to the United States (USDHS 2007).

US immigration policy is organized around several principles. First, the policy gives priority to close relatives of US citizens and, to a lesser extent, to relatives of immigrants already in the country. Second, the policy gives priority to persons with occupations, skills, and capital that will benefit the US economy. This dimension of US immigration policy may contribute to the loss of highly skilled and professionally trained persons from developing countries (referred to as "brain drain"). Third, immigrants to the

United States must be admissible on the basis of a long list of personal characteristics (for example, good health, lack of criminal background, and sufficient economic resources). Fourth, there are annual numerical limits on the major categories of immigration to the United States.

Refugees are resettled in the United States if they meet the criteria of the international definition provided previously. The numbers of persons admitted as refugees reflect international need as identified by the US State Department in consultation with Congress. In the mid-1990s, an average of 80,000 refugees were admitted annually, largely from Bosnia-Herzegovina, countries of the former Soviet Union, and Vietnam (USDHS 2003). Since the September 11, 2001, terrorist attacks, refugee admissions to the United States have been sharply curtailed.

The United States also issues hundreds of thousands of "nonimmigrant" visas each year to persons visiting the country for specific purposes such as tourism and business, university study, consulting, and temporary employment. One likely effect of the September 11 terrorist attacks will be more restriction of temporary visas and a generally greater scrutiny of all visa applications. The relationship between international migration and homeland security has become a significant dimension of the US immigration debate.

Conclusion: Transitional Demographies

Current demographic trends have led to distinctive shifts in social norms and trends on a global scale. A rapid period of population growth followed closely by a swift decrease in fertility rates has contributed to unusual changes in social structure, and will in turn lend itself to continued phases of transition in the future. As indicated by John Weeks:

> The past was predominantly rural and the present is predominantly urban . . . the past was predominantly pedestrian . . . and the present is heavily dependent on the automobile. . . . The past was young . . . whereas the present is older. . . . In the past, people lived in households with more people than today . . . and the average person in each household was considerably less well educated than today, with only 10 percent of those in 1900 achieving a high school education, compared to 80 percent now. (2005: 6)

Put simply, the world is changing quickly—a transformation both influenced by and influencing demographic change.

Population trends and patterns within countries and regions hold fundamental and inescapable implications for the full spectrum of global issues addressed in this book. While the annual rate of world population growth has been declining in recent decades, significant increases in population size, particularly in countries in the developing world, will continue into the near future. Understanding the sources of population change, specifically declines in fertility and patterns of migration, is a critical dimension of efforts to attain sustainable development, reduce poverty, and protect the global environment. Population scientists are directing world attention to the specter of HIV/AIDS and other infectious disease, and also to the rapid urbanization under way throughout many regions of the developing world.

Present trends in fertility, mortality, and migration hold the key for the future trajectory of world population growth. Population policy decisions made today—locally, nationally, and internationally—will influence that trajectory and ultimately the size and distribution of global population.

Discussion Questions

1. Is population growth a major global problem?
2. Do you agree more with the Cornucopians or the neo-Malthusians regarding population growth?
3. What population problems arise from rapidly growing cities?
4. Should countries open their borders to refugees fleeing political persecution and those seeking economic opportunity?
5. Was the Chinese government's one-child policy justified? Should governments be involved in population control?
6. Should the US government give foreign aid to reduce world population growth? Should the aid be conditional?
7. To what groups of immigrants should countries give preference?

Suggested Readings

Arizpe, Lourdes M., Priscilla Stone, and David C. Major, eds. (1994) *Population and Environment: Rethinking the Debate*. Boulder: Westview.

Ashford, Lori S. (1995) "New Perspectives on Population: Lessons from Cairo." *Population Bulletin* 50, no. 2.

Bailey, Adrian (2005) *Making Population Geography*. New York: Oxford University Press.

Castles, Stephen, and Mark J. Miller (2003) *The Age of Migration: International Population in the Modern World*. 3rd ed. New York: Guilford.

Commoner, Barry (1992) *Making Peace with the Planet*. New York: New Press.

Demeny, Paul, and Geoffrey McNicoll (2003) *Encyclopedia of Population*. Vols. 1–2. New York: Macmillan Reference.

Harper, Charles L. (1995) *Environment and Society: Human Perspectives on Environmental Issues*. Upper Saddle River, NJ: Prentice Hall.

McFalls, Joseph, Jr. (1995) "Population: A Lively Introduction." *Population Bulletin* 46, no. 2.

Moffett, George D. (1994) *Critical Masses: The Global Population Challenge.* New York: Penguin.

Myers, Norman, and Julian L. Simon (1992) *Scarcity or Abundance: A Debate on the Environment.* New York: Norton.

United Nations High Commissioner for Refugees (2007) *UNHCR Statistical Yearbook 2006.* Available at http://www.unhcr.org.

United Nations Population Division (2003) *World Population Prospects: The 2002 Revision—Highlights.* New York.

——— (2007) *World Population Prospects: The 2006 Revision—Highlights.* New York.

——— (2008) *World Urbanization Prospects: The 2007 Revision—Highlights.* New York.

US Department of Homeland Security (2007) *2006 Yearbook of Immigration Statistics.* Washington, DC: US Government Printing Office.

10

Women and Development

Elise Boulding and Heather Parker

THIS CHAPTER USES THE TERM *DEVELOPMENT* TO REFER TO social, economic, and political structures and processes that enable all members of a society to share in opportunities for education, employment, civic participation, and social and cultural fulfillment as human beings, in the context of a fair distribution of the society's resources among all its citizenry. The United Nations bound itself in its Charter "to achieve international cooperation in solving international problems of an economic, social, cultural, or humanitarian character." In other words, the more-industrialized countries of the North agreed to help the less-industrialized countries of the South to reach the higher economic and social level already achieved in the North. The first thing to note about this development planning is that it has been done almost entirely by men, for men, with women and children as a residual category. How could half the human race be invisible to development planners despite Mao Tse-tung's well-known saying that "women hold up half the sky"? This chapter will review how this situation evolved, how it has hampered "real development," and what women—and men—are doing about it.

From Partnership to Patriarchy

In the early days of the human species, men, women, and children moved about in small hunting and gathering bands, sharing the same terrain and exploring the same spaces. Role differentiation was minimal, though childbearing restricted women's movements somewhat, and men ranged farther in hunting prey. Even so, it is estimated that women, as roving gatherers, supplied up to 80 percent of a band's diet by weight and therefore carried an important part of society's ecological knowledge in their heads. Based on observation of small hunting-gathering tribes found in Africa, Australia, and South America today, it would seem that old women as well as old men took the role of tribal elders and carried out rituals important to the social life of the band. Around 12,000 B.C.E. a combination of events brought about a major change in the human condition. Improved hunting techniques resulted in a dwindling supply of animals, and women discovered from their plant-gathering activities that seeds spilled by chance near the previous year's campsite would sprout into wheat the following year, creating a convenient, nearby source of food. This resulted in the deliberate planting of seeds to grow food, and agriculture came into being. This changed everything. Since working with seeds was women's work, women became the farmers, and men went farther and farther afield in search of scarce game, thus discovering exciting new terrain that women knew nothing about. The shared-experience worlds of women and men were now differentiated. Women knew about everything that was close to home, and men were exploring the world "out there."

While a lot has happened to humankind since 12,000 B.C.E., bearing children, growing and processing food, and feeding families (also creating shelters and clothing) have continued to be women's work for most of the world's population. Moving from the earliest farming settlements to villages, from villages to towns, and from towns to cities and civilizations took another 8,000 to 10,000 years, but cities never returned to women that freedom of movement they had enjoyed as gatherers. Rather, cities heralded the rise of the rule of men, or patriarchy, to replace an earlier partnership. What cities and civilization brought were concentrations of wealth and power—and houses that enclosed women and shut them off from the outside world. Only poor women were "free"—to scurry about the streets providing services for the rich. With industrialization and major population movements to the cities, where the factories were located, fewer and fewer women enjoyed the relative freedom of farming and craftwork, which families had often carried out as a team in the period known as the Middle Ages. Now they were either shut up in the home or shut up in factories.

The reality for individual women was much more complex than these major trends suggest, and in every age women found ways to be creative and to improve the circumstances in which they lived. While most of women's creativity was invisible, we do find women—queens and saints and philosophers and poets—in the history books. We also find precursors, from the 1500s through the 1700s, of the contemporary women's movement. However, by the nineteenth century, the pressures of urbanization and industrialization in Europe and North America gave rise to a new social sector of educated middle-class women with free time and a growing awareness of the world around them, including an awareness of migrant women from rural areas, chained with their children to factory and slum. This produced a small group of female radicals and revolutionaries and a larger group of liberal reformers and concerned traditionalists who translated their sense of family responsibility into responsibility for the community. These women quickly discovered, when they intruded into the political arena, that as women they had no civic or legal identity, no political rights, and no economic power. The realization that they needed civic rights to get on with reform led to an exciting century and a half of mobilization of women in the public arena. Their concerns about economic conditions for the poor soon spilled over to concern about the frequent wars that rolled over Europe, which they saw as directly related to poverty and suffering. The suffrage movement came into being because women realized that they needed political power in order to fight the social evils they saw.

The patriarchal model pervades society, beginning in the family with the rule of the male head of household over wife or wives and children; it has served as a template for all other social institutions, including education, economic life, civic and cultural life, and governance and defense of the state. Because so many generations of humans have been socialized into the patriarchal model, the struggle to replace it with partnership between women and men will be a long one, and it has barely begun. Now, however, the survival of the planet is at stake. Women's knowledge of their social and physical environment, of human needs, of how children learn, of how conflicts can be managed without violence and values protected without war—as well as their skill in managing households with scarce resources—are urgently needed wherever planning takes place and policies are made. Yet these are precisely the places where women are absent.

The Development Decades

As Western colonial empires began to dissolve during the UN's first decade, the 1950s, it became clear that the ever more numerous new, poor nations that came to be called the "third world" would need a lot of assistance from the founding states of the UN in order to work their way out of poverty. The popular "trickle-down" theory, which asserts that financial benefits given to big business and upper-class sectors will in turn pass down to smaller businesses, consumers, and lower-class sectors, led to the encouragement of capital-intensive heavy industry in poor countries. This type of industry requires large expenditures of capital. The result is a loss of labor-intensive light industries that might have employed the unemployed and the unschooled. The only aid given to agriculture was in the form of capital-intensive equipment that required large acreage and pushed subsistence farmers onto ever more marginal lands with rapidly deteriorating soils. This approach, labeled "economic dualism," leaves the bulk of a country's population without the skills and tools needed to become more productive.

By the end of the 1950s, it was evident that "international cooperation" was not helping poorer countries at all. Then the UN General Assembly declared the 1960s a "Development Decade," a time of catching up for all the societies left behind in the twentieth-century march of progress. Yet the policies based on economic dualism remained unchanged, despite valiant efforts by the United Nations Development Programme (UNDP), established in 1965, to undertake a more diversified approach. By the end of the 1960s, gross national product growth rates were negative in a number of third world countries, even in some that had "good" development prospects.

A second Development Decade was launched in 1970 to try to do what the first had failed to accomplish, but by then it had become clear to many third world countries that the development strategies recommended by "first world" countries were not working. The third world countries formed their own body, the Group of 77 (which eventually grew to include 118 states), to promote a program of action for establishing a new international economic order. This called for more aid from first world countries, debt moratoriums, and other strategies to halt the continuing increase in the economic gap between North and South. A monitoring system for the conduct of multinational corporations was also demanded.

Unfortunately, the North had no serious interest in acting on these proposals, and the poor nations kept getting poorer. In 1980 the UN launched yet another Development Decade, still pursuing failed policies. One problem was the World Bank's structural adjustment programs, which compelled countries of the South to focus on producing cash crops for the international market to reduce their indebtedness, and to spend less money on human services. This intensified the dualism between a low-productivity agricultural sector and a high-productivity agribusiness and industrial sector. People in the South went hungry and without schooling while food was exported to the North (where more money could be made), and countries sank still deeper into poverty.

The goal of all of this Development Decade activity, planned and carried out by men, was to increase the productivity of the male worker. But one quiet day in the late 1960s, a Danish economist, Ester Boserup (1970), decided to study what was actually happening in that intractable subsistence sector of agriculture. She was able to point out that the majority of the food producers were women, not men, but that all agricultural aid, including credit and tools, was given to men. This aid not only failed to reach women, but also encouraged men to grow cash crops (which needed irrigation) for export. This added to women's workload, as the men demanded that their wives tend the new fields in addition to the land that fed the family and provided a modest surplus for the local market.

Now an entirely new picture began to emerge. Boserup focused her studies on Africa, which was bearing the cruelest load of suffering in the food-supply crisis. She found that roughly 75 percent of food producers were women, often either sole heads of households or wives of migrant husbands who lived and worked elsewhere. Their workday, on average, lasted fifteen hours. They worked with babies on their backs, and their only helpers were the children they bore. These farmers never received advice on improved methods of growing food, tools to replace their digging sticks, wells to make water more available, credit to aid them during the years of bad crops, or assistance in marketing. Long hours of walking to procure water and to travel to market (with their produce carried on their heads), in addition to the hard work in the fields, made their lives a heavy struggle.

Yet emphasizing the hardship women experience should not obscure the importance of their special knowledge stock, which is basic to community survival in any country, in both rural and urban settings. In a future world with more balanced partnering between women and men, knowledge will not be so gender-linked; for now, it is imperative that it be recorded, assessed, and used. Women's special knowledge stock relates particularly to six areas: (1) lifespan health maintenance, including care of children and the elderly; (2) food production, storage, and short- and long-term processing; (3) maintenance and use of water and fuel resources; (4) production of household equipment, often including housing construction; (5) maintenance of interhousehold barter systems; and (6) maintenance of kin networks and ceremonials for handling regularly recurring major family events as well as crisis situations. While only the second area relates directly to food, all six factors contribute to the adequate nutrition of a community; the sixth area is particularly important in ensuring food-sharing over great distances in times of shortage and famine. Unfortunately, the extended-kin and ceremonial complex is one of the first resources to be destroyed with modernization.

The story of the UN's third Development Decade could have been different if policy planners had had this information at their fingertips. But now there was a women's movement ready to hear what Ester Boserup and a growing group of female development professionals had to say about women's roles in the development process. The UN came to play an important part in giving them a platform.

The Women's Decade

In the mid-1960s, women began forming traveling teams to explore every continent, and began seeing with their own eyes in the rural areas of Africa,

Asia, and Latin America what Ester Boserup was documenting as an economist. They began to see women as the workers of the world, invisible to census enumerators because much of their productive labor took place in the world of the home: the kitchen, the garden, and the nursery. "Employed persons" were supposed to have identifiable outside jobs and wages. The concepts of unwaged labor (such as childcare and housework) and labor in what is called the informal economy (labor that is not reported in the economy's record books, such as street vending, prostitution, and illegal drug sales) were not taken seriously by economists and statisticians.

As women began to see homemakers as workers, they redefined their own role. They liked to be called "housekeepers of the world." In 1972, "international housekeeping" brought them to the UN Conference on the Human Environment in Stockholm, and in 1974 to the World Population Conference in Bucharest and the World Food Conference in Rome. Each time, they came in larger numbers and with better documentation on how the conference subject was relevant to women. They also became increasingly aware of how blind most of their male colleagues were to the importance of women in economic production and social welfare. Since governments appointed few women to international conferences, the knowledge and wisdom of this growing group of observers went unheard. They were outsiders, petitioners, protesters.

Nevertheless, the UN Economic and Social Council had established its Commission on the Status of Women in 1946, and by 1948 its Declaration on Human Rights included gender as a category. In fact, over the decades, the UN adopted twenty-one conventions on the rights of women, although these remained largely "paper rights" because member states did not provide for their implementation. It was still a man's world, but women were learning strategies for working within it. The Commission on the Status of Women, in particular, was building experience working within the UN system, and began to assert itself. Its members saw to it that the phrase "integrating women into development" was included in the action program for the second Development Decade. Furthermore, in 1972, the UN appointed Helvi Sipila of Finland as the first woman to serve as Assistant Secretary-General for Social Development and Humanitarian Affairs. Empowered by Sipila's support, the Commission on the Status of Women worked closely with older women's organizations and newer networks to create a women's agenda for the United Nations.

The first real breakthrough came when the UN General Assembly declared that 1975 would be "International Women's Year." A global action plan was drafted in 1975 by the attendees of the first World Conference on Women, in Mexico City, defining status in terms of the degree of control women had over the conditions of their lives—a key theme that continues to this day in the women's movement. International Women's Year became the UN International Decade for Women (or Women's Decade), and follow-up conferences were held in 1980 in Copenhagen, in 1985 in Nairobi, and in 1995 in Beijing. These UN-sponsored women's conferences have represented a growing voice for women as participants and co-shapers of the world in which they live. Each of these gatherings, whose attendance swelled from 6,000 women in Mexico City to 14,000 in Nairobi to 50,000 in Beijing, has advanced the same trio of themes: equality, development, and peace. The interrelationship of these three themes in the lives of women, and in the life of every society, represents an important breakthrough in the conceptualization of development.

The guidelines for international action laid down for the Women's Decade are as relevant today as they were over thirty years ago:

- Involve women in the strengthening of international security and peace through participation at all relevant levels in national, intergovernmental, and UN bodies.
- Further the political participation of women in national societies at every level.
- Strengthen educational and training programs for women.
- Integrate women workers into the labor force of every country at every level, according to accepted international standards.
- Distribute health and nutrition services more equitably, to account for the responsibilities of women everywhere for the health and feeding of their families.
- Increase governmental assistance for the family unit.
- Involve women directly, as the primary producers of population, in the development of population programs and other programs affecting the quality of life of individuals of all ages, in family groups and outside them, including housing and social services of every kind.

Immediate outcomes of the 1975 conference included establishment of two new UN bodies, the United Nations International Research and Training Institute for the Advancement of Women (INSTRAW) and the United Nations Development Fund for Women (UNIFEM), as well as a modest working relationship between the two new organizations and the UN Development Programme. With few resources and very small staffs, INSTRAW and UNIFEM have nevertheless played an important part in the gradual recognition of women as actors, not only as subjects needing protection.

Since the UN declared the Women's Decade, some promising international progress has been achieved concerning the role of women in development. For example, the World Bank has developed and implemented a poverty reduction strategy for integrating gender considerations into its development assistance work. And in 1997, the UN's Division for the Advancement of Women (DAW), together with UNIFEM and INSTRAW, founded WomenWatch to monitor the results of the 1995 Beijing conference as well as to create Internet space devoted to global women's issues. The fourth World Conference on Women, and its two subsequent special sessions, Beijing +5 and Beijing +10, identified twelve critical areas of concern to women, ranging from poverty to power and decisionmaking.

As a result of these conferences and global policy decisions, many countries have changed their laws concerning women, particularly regarding discrimination: in Mexico, a law that required women to wait 300 days after a divorce to remarry was rescinded; in Costa Rica, a law that excepted rapists from punishment if they married their victims was repealed; and in France, prohibitions that kept women from working at night were removed. Further, many countries and organizations have sought to enhance women's economic independence by increasing access to credit, land, inheritance, markets, employment, and public transportation. For instance, in 2004, Bangladesh aimed to improve women's use of public transportation by creating "gender-friendly" bus services.

In addition, in 2000 the UN General Assembly adopted the Millennium Declaration. This resolution led to the creation of the Millennium Development Goals, a group of eight goals and eighteen targets that form a development blueprint for the world. All the world's countries and leading development organizations have agreed to these goals, which range from halving extreme poverty, to halting the spread of HIV/AIDS, to increasing women's and girls' access to education, all by 2015. Though the Millennium Development Goals mirror the critical targets noted by the Beijing conferences, and advance the objectives of the Convention on the Elimination of All Forms of Discrimination Against Women and other international treaties and conventions, they are different in that they require concrete, time-bound, quantitative action.

In short, women are no longer to be treated as invisible or as subject to patriarchal rule, but as persons, citizens, and actors on the local and global scene. The importance of the United Nations in providing an international platform for women to be seen, heard, and listened to with respect cannot be underestimated, even though the UN still has a long way to go in making senior posts available to women. Now the concepts are here, in public international discourse, and the UN and its member states cannot completely ignore them, however much they drag their feet. This change in visibility could not have happened without the strong involvement of women from every continent.

Women's Networks Redefine the Meaning of Development

As mentioned, women were already active internationally on behalf of the oppressed poor by the middle of the nineteenth century. They came to know each other across continents by meeting at the world's great fairs—London in 1851, Paris in 1855 and 1867, and Chicago in 1893. These were women of the upper middle classes who could travel, of course. Many innovations in educational, welfare, and home services for working women, as well as for children, prisoners, and migrants, emerged from those meetings. By 1930, there were thirty-one international nongovernmental women's organizations, and many more of them had working-class members. However, it was not until the United Nations brought together grassroots women's groups from each continent that there emerged a deeper understanding of the problems that women faced in countries formerly colonized by the West. While every country experienced gender-based occupational dualism (women's-only and men's-only jobs), only in the third world was this dualism linked to an economic dualism that trapped women in the subsistence sector. In other words, women were left without the skills and opportunities for economic improvement.

While the cultural specifics of their situations differed among countries and continents, all women had a common base of experience linked to their ability and expectation to bear and raise children, and their responsibility for the nourishment, health, and well-being of family members. Thus, while men measured development in terms of gross national product growth rates, women thought in terms of human and social development. What would make life better for individual human beings, and what would make societies more humane and joyful? These were the questions women were asking as they formed new organizations committed to human rights, development, and peace. The same questions were being asked by the new women's networks that began forming during preparations for the first International Women's Year.

The first of the new networks was formed at the International Tribunal on Crimes Against Women, which met in Brussels in 1976. From that tribunal came the first major international statement about patriarchal power as violence against women per se, apart from specific abusive acts. After listening to testimonials of the horrifying array of violence experienced by women from all classes, on all continents, at the hands of men in their families, communities, and places of work, the tribunal's participants joined together to establish the International Feminist Network. Pressure from this network and others finally pushed the UN to recognize and act on another form of violence—sexual assault of women during times of conflict—

during a later tribunal on war crimes in the former Yugoslavia in 1993. With this growing awareness of the obstacles to achieving a better life, not only for themselves but also for society as a whole, women began to seek full participation in the development process, confronting violence and other human rights abuses, including war itself.

Because being able to work with men as equals is a precondition for women's voices to be heard, and because many laws precluded women from obtaining equal employment opportunities, political equality as a means to change these laws has been a goal at least since the beginning of the suffrage movement. But gaining the vote, with some associated legal and property rights, did not greatly improve women's economic situation, nor did it bring sexual discrimination to an end, or place women in policy-making positions equally with men. Therefore, the equal rights movement has devoted painstaking attention to legal rights in the areas of family, employment, education and training, and access to opportunity in general. This is important for women everywhere, but especially in the poorest countries, where men's literacy rate is barely 70 percent and women's is less than half of that. As opportunities for advanced study become more available to women, and increased numbers of women attain degrees in law, engineering, and the sciences, both physical and social, the sisterhood for social change will be able to draw on increasing competence and expertise in its work for equality on all continents.

The equal rights movement has taken a more direct political track as increasing numbers of women seek to become involved in the lawmaking process as elected officials. Since 1998, women's representation in national parliaments has increased in every world region. Unfortunately, this increase has been relatively slow (see Table 10.1). Women are still greatly underrepresented in national parliaments throughout the world.

In addition, women are underrepresented in executive branches of government around the world (including in presidential, ministerial, and cabi-

Table 10.1 Representation of Women in National Parliaments by Region, 1998–2007 (percentages)

	1998	2001	2004	2007
World (average)	12.7	14.0	15.7	17.7
Sub-Saharan Africa	11.5	12.5	14.8	17.7
Asia	13.7	15.4	15.0	16.7
Arab states	3.3	4.3	6.5	9.0
Europe	14.4	16.0	18.4	20.3
Americas	15.2	15.1	18.6	20.6

Source: Inter-Parliamentary Union, "Women in Parliament: World Classification" (December 2007), available at http://www.ipu.org/wmn-e/world.htm.

net positions). Since 1974, only eighteen countries have elected a woman president, and since 1960, women have served as prime ministers in only thirty-two countries. Yet around the world there is evidence that women's representation is slowly increasing. In 1998, twenty-eight countries had women in at least 15 percent of ministerial positions, versus fifty-eight countries in 2001; by 2005, the number of countries with women in these posts had increased to sixty-eight. And while women held no ministerial positions in fifty-four countries in 1994, by 2005 this number had dropped to nineteen countries.

Women's political participation may be thought of as a pyramid, with the bulk of women found at the bottom, at the local level, and the fewest at the top, at the diplomatic level. In general, there is a tipping-point phenomenon for women in elected or appointed office. When there are very few "token" women in office, they tend to confine themselves to what are thought of as traditional women's issues, notably the well-being of families and children. As women grow in numbers and self-confidence, however, they become empowered to address broader systemic issues that affect all sectors of society.

As women enter into the legislative process, they also bring with them alternative development expertise. Instead of investing in high-tech industrial enclaves and agribusinesses that rob the poor of their intensive hand labor, alternative development means investing in tools and water supplies for farmers and craftworkers; making credit available for small improvements, including equipment for small local factories; and building local roads, schools, and community centers that can serve many needs, including daycare for young children and health services. While women have been leaders in all of these efforts to achieve human and social development, not simply economic development, they have certainly not been alone. Ever since E. F. Schumacher (1993) wrote *Small Is Beautiful: Economics As If People Mattered,* there have been creative and humanistically oriented male development professionals working both alone and together with women to further these broader goals. They, too, have helped to strengthen the women's networks.

Protecting the environment—such as from logging that destroys entire forests and leaves fragile soils to erode, mining that destroys local farmland and waterways, and dams that destroy vast acreage and increase homelessness and unemployment—has also become an important agenda for some women's networks, who argue that a relationship exists between the oppression of women and the degradation of nature. These scholars and activists are proud to call themselves "ecofeminists" (it was a woman, Ellen Swallow, who coined the term *ecology* in 1892, and it is a woman, Rachel Carson, who is credited with initiating the birth of the modern environmental movement with her book *Silent Spring,* published in 1962).

From Patriarchy to Partnership

The world is midstream in a long, slow process of transformation. As described more fully in *The Underside of History: A View of Women Through Time* (Boulding 1992), in one sense the institutions of society continue to be stacked against women. There are strong expectations of subservient behavior on the part of women, reinforced by upbringing, teachings of church and school, continuing inequality under the law, underrepresentation in government, and media portrayals of women as consumer queens. However, we must not underestimate the fact that there has always been a women's culture inside the patriarchal culture, one that modifies and changes patriarchal institutions. We can therefore think of women as shapers, not only as victims, of society. However, individual initiative alone can be weak and ineffective. The support systems found in women's organizations and networks act as great multipliers of individual effort.

Organizational Examples of the Women-Development Relationship

Environmentalism

The Green Belt movement is responding to the environmental crisis of deforestation. The trees of the Rift Valley in Kenya are already gone, and the land is undergoing soil erosion and desertification on a large scale. The Green Belt movement brings together the women of local villages for tree-planting activities. The plan is simple: reforest the land to provide renewable food and fuelwood resources. The women of each village plant a community woodland of at least a thousand trees, a "green belt." They prepare the ground, dig holes, provide manure, and then help take care of the young trees.

The inspiration for this movement came from a female Kenyan biologist, Wangari Maathai, who saw that the natural ecosystems of her country were being destroyed. As she has pointed out, the resulting poverty was something men could run away from, into the cities; however, women and children remained behind in the rural areas, hard-pressed to provide food and water for hungry families. Working with Kenya's National Council of Women, Maathai went from village to village to persuade women to join the Green Belt movement, which to date has involved more than 100,000 women and half a million schoolchildren in establishing over 6,000 local tree nurseries and planting more than 30 million trees in community woodlands, with a tree survival rate of 70 to 80 percent. This revival of local ecosystems has made fuelwood available, and family gardens are once more viable. Furthermore, women have been empowered to develop more ways of processing and storing foodstuffs that were formerly vulnerable to spoilage. The movement's success has spread throughout the nations of Africa and onto other continents as well.

In Kenya, educated urban women were in close enough contact with their village sisters to be aware of their needs. The basic development issue at stake is food security, without which no country can achieve any meaningful development. Women helped introduce the food security concept to development professionals. As a result of her work, Maathai was awarded the Nobel Peace Prize in 2004 (the first African woman to receive the prestigious award), and in 2005 both *Time* and *Forbes* magazines named her as one of the world's hundred most influential people.

Microfinance

Long before the days of colonialism, the traditional practice of forming local credit associations tided over many women in hard times. The new situation of a growing cash economy in both rural and urban areas has greatly increased the need for credit, but credit is rarely available to women through local banks. In Bangladesh, the Grameen Bank was started specifically to aid the poorest sector of the population; as of March 2007, it has over 7 million borrowers throughout the country, 97 percent of whom are women. A network of thousands of "bankers on bicycles," each trained by the Grameen organization, covers many areas of Bangladesh, and the Grameen principle has now spread to other Asian countries, to Africa, and to Latin America. Further loans to a borrower group depend on repayment of previous loans, which averages about 97 percent, with women having consistently higher repayment rates than men. Once a borrower group has successfully completed its first round of loans for farming land, livestock, or tools, it often expands its activities to include building schools, clinics, and needed local production facilities. Studies of this program find that the incomes of borrowers in Grameen villages are 50 percent higher than those of borrowers in non-Grameen villages. To continue the steady move away from poverty, the Grameen Bank now also awards scholarships and educational loans to students. In 2006, as a result of these achievements, the Grameen Bank and its founder, Muhammed Yunus, were jointly awarded the Nobel Peace Prize. This is a tribute both to the Grameen method and to the business acumen of women borrower groups.

Similar microcredit lenders, such as Kiva, have achieved comparable success. Established in 2005, Kiva is a nonprofit organization that permits individual donors to help sponsor a business in the developing world for as little as $25. The process begins with a lender choosing a small business entrepreneur through Kiva's website. The borrower then receives 100 percent of the lender's money, through Kiva, to help start or expand their business. During the course of the loan period (six to twelve months), the lender receives e-mail journals, including progress reports, from the funded business. After the loan is repaid, the lender can either lend the money again or withdraw it from the program. During its short existence, Kiva has helped to raise loan funds for more than 4,000 borrowers, nearly 70 percent of whom are women, and has maintained 100 percent loan repayment (for more information, see http://www.kiva.org).

Assessing Where We Are

During the Development Decades, women were largely absent from the planning process. Male planners' lack of knowledge of the importance of women's agricultural labor, and of their economic and social contributions to a national standard of living, led development professionals and the UN development institutions, including the World Bank, to concentrate on a capital-intensive type of development. This included agricultural development that remained in special enclaves. Nothing "trickled down" to poor farmers, mostly women, and the countries of the South grew poorer as resources were diverted from the areas of greatest need.

The international women's movement, with support from the UN, has set about addressing this ignorance of male planners about women's work. At the same time, women's efforts to promote equality of participation in development, starting with the Women's Decade, have led to a growing awareness that overcoming poverty and improving the quality of life for all requires a different relationship between women and men, moving from patriarchy to partnership. When women and men can share their experience, resources of the World Bank and UN member states will be better used at local levels.

Fortunately, this is already happening. In the spring of 1997, a noteworthy global campaign—the Microcredit Summit Campaign—was launched by nongovernmental organizations and private-sector antipoverty groups, with the backing of the World Bank, the UN Development Programme, national leaders, international aid agencies, and other foundations and corporations, to make small loans to 100 million of the world's poorest families by 2005. The loans financed small-scale farming and trade, with a special focus on female farmers and entrepreneurs. In 2006 the campaign was relaunched, with two new goals to reach by 2015: (1) ensure that 175 million of the world's poorest families, particularly the women of these families, are receiving credit for self-employment and other financial or business services; and (2) raise 100 million families above the $1-per-day threshold. The Women's World Banking group has been a leading promoter of this new campaign, which has resulted in a tremendous multiplication of Grameen Bank–type projects. This represents an important step toward understanding development as human and social development, opening up more possibilities for diverse approaches to human betterment, and recognizing that, as Schumacher said many years ago, "small is beautiful." Most of all, it points to a growing partnership between the women and men who will be working for that more diversified, more earth-loving, more local, and yet more connected world of the future.

Discussion Questions

1. After reading this chapter, what does development mean to you?
2. The international women's conferences of 1975–1995 each advanced the same three themes: equality, development, and peace. Do you agree that these themes belong together?
3. This chapter focuses primarily on development problems in the South. Do you think the United States has development problems?
4. What does the phrase "from patriarchy to partnership" mean? Where is the United States in terms of this transformation?
5. What skills are needed to solve the world's current and future economic and environmental problems?

Suggested Readings

Boserup, Ester (1970) *Women's Role in Economic Development.* New York: St. Martin's.

Boulding, Elise (1992) *The Underside of History: A View of Women Through Time.* Rev. ed. Newbury Park, CA: Sage. (See vol. 1, chap. 4, "From Gatherers to Planters"; and vol. 2, chap. 5, "The Journey from the Underside: Women's Movements Enter Public Spaces.")

Caldecott, Leonia, and Stephanie Leland, eds. (1983) *Reclaim the Earth: Women Speak Out for Life on Earth.* London: Women's Press.

Fisher, Julie (1993) *The Road from Rio: Sustainable Development and the Nongovernmental Movement in the Third World.* Westport: Praeger.

Masini, Eleanora, and Susan Stratigos, eds. (1991) *Women, Households, and Change.* Tokyo: United Nations University Press.

Rehn, Elisabeth, and Ellen Johnson Sirleaf (2003) *Progress of the World's Women, 2002.* New York: United Nations Development Fund for Women.

Shiva, Vandana (1989) *Staying Alive: Women, Ecology, and Development.* London: Zed.

Turpin, Jennifer, and Lois Ann Lorentzen (1996) *The Gendered New World Order: Militarism, Development, and the Environment.* New York: Routledge.

United Nations (1995) *Report of the Fourth World Conference on Women.* UN Doc. A/Conf.177/20. October 17. New York.

United Nations Statistical Division (2006) *The World's Women 2005: Trends and Statistics*. New York.
Waring, Marilyn (1988) *If Women Counted: A New Feminist Economics*. San Francisco: Harper and Row.

11

Children

George Kent

WORLDWIDE, MANY CHILDREN LIVE IN WRETCHED CONDITIONS, suffering from malnutrition and disease, laboring in abusive work situations, and suffering exploitation of the most grotesque forms. The gravest problems are found in poorer countries, but many children are severely disadvantaged even in the richer countries. In the United States, for example, fully one-fifth of the nation's children live below the official poverty line. The situation of children is not merely a series of unconnected localized and private problems, but a series of systemic problems of public policy requiring attention at the highest levels of national and international governance.

Increasing attention by policymakers has resulted in real progress in improving the quality of children's lives. The advances are documented every year in reports of the United Nations Children's Fund (UNICEF). However, satisfaction with such successes must be tempered with appreciation of the great distance still to be traveled if all children are to live a life of decency. Perhaps the clearest lesson learned in recent years is that significant gains in children's well-being do not result from economic growth alone. They also require progressive social policy based on a sustained commitment to improvements in the well-being of the poor in general and children in particular.

While many different kinds of programs have been developed over the years to address the situation of children, most have been inadequate to the task. There is now new hope in the rapidly advancing recognition of children's rights, based on the acknowledgment that every single child has the right to live in dignity. The legal obligation for the fulfillment of children's rights falls primarily on national governments, but for large-scale global issues, there are also global obligations that need to be clarified (Kent 2008).

Child Labor

Children work all over the world, in rich as well as poor countries. They do chores for their families, and many work in fields and factories to earn modest amounts of money. Children's work can be an important part of their education, and it can make an important contribution to their own and their families' sustenance. There can be no quarrel with that. The concern here, however, is with child *labor.* Child labor can be defined as children working in conditions that are excessively abusive and exploitative. In some cases the excesses are plainly evident, as in reports of a scandal "that involves the kidnapping in central China of hundreds of children, and perhaps more, some reportedly as young as 8, who have been forced to work under brutal conditions—scantily clothed, unpaid and often fed little more than water and steamed buns—in the brick kilns of Shanxi Province" (French 2007). It is not clear where exactly the boundary between acceptable child work and unacceptable child labor should be located, but there are many situations that no doubt cross the line:

- Conditions of child labor range from that of four-year-olds tied to rug looms to keep them from running away, to seventeen-year-olds helping out on the family farm. . . .
- Children who work long hours, often in dangerous and unhealthy conditions, are exposed to lasting physical and psychological harm. Working at looms, for example, has left children disabled with eye damage, lung disease, stunted growth, and a susceptibility to arthritis as they grow older.
- Denied an education and a normal childhood, some are confined and beaten, reduced to slavery. Some are denied freedom of movement—the right to leave the workplace and go home to their families. Some are even abducted and forced to work.
- Bonded labor takes place when a family receives an advance payment (sometimes as little as US$15) to hand a child—boy or girl—over to an employer. In most cases the child cannot work off the debt, nor can the family raise enough money to buy the child back. The workplace is often structured so that "expenses" and/or "interest" are deducted from a child's earnings in such amounts that it is almost impossible for a child to repay the debt. In some cases, the labor is generational—that is, a child's grandfather or great-grandfather was promised to an employer many years earlier, with the understanding that each generation would provide the employer with a new worker—often with no pay at all.
- Millions of children work as bonded child laborers in countries around the world, 15 million in India alone. (HRW 2007)

The International Labour Organization (ILO) explains how bonded labor works:

> The employer typically entraps a "bonded" labourer by offering an advance which she or he has to pay off from future earnings. But since the

International Labour Office/J.M. Derrien

Young brick carriers in Madagascar

> employer generally pays very low wages, may charge the worker for tools or accommodation, and will often levy fines for unsatisfactory work, the debt can never be repaid; indeed it commonly increases. Even the death of the original debtor offers no escape; the employer may insist that the debt be passed from parent to child, or grandchild. Cases have been found of people slaving to pay off debts eight generations old. (ILO 1993: 11)

Paradoxically, the acceptance of child labor tends to be higher where there are higher surpluses of adult labor. The addition of children to the labor force helps to bring down wage rates, which in turn makes it more necessary to have all family members employed. The widespread employment of children keeps them out of school and thus prevents the buildup of human capital that is required if poor nations are to develop.

In India there is active trafficking in child laborers: "Child trafficking has become an endemic problem in the poorest villages of India. . . .

Children, who are cheaper than animals, are sold by their families to work as domestic labourers, in the carpet industry, on farms or as sex workers. In fact, whilst buffaloes may cost up to Rs 15,000 [US$360], children are sold at prices between Rs 500 [US$12] and Rs 2,000 [US$48]" ("Sold As Slaves" 2007). A recent study showed that currently there are hundreds of thousands of children under the age of eighteen working in cotton production in India, many of them as bonded laborers (India Committee of the Netherlands 2007).

Children work in rich countries as well. In 1998 the US government reported that 300,000 youths aged fifteen to seventeen were working in agriculture. This is not surprising, given that more than half the farmworkers in the United States earned less than $12,500 per year in 2002 (American Federation of Teachers 2007). In the United States, just as in poor countries, when adults cannot earn a living wage, their children often work. Enforcement of child labor laws has been weak in many states, apparently due to the greater concern with protecting the interests of employers. The federal rules regarding teen labor in the United States are readily available (USDOL 2007).

Human Rights Watch, a nongovernmental organization, has an active program on the elimination of child labor, as does the International Labour Organization; since 2003 it has designated June 12 of every year as "World Day Against Child Labour." Despite the ILO's optimism regarding the prospects for ending child labor (ILO 2006), much work remains to be done:

> While most international attention during the 1990s was focused on the formal and export sectors, only 5 per cent of child labour is found there, and an estimated 70 per cent of children in developing countries work far from public scrutiny in agriculture and the informal sectors. The invisibility of the bulk of child labour—including work in the informal sector or in the family, represents a serious challenge and is compounded by the clandestine nature of such practices as trafficking. (UNGA 2001: 92)

Despite extensive efforts to control child labor, there are still millions of children who work, many under grossly exploitative conditions.

Child Prostitution

Child prostitution refers to situations in which children engage in regularized sexual activity for material benefits for themselves or others. These are institutionalized arrangements—sustained, patterned social structures—in which children are used sexually for profit. Prostitution is an extreme form of sexual abuse of children and an especially intense form of exploitative child labor. Most prostitution is exploitative, but for mature men and women there may be some element of volition, some consent. The assumption here is that young children do not have the capacity to give valid, informed consent on such matters.

In some places, such as India and Thailand, child prostitution was deeply ingrained as part of the culture well before foreign soldiers or tourists appeared in large numbers. There are many local customers. Some Japanese and other tourists may use the child prostitutes in the "tea houses" in the Yaowarat district of Bangkok, but traditionally most of their customers have been locals, especially local Chinese. Similarly, in the sex trade near the US military bases in the Philippines before they closed down, more than half the customers were locals. There is big money associated with the foreign child sex trade, but there are bigger numbers in the local trade. According to a UN report on the status of children:

> Sexual abuse occurs in the home, in communities and across societies. It is compounded when abuse takes place in a commercial setting. The worst forms of exploitation include commercial prostitution and child slavery, quite often in the guise of household domestic work. The trafficking of children, as well as women, for sexual exploitation, has reached alarming levels. An estimated 30 million children are now victimized by traffickers, so far largely with impunity. (UNGA 2001: 22)

Armed Conflict

Armed conflicts hurt children in many ways. Wars kill and maim children through direct violence. Children are killed in attacks on civilian populations, as in Hiroshima and Nagasaki. In Nicaragua, many children were maimed or killed by mines. The wars in Afghanistan in the 1980s and in Bosnia in 1993 were especially lethal to children. Many children have been brutally killed in the genocide in Darfur, Sudan. In September 2004, a siege by Chechnyan rebels at a school in Beslan, Russia, led to the killing of several hundred children. Wars now kill more civilians than soldiers, and many of these civilians are children. Children have been counted among the casualties of warfare at a steadily increasing rate over the past century. Historically, conflicts involving set-piece battles in war zones away from major population centers killed very few children. However, wars are changing form, moving out of the classic theaters of combat and into residential areas, where civilians are more exposed.

A great deal of violence is also perpetrated against children under repressive conditions short of active warfare. Death squads in Latin American countries have killed thousands of street children with impunity. As well, children are frequently hurt in the aftermath of warfare by land-

Small Boys Unit at a checkpoint near Lunsar, Sierra Leone

mines. Given current technology, clearing these mines would take many years, and it may be impossible to clear them completely.

Often children are pressed to participate directly in armed combat, which harms them psychologically as well as physically. Children can be the agents as well as the victims of violence. Increasingly, older children (ten to eighteen years old) are engaged not simply as innocent bystanders but as active participants in warfare.

Dorothea Woods, associated with the Quaker United Nations Office in Geneva, dedicated herself to chronicling the plight of child soldiers in a monthly survey of world press reports titled *Children Bearing Military Arms.* She cited these cases:

- *Afghanistan:* "Hundreds of thousands of youth . . . were being raised to hate and fight a 'holy war.' . . . Many of those children are now with the Taliban army."
- *Burma/Myanmar:* "A Shan boy . . . had been a porter-slave to carry heavy things to the place of fighting. . . . He fell down and was kicked by a Burmese soldier . . . until his leg broke like a stick in three places."
- *Chechnya:* "Government security forces have often detained young males between the ages of 14 and 18 as potential combatants in order to prevent them from joining the rebel forces."
- *Guatemala:* "Forcing the under 18's from the indigenous communities to enroll in the army practically severs and destroys the future of these communities."

- *Liberia:* "Because of the socio-economic crisis a part of the youth population is inclined to join one of the factions. The possession of a Kalachnikov gives the means to live by pillage and racketeering if necessary. . . . Various estimates have put the total number of Liberian soldiers below the age of 15 at around 6,000."
- *Mozambique:* "For the 10,000 children who took part in the civil war, the war is not over; it has been replaced by a multitude of small wars in their heads."
- *Sierra Leone:* "After the outbreak of the civil war in 1991 some five thousand youngsters joined either the governmental army or the rebel Revolutionary United Front."
- *Uganda:* "Some 3,000 children have been kidnapped in the northern part of Uganda in the last four years according to UNICEF. The guerrillas who took these children have enrolled the boys in their army and have forced the girls to 'marry' the soldiers."

More recently, Human Rights Watch reported that the armed forces in Myanmar were luring children as young as ten into military service, and even sending them into combat zones. One boy was reported to be just over four feet tall, weighing only seventy pounds (CNN.com/asia 2007).

Wars sometimes harm children indirectly, through their interference with normal patterns of food supply and healthcare. Many children died of starvation during the wars under the Lon Nol and Pol Pot regimes in Kampuchea (Cambodia) in the 1970s. From 1980 to 1986 in Angola and Mozambique, about half a million more children under age five died than would have in the absence of warfare. In 1986 alone, 84,000 child deaths in Mozambique were attributed to warfare and to South Africa's destabilization efforts. The famines in Ethiopia in the mid-1980s and again later in that decade would not have been so devastating had it not been for the civil wars involving Tigre, Eritrea, and other provinces of Ethiopia. Civil war has also helped to create and sustain famine in Sudan.

The interference with food supplies and health services is often an unintended byproduct of warfare, but in many cases it has been very deliberate. In some cases, the disruption of infrastructure can have deadly effects well beyond the conclusion of the war. One example is the trade sanctions imposed on Iraq after the Gulf War of 1991. That war led to more deaths *after* the war than during it. In July 2000, the United Nations Children's Fund (UNICEF) estimated that if trends of the 1980s had not been interrupted by the war and the subsequent sanctions, there would have been half a million fewer deaths of children under age five in Iraq in the 1990s (Hijah 2003).

Children also suffer serious psychological stress as a result of their exposure to warfare, as has been the case in Iraq during the ongoing conflict there:

> Parents, teachers and doctors . . . cite a litany of distress signals sent out by young people in their care—from nightmares and bedwetting to with-

> drawal, muteness, panic attacks and violence toward other children, sometimes even to their own parents.
>
> Amid the statistical haze that enshrouds civilian casualties, no one is sure how many children have been killed or maimed in Iraq. But psychologists and aid organisations warn that while the physical scars of the conflict are all too visible—in hospitals and mortuaries and on television screens—the mental and emotional turmoil experienced by Iraq's young is going largely unmonitored and untreated. (French 2007)

Malnutrition

There are many different kinds of malnutrition. One of the most important, protein-energy malnutrition, is usually indicated in children by growth retardation. It is widely accepted that a malnourished child is one whose weight is more than two standard deviations (about 20 percent) below the normal reference weight for his or her age. By this standard, an enormous number of the world's children are malnourished. The number is decreasing, but much too slowly, and progress has been uneven.

> In 1990, 177 million children under five years of age in developing countries were malnourished, as indicated by low weight-for-age. Estimates suggest that 149 million children were malnourished in 2000. The prevalence of under-five malnutrition in developing countries as a whole decreased from 32 per cent to 27 per cent. . . . The most remarkable progress has been in South America, which registered a decrease in child malnutrition rates from 8 to 3 per cent. Progress was more modest in Asia, where rates decreased from 36 to 29 per cent and the number of underweight children under five years of age fell by some 33 million. Even this relatively limited achievement probably had a significant positive impact on child survival and development. Still, more than two thirds of the world's malnourished children—some 108 million—now live in Asia. . . . In sub-Saharan Africa, the absolute number of malnourished children has increased despite progress achieved in a few countries. (UNGA 2001: 37)
>
> Throughout the developing world, one out of every four children—roughly 146 million children—under the age of five is underweight. Among developing regions, child undernutrition is most severe in South Asia and, to a lesser extent, sub-Saharan Africa. For children whose nutritional status is deficient, common childhood ailments such as diarrhoea and respiratory infections can be fatal. Undernourished children who survive the early years of childhood often have low levels of iodine, iron, protein and energy, which can contribute to chronic sickness, stunting or reduced height for age, and impaired social and cognitive development. (UNICEF 2007b)

Contrary to the common belief that the problem is most widespread in Africa, there are far more malnourished children in Asia. More than half the developing world's underweight children live in South Asia alone (UNACC-SCN 1997). In developing countries, almost a third of all children under the age of five are malnourished, half of whom die as a result (UNICEF 1996).

Child Mortality

Nothing conveys the plight of children worldwide as clearly as their massive mortality rates. Estimates of the number of global under-five deaths for 1960–2005 are shown in Table 11.1. In 2006 the number fell below 10 million for the first time, to 9.7 million (McNeil 2007a). However, critics point out that this number is still far too high and that the progress is far too slow (Murray et al. 2007).

Children's deaths account for about one-third of all deaths worldwide. In northern Europe and in the United States, children account for only 2–3 percent of all deaths. But in many less-developed countries, more than half of all deaths are deaths of children. The child mortality rate for any given region is the number of children who die before their fifth birthday for every thousand

Table 11.1 Global Annual Child Deaths, 1960–2006

	Under-Five Deaths (millions)
1960	18.9
1970	17.4
1980	14.7
1990	12.7
1991	12.8
1992	13.2
1993	13.3
1994	12.6
1995	12.5
1996	11.7
1997	11.6
1998	11.1
1999	10.6
2000	10.9
2001	10.8
2002	10.9
2003	10.6
2004	10.5
2005	10.1
2006	9.7

Source: United Nations Children's Fund, *The State of the World's Children* (New York: Oxford University Press, annual).

born. As indicated in Figure 11.1, the rate at which children are dying each year has been declining. However, the numbers are still enormous.

The number of children who die each year, currently about 10 million, can be made more meaningful by comparing it with mortality due to warfare, which results in a long-term average of about 300,000–400,000 fatalities per year (Kent 1995). The most lethal war in all of human history was World War II, which resulted in about 15 million battle deaths. If civilian deaths are added to this total, including deaths due to genocide and other forms of mass murder, then World War II resulted in about 51 million deaths, or about 8.6 million deaths a year over six years—compared to under-five child deaths of well over 20 million per year. Thus the most lethal war in history resulted in a lower death rate, over a very limited period, than results from children's mortality year in and year out.

There are about 58,000 names on the Vietnam Veterans Memorial in Washington, DC. Significant to be sure, but this is less than the number of children under the age of five who die every two days throughout the

Figure 11.1 Under-Five Mortality Rate, 1990 vs. 2006

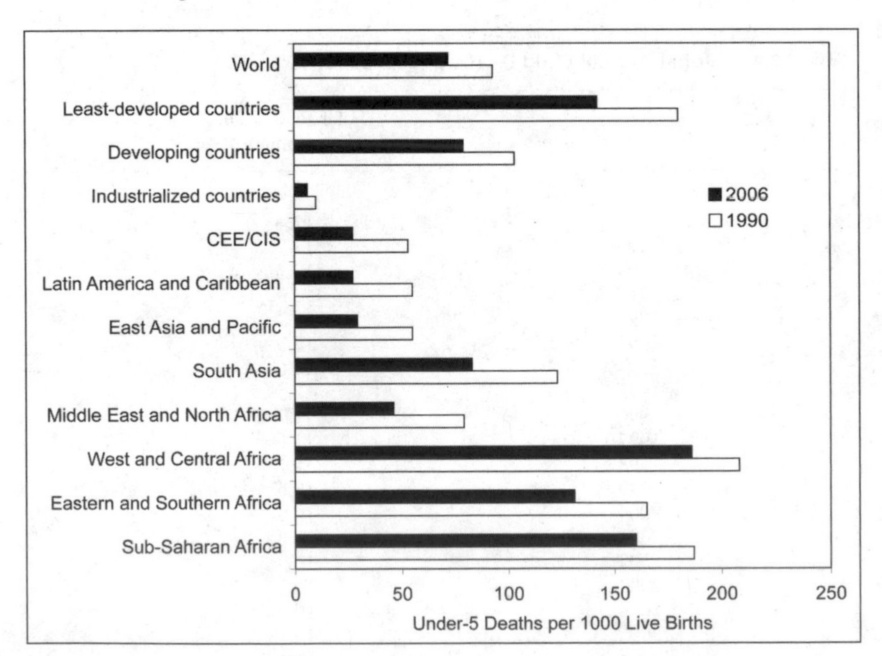

Source: United Nations, *The State of the World's Children, 2008* (March 25), available at http://www.unicef.org/sowc08/docs/Figure-1.4.pdf.

Note: CEE/CIS = Central and Eastern Europe/Commonwealth of Independent States.

world. For a memorial to list the names of all the children who die worldwide each year, it would need to be more than 250 times as long as the Vietnam Veterans Memorial, and a new such memorial would be needed every year.

Children in the United States

On the whole, children in the United States are far better off than most children in the rest of the world. However, there is a dark side that many in the United States never experience directly, one that is rarely reported by the US government or media. Although the Chinese may not be any more objective than Americans themselves about the situation of children in the United States, their views may provide a helpful corrective to the usual complacence of Americans. The following excerpt from a report by the government of China is based entirely on data from US sources:

> Child poverty was a serious problem. . . . [T]he number of children in poverty climbed from 12.1 million in 2002 to 12.9 million in 2003, a year-on-year increase of 0.9 percent. About 20 million children lived in "low-income working families"—with barely enough money to cover basic needs. . . . In California, one in every six children did not have medical insurance [and] in the metropolitan area the number of homeless children found wandering on the streets at nights numbered 8,000, which had stretched the 2,500-bed government-run emergency shelter system well beyond capacity. Poverty deprived many children the opportunity to obtain higher education. In the 146 renowned institutions of higher learning, only 3 percent of the students came from the low-income class, while 74 percent of them were from the high-income class.
>
> Children were victims of sex crimes. Every year about 400,000 children in the US were forced to engage in prostitution or other sexual dealings on the streets. Home-deserting or homeless children were the most likely to fall victims of sexual abuse. Reports on children sexually exploited . . . soared from 4,573 cases in 1998 to 81,987 cases in 2003. . . .
>
> Violent crimes occurred frequently. . . . [N]early 20 percent of US juveniles lived in families that possessed guns. In Washington, D.C., 24 people younger than 18 were killed in 2004, twice as many as in 2003. . . . In Baltimore, 29 juveniles were killed from Jan. 1 to Sept. 27 in 2004. In 2003, 35 were killed. . . . [A]bout 9 percent of school kids aged 9 to 12 admitted being threatened with injury or having suffered an injury from a weapon while at school in 2003. After the massacres at Columbine and Virginia Tech and many other places, shooting rampages at high schools and colleges in the United States now seem almost commonplace.
>
> Every year, 1.98 out of every 100,000 American children were killed by their parents or guardians. In Maryland, the rate was as high as 2.4 per 100,000. . . . [I]n Texas, each staff of local government departments responsible for protecting children's rights handled 50 child abuse cases every month.

> Two-thirds of juvenile detention facilities in the United States lock up mentally ill youth; every day, about 2,000 youth were incarcerated simply because community mental health services were unavailable. In 33 states, juvenile detention centers held youth with mental illness without any specific charges against them. . . .[B]etween Jan. 1 and June 30 of 2003, 15,000 youth detained in US youth detention centers were awaiting mental health services, while children at the age of 10 or younger were locked up in 117 youth detention centers. The detention centers totally ignored human rights and personal safety with excessive use of drugs and force, and failed to take care of inmates with mental problems in a proper way. They even locked up prisoners in cages. There were reports about scandals involving correctional authorities in California, where two juvenile inmates hanged themselves after they were badly beaten by jail police. (CIOSC 2005)

Children's Rights

Many different kinds of service programs are offered by both governmental and nongovernmental agencies to address children's concerns, and many of them have been very effective. However, the coverage is uneven, largely a matter of charity and chance. There is now an evolving understanding that if children everywhere are to be treated well, it must be recognized that they have specific rights to good treatment. Thus there is now a vigorous movement to recognize and ensure the realization of children's rights.

Children's rights have been addressed in many different international instruments. On February 23, 1923, the General Council of the Union for Child Welfare adopted the Declaration of Geneva on the Rights of the Child, which on September 26, 1924, was also adopted by the League of Nations. It was then revised and became the basis of the Declaration of the Rights of the Child, which was adopted without dissent by the UN General Assembly in 1959. The declaration enumerates ten principles regarding the rights of the child. As a nonbinding declaration, it does not provide any basis for implementation of those principles.

The UN General Assembly approved the Universal Declaration of Human Rights without opposition in 1948. It was given effect in the International Covenant on Civil and Political Rights and the International Covenant on Economic, Social, and Cultural Rights. The two covenants were adopted in 1966 and entered into force in 1976. The covenants include specific references to children's rights.

On November 20, 1989, after ten years of hard negotiations, the UN General Assembly adopted the Convention on the Rights of the Child. It came into force on September 2, 1990, when it was ratified by the twentieth nation. Weaving together the scattered threads of earlier international statements of the rights of children, the convention's articles cover civil, political, economic, social, and cultural rights, including not only rights to basic survival requirements such as food, clean water, and healthcare, but also rights of protection against abuse, neglect, and exploitation, and the rights to education and to participation in social, religious, political, and economic activities.

The convention is a comprehensive legal instrument, legally binding on all nations that accept it. The articles specify the actions that states are obligated to take under different conditions. National governments that agree to be bound by the convention have the major responsibility for its implementation. To provide added international pressure for responsible implementation, Article 43 calls for the creation of a Committee on the Rights of the Child, to comprise experts whose main functions are to receive and transmit reports on the status of children's rights. Article 44 requires states to submit "reports on the measures they have adopted which give effect to the rights recognized herein and on the progress made on the enjoyment of those rights." Article 46 entitles UNICEF and other agencies to work with the committee within the scope of their mandates.

All countries except Somalia and the United States have ratified or otherwise acceded to the Convention on the Rights of the Child. Somalia has not ratified because it does not have a functional government. The reasons for the US failure to ratify are not so clear. Both Bill and Hillary Clinton were known as strong advocates of children's rights, so it was a serious disappointment when the convention was not quickly signed and ratified following Bill Clinton's accession to office in 1993. The United States finally did sign the convention in February 1995. That signing, handled very quietly, apparently was done to fulfill a deathbed promise to James Grant, who had been executive director of UNICEF. However, the convention does not become binding on the United States until it is ratified through the advice and consent of the Senate. The convention still has not been forwarded to the Senate.

The US government has never offered any official explanation for its reluctance to ratify the Convention on the Rights of the Child. However, three major concerns have been voiced. The first relates to states' rights. Because the historical struggle to find an appropriate balance between the powers of the states and the powers of the national government has not been fully resolved, there is a fear that the US government, through its power to make international agreements, might federalize issues that previously had been addressed only in state law.

The second concern relates to capital punishment. Article 37 of the convention states that "neither capital punishment nor life imprisonment without possibility of release shall be imposed for offenses committed by persons below eighteen years of age." Until recently, the United States (along with Pakistan, Iran, Iraq, and Bangladesh) was among the few countries that still execute people for crimes committed before their eighteenth birthday. This argument was tied in with the argument that capital punishment should be a matter of state rather than federal policy. However, in March 2005 the US Supreme Court ruled this practice unconstitutional. This Supreme Court decision brought the United States into line with international law that prohibits such executions for crimes committed under the age of eighteen. According to Amnesty International (2006), twenty-two child offenders were executed in the United States between 1977 and 2005.

The third concern relates to abortion. The preamble of the convention states, in reference to the Declaration of the Rights of the Child, that "the child, by reason of his physical and mental immaturity, needs special safeguards and care, including appropriate legal protection before as well as after birth." The last six words conform to the pro-life, antiabortion position. However, because of the divisiveness of the abortion issue, the drafters of the convention chose not to elaborate the theme. For pro-life activists, the convention is not explicit enough regarding safeguards before birth and thus is not acceptable.

Despite these various objections, the United States could ratify the convention with reservations regarding the provisions it finds unacceptable, in order to support the many other provisions that it does favor. So perhaps the most serious obstacle to US ratification is ideological. The United States tends to support civil and political rights but not economic and social rights. The Convention on the Rights of the Child asserts economic rights such as "enjoyment of the highest attainable standard of health," including universal provision of adequate food and medical care. Economic rights of this sort trouble many officials in the US government. The government is willing to provide a broad array of social service programs, but it balks at the idea that people have a right, an entitlement, to them (Kent 2005).

Conservative elements in the United States have organized systematic campaigns of opposition to the Convention on the Rights of the Child based on the unfounded argument that ratification would undermine the family and take away parents' rights to raise their own children as they see fit.

International Obligations

The rights enumerated in the Convention on the Rights of the Child were further clarified with the Optional Protocol on the Sale of Children, Child Prostitution, and Child Pornography, which entered into force on January 18, 2002, and by the Optional Protocol on the Involvement of Children in Armed Conflict, which entered into force on February 12, 2002. With few exceptions, children also have all other human rights.

In some cases, other agreements give explicit attention to children. For example, the Convention on the Elimination of All Forms of Discrimination Against Women makes frequent reference to children. Several of the agreements developed through the International Labour Organization refer to children and particularly to the need to limit child labor. On June 17, 1999, the International Labour Organization adopted the Convention on the Worst Forms of Child Labor, which entered into force in November 2000.

The articulation of these human rights in international instruments represents an important advance, but there is still much more to be done to ensure that these rights are fully realized. Although technically binding on the states that ratify these international agreements, the human rights claims therein are not precisely specified. Latitude for interpretation is provided, because it is left to the national governments, representing the states that are party to the agreements, to concretize them in ways appropriate to their particular circumstances. As well, there remains the question of the international community's obligations, especially where national governments are unwilling or unable to do what needs to be done to ensure that children's human rights are fully realized.

There are many international organizations, governmental and nongovernmental, that work to alleviate suffering, and development and foreign aid programs do a good deal to improve quality of life. But these efforts are largely matters of politics and charity. There may be a sense of moral responsibility, but there is no sense of legal obligation, no sense that those who receive assistance are entitled to it and that those who provide it owe it. Historically, the idea of a duty to provide social services and to look after the weakest elements in society has been understood as something undertaken at the national and local levels, not as something that ought to be undertaken globally. Indeed, the only major market economy in which there is no clearly acknowledged responsibility of the strong with respect to the weak is the global economy.

Within nations, citizens may grumble when they are taxed to pay for food stamps for their poor, but they pay. Globally, there is nothing like a regular tax obligation through which the rich provide sustenance to the poor in other nations. The humanitarian instinct and sense of responsibility is extending worldwide, but there is still little clarity as to where duties lie. There is no firm sense of sustained obligation at the global level.

Most current discussions of global governance focus on security issues, the major preoccupation of the powerful, and give too little attention to the need to ensure the well-being of ordinary people. Just as there should be clear legal obligations to assist the weak in society at the local and national levels, those sorts of obligations should be recognized at the global level as well. Discussion of this idea has begun in the United Nations, but just barely.

There is much discussion of international protection of human rights, but what does this mean? If one party has a right to something, some other party must have the duty to provide it. Children's rights would really be international only if, upon failure of a national government to do what was necessary to fulfill those rights, the international community was *obligated* to step in to do what needed to be done—with no excuses. There is now no mechanism, nor any commitment, to do this. The international community provides humanitarian assistance in many different circumstances, but it is not required to do so. International law does not now require any nation to respond to requests for assistance.

There should be clear global obligations, codified in explicit law, to sustain and protect those who are most in need. The exact nature of those obligations and their magnitude and form will have to be debated, but the debate must begin with the principle that international humanitarian assistance should be regularized through the systematic articulation of international rights and obligations regarding assistance. Regularization can begin with the formulation of guidelines and basic principles and then perhaps codes of conduct. These can be viewed as possible precursors of law.

The nations of the world could collectively agree that certain kinds of international assistance programs *must* be provided, say, to children in nations where their mortality rates exceed a certain level. This international obligation to provide assistance should stand unconditionally where national governments or, more generally, those in power, consent to receiving the assistance. The obligation must be mitigated, however, where those in power refuse the assistance or where delivering the assistance would require facing extraordinary risks.

Part of the effort could focus on helping nations ensure that their children's nutrition rights are realized. The most prominent intergovernmental organizations concerned with nutrition are the Food and Agriculture Organization of the United Nations, the World Food Programme, the International Fund for Agricultural Development, the World Health Organization, and the United Nations Children's Fund. They are governed by boards composed of UN member states. Responsibility for coordinating nutrition activities among these and other intergovernmental organizations within the UN system rests with the United Nations System Standing Committee on Nutrition. Representatives of bilateral donor agencies, such as the Swedish International Development Agency and the US Agency for International Development, also participate in activities of the Standing Committee on Nutrition, as do numerous international nongovernmental organizations.

The main role of intergovernmental organizations is not to feed people directly, but to help nations use their own resources more effectively. A new regime of international nutrition rights would not involve massive international transfers of food. Its main function would be to press and help national governments address the problem of malnutrition among their own peoples, using domestically available food, care, and health resources. There may always be a need for a global food facility to help in emergency situations that are beyond the capacity of individual nations, but a different kind of design is needed for dealing with chronic malnutrition. Moreover, if chronic malnutrition were addressed more effectively, nations would increase their capacity for dealing with emergency situations on their own. Over time, the need for emergency assistance from the outside would decline.

Intergovernmental organizations could be especially generous in providing assistance to those nations that create effective national laws and national agencies devoted to implementing nutrition rights. Poor nations that are relieved of some of the burden of providing material resources would be more willing to create programs for recognizing nutrition rights. Such pledges by international agencies could be viewed as a precursor to recognition of a genuine international duty to effectively implement rights to adequate nutrition.

Of course, the objective of ending children's malnutrition in the world by establishing a regime of hard international nutrition rights is idealistic. Nevertheless, the idea can be useful in setting the direction of action. We can think of the intergovernmental organizations as having specific duties with regard to the fulfillment of nutrition rights. We can move progressively toward the ideal by inviting these organizations to establish clear rules and procedures that they would follow *as if* they were firm duties.

Efforts are under way to clarify and concretize global obligations with regard to human rights (Kent 2008). The core premise here is that a child may be born into a poor country, but that child is not born into a poor world. That child has claims not only on its own country and its own people, but also on the entire world. If human rights are meaningful, they must be seen as universal, and not merely local. Neither rights nor obligations end at national borders. While national governments have primary responsibility for ensuring realization of the right to food for people under their jurisdiction, all of us are responsible for all of us, in some measure. The task is to work out the nature and the depth of those global obligations.

Conclusion

Within nations, through democratic processes managed by the state, some moral responsibilities become legal obligations. A similar process is needed at the global level. Internationally recognized and implemented rights and obligations should not, and realistically cannot, be imposed. They should be established democratically, through agreement among the nations of the world. Reaching such agreement would be an action not against sovereignty but against global anarchy. It is important to move toward a global rule of law.

Regularized assistance to the needy under the law is a mark of civilization *within* nations. If we are to civilize relations *among* nations, international humanitarian assistance also should be governed by the rule of law. Looking after our children internationally could become the leading edge of the project of civilizing the world order.

The 1990 World Summit for Children produced ringing declarations and a promising plan of action to improve the conditions of children worldwide. In May 2002, a special session of the UN General Assembly was held to review the progress that had been made in the intervening decade and to make new plans and new commitments for the future. The Secretary-General's review, prepared for the special session, showed that substantial progress had been made on many of the issues of concern (UNGA 2001). It also showed that much more remained to be done.

Like the 1990 summit, the May 2002 special session concluded with impressive declarations and plans of action. Some will say that we can only wait and see if these commitments will be taken seriously. They are mistaken. It is important for people everywhere, in all walks of life, to actively and persistently insist that governments and international agencies at every level honor their commitments to the children of the world.

Discussion Questions

1. Should countries be concerned with the treatment of children within other countries? Should the United Nations be concerned? Why or why not?

2. Do you see many children (under eighteen) working in your community? Are they working in violation of state or federal labor laws?

3. Should corporations be allowed to benefit from exploitative child labor in other countries? Why or why not?

4. Under what conditions do you think it is acceptable for children to work?

5. Should the United States ratify the Convention on the Rights of the Child? Why or why not?

6. Do you think that the international community should be obligated to uphold children's rights?

Suggested Readings

America Federation of Teachers (2007) "In Our Own Backyard: The Hidden Problem of Farmworkers in America." Available at http://www.ourownbackyard.org.

Freeman, Michael, and Philip Veerman, eds. (1992) *The Ideologies of Children's Rights*. Dordrecht: Martinus Nijhoff.

Howard, Michael (2007) "Children of War: The Generation Traumatized by Violence in Iraq." *The Guardian* (February 6). Available at http://www.commondreams.org/headlines07/0206-05.htm.

Human Rights Watch (2007) "Child Labor." Available at http://www.hrw.org/about/projects/crd/child-labor.htm.

International Journal of Children's Rights (quarterly).

International Labour Organization (2006) "The End of Child Labour: Within Reach." Available at http://www.ilo.org/dyn/declaris/declarationweb.download_blob?var_documentid=6176.

Kent, George (1991) *The Politics of Children's Survival.* New York: Praeger.

——— (1995) *Children in the International Political Economy.* London: Macmillan.

——— (2005) *Freedom from Want: The Human Right to Adequate Food.* Washington, DC: Georgetown University Press.

——— (2008) *Global Obligations for the Right to Food.* Lanham: Rowman and Littlefield.

LeBlanc, Lawrence J. (1995) *The Convention on the Rights of the Child: United Nations Lawmaking on Human Rights.* Lincoln: University of Nebraska Press.

Sawyer, Roger (1988) *Children Enslaved.* London: Routledge.

"Sold As Slaves, Children Are Cheaper Than Animals" (2007) *Asia News* (April 6). Available at http://www.asianews.it/index.php?l=en&art=8946&size=a.

United Nations (1992) *Child Mortality Since the 1960s: A Database for Developing Countries.* New York.

United Nations Children's Fund (annual) *The Progress of Nations.* New York.

——— (annual) *The State of the World's Children.* New York: Oxford University Press.

——— (1993) *Food, Health, and Care: The UNICEF Vision and Strategy for a World Free from Hunger and Malnutrition.* New York.

United Nations General Assembly (2001) *We the Children: End-Decade Review of the Follow-Up to the World Summit for Children—Report of the Secretary General.* UN Doc. A/S-27/3. Available at http://www.unicef.org/specialsession/documentation/index.html.

US Department of Labor (2007) "Youth Rules." Available at http://www.youthrules.dol.gov.

Veerman, Philip E. (1992) *The Rights of the Child and the Changing Image of Childhood.* Dordrecht: Martinus Nijhoff.

12

Health

Lori Heninger and Kelsey M. Swindler

WHAT IS HEALTH? WHEN WE SAY "HEALTHCARE," DOES IT mean treatment for sickness, or the prevention of disease? Does it pertain to the individual, or the society? Does it imply that health equals happiness?

The World Health Organization defines health as "a complete state of physical, mental and social well-being, and not merely the absence of disease or infirmity." For Australian Aboriginal people, "Health does not mean just the physical well-being of the individual, but refers to the social, emotional, spiritual and cultural well-being of the whole community" (WHO 2007c).

To most accurately define health and nonhealth in a given society, one must look at multiple factors, including what people in that society understand health and illness to be; the culture, such as traditional beliefs, religion, and social structures; medical and scientific knowledge; history; and political structures. For this chapter, health and illness are viewed as two sides of the same coin; both exist in each of us and in the communities and countries in which we live. We include intellectual, mental, physical, social, spiritual, environmental, and occupational concerns in the overall definition of health. We also approach health, and access to quality, appropriate healthcare, as a human right; Article 25 in the Universal Declaration of Human Rights states:

> Everyone has the right to a standard of living adequate for the health and well-being of himself and of his family, including food, clothing, housing and medical care and necessary social services, and the right to security in the event of unemployment, sickness, disability, widowhood, old age, or other lack of livelihood in circumstances beyond his control. . . . Motherhood and childhood are entitled to special care and assistance. All

children, whether born in or out of wedlock, shall enjoy the same social protection.

In 2000, world leaders gathered at the United Nations in New York and adopted the Millennium Development Goals, eight goals to be reached by 2015. Each of the goals has a relation to health/illness; however, five are specifically directed to the global eradication of health problems:

- Reduce the rate of child mortality among children under five by two-thirds.
- Reduce maternal mortality by three-quarters.
- Halt and begin to reverse the spread of HIV/AIDS and the incidence of malaria and other major diseases.
- Reduce by half the proportion of people without access to safe drinking water, and achieve significant improvement in the lives of at least 100 million slum dwellers (the latter by 2020).
- Develop a global partnership for development, one part of which calls for providing access to affordable essential drugs in developing countries.

How illness is caused varies greatly, and many types and causes of ill health are geographically and culturally specific. Malaria is not present in northern and western countries, because either the climate is too cold to sustain the parasite-bearing mosquito, or the mosquito has been eradicated. Japan and Nigeria have very low incidences of colon cancer; this may be due to a higher-fiber diet. River blindness, caused by a parasite, occurs mainly in Africa, South and Central America, and Yemen; the parasite does not occur in other areas. Migraines are much more prevalent in developed countries than in developing ones—cause unknown.

War zones, as can be imagined, are particularly unhealthy, more so for women and children than for fighters. In modern warfare, women, youth, and children suffer the greatest number of casualties; many more die during conflicts than those actually fighting the war. In addition, rape has been and continues to be used by some fighting factions, as in the Balkan conflict and by the Janjaweed in Darfur, as a weapon of war. Almost 80 percent of those displaced during conflicts are women, children, and youth; displacement can lead to malnutrition, lowered resistance to disease, mental health problems, environmental degradation and unclean or unavailable water, and in some cases, employment in unsafe conditions or sex work—the need for money drives women, youth, and children to work in situations in which they can be severely exploited (Amnesty International 2004).

The availability of medical services and treatment also varies widely from country to country, from a low of 0.02 doctors per 1,000 people in

Malawi, Niger, and Tanzania, to a high of 5.91 per 1,000 in Cuba (WHO 2007f). In many developing countries, nurses and community health workers supply most of the medical treatment and care received by people. Transportation to a clinic or medical office, or lack of ability to pay for treatment, often stops people from seeking help; this is true in developed as well as developing countries. When people are not able to access regular healthcare, disease often progresses to the point that radical, expensive intervention is required, if it can still be treated at all. These realities have an important effect on life expectancies throughout the world (see Table 12.1).

There are two main categories of ill health or disease, infectious and noninfectious. Infectious diseases are those that are spread from person to person, as with the flu; from animal to human, as with rabies; or from the environment to human, as with tetanus. Noninfectious diseases are those that are not spread; they either originate within one's body, such as cancer or lupus, or result from an external event, such as assault.

Due to the vastness of the subject, this chapter focuses on physical and mental health, including infectious and noninfectious health concerns; even with this limitation, we can barely begin to scratch the surface of global health. Other types of health not addressed here include intellectual, social, spiritual, environmental, and occupational health ("Dimensions of Health" 2007).

Table 12.1 Life Expectancy at Birth, Selected Countries, 2005

	Life Expectancy (years)
Australia	81.4
Canada	80.5
Austria	79.6
South Korea	78.5
United States	77.9
Panama	75.9
Ecuador	72.6
Jamaica	72.4
Malaysia	71.8
Fiji	68.9
Russia	65.2
Bangladesh	62.6
Mauritania	57.4
Haiti	54.7
Cambodia	53.8
Afghanistan	41.9
Angola	40.0
Sierra Leone	38.6

Source: World Health Organization, "Core Health Indicators" (2007), available at http://www.who.int/whosis/database/core/core_select.cfm.

Nutritional Health

The World Health Organization states that "nutrition is an input to and a foundation for health and development" (WHO 2006a). It is this broad definition that leads one to wonder what constitutes "good" nutrition and what factors affect the nutritional health of an individual or a community. The connotation of the word *nutrition* itself differs greatly when comparing developed nations to developing nations. In a rich Northern nation, the term conjures images of low-carbohydrate diets, fortified cereal, and foods that are free of trans-fats. In the context of the poverty, disease, and extreme weather conditions that affect so many developing nations, different pictures arise, such as the swollen belly of an undernourished child in Africa, or the goiter on a South American woman who is iodine-deficient.

For the purposes of this chapter, good nutrition signifies the presence of adequate calories, protein, fats, vitamins, and minerals in the diet to ensure proper growth and health. The absence of any one of these factors has a wide array of effects on the body of the individual and on the community as a whole. In order to better understand nutritional health and its effect on the global community, one must first focus on those factors that are a direct influence, such as availability of land, population of a region, food production, and climate change.

Food Production

The term *food security* describes the ability of a family or household to obtain enough food to sustain its members, and is dependent on both the amount of land available for agricultural use and the population of the region or nation. There are many other variables—but for simplicity we focus on these two primary influences.

Table 12.2 compares food security across five nations. The United States represents the Northern world (rich and developed); China and India represent two sides of a rapidly developing Asia; Zambia is an African nation fighting infectious disease and poverty; and Nicaragua is a Central American country plagued by poverty. As can be seen, although a developing nation may use a large percentage of its land for agriculture, it is unable to adequately feed its people. This results in undernourished children and a weakened population. India is a primary example: while close to 49 percent of its land is arable land (capable of producing crops), 47 percent of its young children (under age five) are underweight. This is due primarily to the overwhelming population, as the number of people has surpassed the availability of arable land.

Table 12.2 Food Security, Selected Countries, 2004

	Total Population	Percentage of Arable Land[a]	Percentage of Under-Five Children Who Are Underweight
China	1,307,989,000	14.7	8.0
India	1,087,124,000	48.8	47.0
Nicaragua	5,376,000	14.8	10.0
United States	295,410,000	18.0	1.1[b]
Zambia	11,479,000	6.9	23.0

Sources: United Nations, *Human Development Reports,* available at http://hdr.undp.org/en; Central Intelligence Agency, *World Factbooks,* available at https://www.cia.gov/library/publications/the-world-factbook/index.html.

Notes: a. Data for 2005.

b. Data for 2002.

Under- and Overnutrition

Undernutrition

In 2004, 5.6 million deaths worldwide of children under five years of age were related to undernutrition. In developing countries, 27 percent of children under age five are considered underweight. In South Asia alone, 45 percent of the total population is considered underweight.

Undernutrition is a form of malnutrition, which simply means "bad nourishment" (WHO 2006a), and is a major concern in most developing nations. It can be divided into two main types: protein and energy malnutrition, and micronutrient deficiency. The first occurs when the body lacks sufficient amounts of protein to sustain bodily functions. Protein is responsible for building tissues in the body, and acts as an enzyme in many life-sustaining biological reactions: the absence of protein is detrimental to health. As stated in *Critical Issues in Global Health,* "Recent data supports the hypothesis that lasting risks for the development of diet-related diseases in later life, including hypertension, diabetes, obesity, and heart disease, are associated with fetal and infant under-nutrition" (Koop, Pearson, and Schwarz 2002: 232).

Micronutrient deficiency is the lack of certain necessary vitamins and minerals in the diet. When speaking of global health, the most commonly referenced micronutrients are iodine, Vitamin A, and iron. Iodine deficiency is the most common cause of brain damage and can cause goiter, a growth on the thyroid. Iodized salt is an easy prevention method, but even with this relatively cheap supplement, people in fifty-four countries remain iodine-deficient. Vitamin A deficiency is a major cause of blindness in developing countries, and anemia, brought on by iron deficiency, affects developed and developing nations alike.

The ramifications of undernutrition are far-reaching and often deadly. Protein and micronutrient deficiencies affect the health and immunity of the individual and subsequently the economic and social well-being of the community at large. At the individual level, poor nutrition increases the rate of death. In those tested positive for HIV (human immunodeficiency virus), nutrient deficiency can quicken the onset of AIDS (acquired immunodeficiency syndrome), and affects the safety and effectiveness of AIDS treatment by increasing the body's susceptibility to opportunistic infections. Inadequate nutrition also increases the probability of death in malaria patients. At the community level, undernutrition affects the individual's ability to contribute to society; it reduces worker productivity by limiting mental and physical capacity, simultaneously increasing the burden on government healthcare systems.

If malnutrition is to be overcome worldwide, a significant focus on preventative, rather than curative, healthcare is vital. As a global community, we need to refocus research and education on preventing the causes of under- and overnutrition, in order to see a significant decrease in chronic diseases. If the Millennium Development Goal of halving the number of under-five underweight children were achieved, this would prevent 50 million deaths annually (UNICEF 2007a).

Overnutrition

Overnutrition has recently become much more prevalent, in the developed world as well as in the developing world. The "global obesity pandemic" is related to increased consumption of high-fat, high-calorie foods that are low in fiber and nutrients. Many factors have contributed to the spread of obesity, including poor diet, decreased physical activity, and globalization and urbanization. The prevalence of "Westernized" diets that are high in saturated fats, sugar, and refined foods has increased worldwide, greatly affecting developing countries in the midst of economic transition. Due to the availability of modern transportation and the widespread use of technology, there has been a severe decrease in physical activity—which means that instead of burning excess calories, the body stores them as fat. As regions become more industrialized, reliance on factory-made foods increases and people move away from more balanced diets based on local foodstuffs. This shift has altered the geographic distribution of obesity; rates of obesity in poorer countries are now more quickly approaching the rate of obesity in the United States.

The ramifications of overnutrition are many: Type 2 diabetes (also known as adult-onset diabetes), cardiovascular disease, stroke, and cancer are just a few of the chronic conditions associated with obesity. These diseases, once considered "diseases of the affluent," are now becoming major health concerns for developing nations. According to the International Diabetes Federation, 70 percent of diabetes cases occur in the developing world (Carmichael 2007). Economic impacts stem from the burden these diseases place on the healthcare system; according to the World Health Organization, obesity accounts for up to 7 percent of total healthcare costs in several developed countries (WHO 2006b).

To define overweight and obese ranges, a weight-to-height ratio is used, called the "body mass index." In adults, a body mass index of 25.0–29.9 is considered overweight, while an index of 30.0 or greater is considered obese. For example, an adult who is 5-foot, 9-inches tall and weighs 169–202 pounds is considered overweight. The United States leads the world in obesity rates, with 140 million overweight adults (age twenty and older); of these, 66 million are obese (AHA 2007).

Infectious Disease

> [It] was always present, filling the churchyard with corpses, tormenting with constant fear all whom it had not yet stricken, leaving on those whose lives it spared the hideous traces of its power, turning the babe into a changeling at which the mother shuddered, and making the eyes and cheeks of the betrothed maiden objects of horror to the lover.
>
> —T. B. Macaulay, British historian, 1848 (Koplow 2003)

"It" is the ominous global scourge smallpox, a highly contagious infectious disease. Smallpox contributed to the fall of the Aztecs, created instability in the Hittite Empire, defied the "protection" of the Great Wall of China, altered the strategy of the American Revolutionary War, and infected pharaohs, tsars, queens, and emperors alike (Koplow 2003).

However, "It" can describe a number of infectious diseases that have plagued the global community since its infancy. Polio, for example, was prevalent during the Egyptian dynasties; stone slabs depicting priests with withered legs have been uncovered. HIV/AIDS, a "young" disease, has claimed over 25 million lives in the more than two decades since it was declared an epidemic. And some scientists believe that malaria has killed one of every two people who have ever lived on the planet (Finkel 2007).

An infectious disease is a disease that invades the body after the infection by a pathogenic microorganism (a microorganism capable of hosting and transferring a disease). This microorganism survives and replicates within the host, disrupting internal balance and destroying oxygen-carrying red blood cells. Infectious disease can be transmitted to humans in a variety of ways: from other persons, from insects and animals, and from contaminated food and water.

Infectious disease does not need a passport to cross borders. Smallpox was carried from country to country by ship; HIV spread swiftly through air travel. Infectious disease is influenced today by the rapid globalization of society, profit-driven pharmaceutical companies, and climate change. The idea of a "global market" with limited barriers on trade has allowed the mass production and distribution of food worldwide, creating an ideal scenario for the spread of food-borne illness. Pharmaceutical companies have also contributed to the unchecked spread of disease: by patenting their drugs and obstructing companies that seek to produce generic versions, they have made it more difficult for the poor to obtain life-saving medications.

Smallpox: A Story of Success

Smallpox attacks the lungs, heart, liver, intestines, eyes, and skin, disfiguring, blinding, and often killing its victims. On average, smallpox proved deadly in 25 percent of cases, killing roughly 500 million in the twentieth century alone. It was still widespread just three decades ago—infecting nearly 15 million annually and killing 2 million (Koplow 2003); the disease is still known for its infamous lesions and pustules.

After the discovery of an effective smallpox vaccine in the late eighteenth century, prevention of the disease became possible. In 1958, Victor M. Zhdanov of the Soviet Union proposed to the eleventh World Health Assembly that smallpox could be completely eradicated, and what became known as the Intensified Smallpox Eradication Program was born. By improving international communication, providing a system for identifying and reporting cases of infection, and developing mass quantities of vaccine, the world did indeed succeed in eradicating smallpox, as officially announced on December 9, 1979 (Koplow 2003).

The success of this campaign rested on the ability of countries to unite against a common enemy. Many of these nations were in the midst of civil war and racial strife, caught between the feuding superpowers of the late twentieth century. The United States and the Soviet Union, despite their entrenchment in the Cold War, and together with over seventy other nations, overcame political sensitivities and contained a virus unhindered by borders (Koplow 2003).

Polio: The Eradication Initiative Continues

The success of the smallpox campaign led the forty-first World Health Assembly to adopt a similar resolution for the worldwide eradication of polio. Poliomyelitis, as it is formally known, is a highly infectious disease that attacks the nervous system and can paralyze its victims in just a few hours. This paralysis is permanent in 1 out of 200 infections, and in some cases affects the breathing muscles, thus proving deadly. Symptoms of the virus include fever, fatigue, headache, nausea, and stiffness in the neck and limbs. While there is no cure for polio, a vaccine administered several times throughout the first year of infancy effectively immunizes a child for life (WHO 2006c). This vaccine has served as the world's main weapon in combating polio.

The Global Polio Eradication Initiative, led by the World Health Organization, Rotary International (a nongovernmental organization), the US Centers for Disease Control and Prevention, and the United Nations Children's Fund, has outlined three central objectives: interrupt transmission of the polio virus, achieve certification of its eradication, and contribute to and strengthen global healthcare systems. The core strategies of the initiative include infant immunization, supplementary immunization of children up to five years of age, increased surveillance of all polio cases among those under fifteen years of age, and "mop-up" campaigns for specified locations where the polio virus has been isolated. This campaign has achieved tremendous results, decreasing polio cases by 99 percent since the initiative was started in 1988 (WHO 2006c).

The job is far from complete, however, "because no country is safe from infectious diseases unless all countries are safe" (Koop, Pearson, and Schwarz 2002: 232). As with smallpox, polio knows no borders, and can even reinfect populations due to international food importations. Further, developing countries, in particular, are in need of adequate funds (an additional $390 million for 2007–2008) if they are to continue fighting the disease (WHO 2006c).

HIV/AIDS: The Epidemic Rages On

While polio captures a relatively small media audience, HIV/AIDS has drawn worldwide attention in the past several years. The human immunodeficiency virus attacks the body's T-cells (which fight off infections), thereby weakening the immune system and allowing the onset of additional, opportunistic infections. For an HIV patient, even an otherwise minor opportunistic infection, such as the common cold, can be extremely hazardous. In virtually all cases, HIV progresses to AIDS. HIV/AIDS has had devastating repercussions throughout the world, in both developed and developing countries. In 2006 there were 4.3 million new HIV infections; 2.8 million (65 percent) of these infections occurred in sub-Saharan Africa (WHO 2008). The parents of 15.2 million children have died from AIDS; 12 million of these children live in sub-Saharan Africa alone.

Transmission. Unlike smallpox and polio, which can be transmitted through simple person-to-person contact, HIV is transmitted solely through bodily fluids. There are three primary routes of transmission: through sexual intercourse, through blood and blood products (e.g., contaminated drug needles, transfusions), and from mother to child (during birth and through breast milk). Sexual transmission is by far the most common, resulting in 85 percent of HIV infections worldwide.

Prevention and treatment. The Millennium Development Goals specifically address HIV/AIDS, seeking to halt and reverse its spread by 2015. This ambitious task has been strengthened through worldwide consent; in June 2001 (the year following the Millennium Summit), heads of state and government representatives from 189 nations convened for the first time at a special session of the United Nations General Assembly on HIV/AIDS. This meeting resulted in the adoption of the Declaration of Commitment on HIV/AIDS. The Joint United Nations Programme on HIV/AIDS (UNAIDS) has united the efforts of many international governmental organizations and nongovernmental organizations in order to effectively combat the effects of this horrifying pandemic (UNAIDS 2007).

While treatment is often the primary focus of scientific research, prevention will prove our greatest ally in combating HIV/AIDS. Because there are more HIV infections each year than there are deaths due to AIDS, thus this specialized UN coalition has developed a framework for prevention: "comprehensive HIV prevention requires a combination of programmatic interventions and policy actions that promote safer behaviors, reduce biological and social vulnerability to transmission, encourage use of key prevention technologies, and promote social norms that favour risk reduction" (UNAIDS 2007). Primary methods of prevention target sexually transmitted HIV and include increased distribution and use of condoms and education concerning safe sex practices, the advantages of abstinence and monogamy, and the advantages of male circumcision. The latter method has recently shown promise: the results of three randomized trials conducted throughout Kenya, Uganda, and South Africa have shown a 60 percent reduction in the risk of heterosexually acquired HIV infection in men (UNAIDS 2007).

Since the introduction of antiretroviral drugs in 1996, HIV/AIDS treatment has become increasingly cheaper and more effective; this improvement has extended life expectancy of those diagnosed with HIV, and has heightened their quality of life. Antiretroviral drugs disrupt the life cycle of HIV by inhibiting replication, halting entry into the host cell, and blocking the release of enzymes critical to its integration into the host cell's genetic material. Antiretrovirals are most effective when administered in a combination "cocktail" (the virus mutates rapidly and has quickly developed resistance to most uniform drugs). This drug combination is known as "highly active antiretroviral therapy" and has proven successful in suppressing the virus when used properly (NIAID 2007).

Global progress. Table 12.3 compares the impact of HIV/AIDS on different regions of the world and the state of the epidemic in 1996 and 2006. While tremendous efforts have been made to combat HIV/AIDS, the epidemic continues to worsen. Despite advances in therapy and the latest antiretroviral treatment, the disease continues to take the lives of millions and is now recognized as the leading cause of death for those between fifteen and fifty-nine years old. In order to slow and eventually halt this devastating disease, prevention education is vital. Sadly, those areas that are the worst affected by the infection are also the least likely to have access to prevention education. Take, for example, the Papua region of Indonesia: 48 percent of the people in this remote, mountainous area do not know about AIDS. This is especially frightening because the Papua region has the highest AIDS prevalence rate in Indonesia (Arga 2007). This is not uncommon, as many of those affected by the disease live in remote locations of the world and not only are unaware of the virus, but also have no knowledge of or access to condoms. If the global community wishes to halt this silent killer, education, accessible medications, and increased government funding must become the top priorities.

A ray of hope. UNAIDS reported in 2006 that about 39.5 million people were infected worldwide; but in November 2007, UNAIDS revised this sta-

Table 12.3 Global Effects of HIV/AIDS, 1996 vs. 2006 (estimates)

	People Living with HIV/AIDS (thousands)		Deaths due to HIV/AIDS (thousands)	
	1996	2006	1996	2006
Sub-Saharan Africa	20,800	24,700	1,800	2,100
South and Southeastern Asia	6,000	7,800	250	590
East Asia and the Pacific (Oceania)	452	831	7	47
Caribbean	310	250	19	19
Latin America	1,300	1,700	81	65
North America	860	1,400	29	18
Europe and Central Asia	680	2,440	15	96
Middle East and North Africa	210	460	13	36
World (millions)	31	40	2	3

Source: Joint United Nations Programme on HIV/AIDS (2007), available at http://www.unaids.org/en/policyandpractice/default.asp.

tistic to 33.2 million, reporting that it had grossly overestimated the number of infections by millions. This miscalculation was primarily caused by faulty surveys that disregarded representative sampling. In July 2007, India's estimated AIDS caseload was halved (from 5.7 million to 2.5 million) after randomized household surveys were implemented to account for the general population (McNeil 2007b). The revised statistics offer hope to the campaign, as they signal a decrease in new infections (since their peak in the 1990s).

Malaria: A Tropical Disease?

As the global community continues to fight polio and works to educate people on HIV transmission and AIDS treatment, another deadly disease is sweeping through the tropical world, infecting millions through the bite of the Anopheles mosquito. This ancient parasite is the cause of malaria, an infectious disease that is currently endemic to 106 nations, kills over 1 million people every year, is Africa's leading cause of under-five mortality, and accounts for 40 percent of public health expenditure in Africa (UNICEF 2007b).

Treatment is difficult, as the parasite has developed a resistance to many drugs. The first widely used individual remedy was quinine. Chloroquinine, a synthetic malaria drug, was discovered in the 1940s; although it is safer than its predecessor, newer mutations of the parasite have developed a resistance. In order to prevent drug-resistant malaria, patients are now given an artemisinin-based combination therapy, which uses several antimalarial compounds mixed with an ancient herbal remedy. While this combinations therapy has proven effective, it is often expensive, sometimes costing up to a dollar a dose (Finkel 2007).

Because eradication of malaria was achieved in the United States and other northern nations, research and funding for malaria prevention and treatment have long been neglected. Pharmaceutical companies, which spend huge sums on research and development of the next groundbreaking drug, are looking for significant return on their investment. Since people with malaria are often poorer than people in the more-developed world, the return on investment is greatly reduced, reducing the incentive to conduct research into antimalarial drugs. As a result of the huge disparity between the health concerns of the developed versus the developing world, 90 percent of pharmaceutical research funds worldwide are going toward the health concerns of 10 percent of the global population (Kawachi and Wamala 2007).

Recently, the disregard for tropical diseases has begun to dissipate. Bill Gates, of Microsoft, is channeling money into malaria research, and the United States is working to deliver on a $1.2 billion pledge to fight the disease. The World Health Organization and other international organizations have renewed their commitment through the Global Malaria Programme (initially abandoned in 1969) and the Roll Back Malaria Partnership. Focus has been placed on prevention, with governments and agencies working to distribute insecticide-treated bed nets. Bed nets have shown great success, Kenya being a prime example. By following the World Health Organization's guidelines for distributing and using long-lasting insecticidal nets, Kenya reduced child deaths due to malaria by 44 percent between 2004 and 2006 (WHO 2007e).

These efforts, while substantial, will not be enough to eradicate malaria. Increased funding is needed. The cost for supporting the minimal set of malaria interventions in those countries that are worst affected is over $3 billion. Sadly, the "spark" for malaria research, and funding for tropical diseases in general, may come only as these diseases move north. Although the United States halted malaria transmission within its borders in 1950, the gradual warming of the atmosphere may reintroduce northern regions to these tropical diseases. Nairobi, Kenya, due to its high altitude, has long been considered an environment unsuitable to the mosquito transmitting malaria. However, malaria is now its most common disease (IRIN 2007).

* * *

Infectious disease will continue to plague our world until we overcome these social, economic, and political obstacles. In order to halt the scourges of the twenty-first century, it is imperative that we begin to see ourselves as a global community rather than a divided world of North and South. As evidenced by the success of the smallpox campaign, we *can* overcome political rivalries and break down the walls of economic and social differences in order to eradicate disease.

Global Climate Change and Health

Many scientists believe that climate change (see Chapter 14) is increasingly impacting global health. Global climate change can be seen from two different perspectives: rising average surface temperatures (land and sea), and increasing frequency and severity of extreme weather events worldwide. These two major categories of climate change are interconnected to a great extent, as an increase in the frequency and severity of extreme weather events has been linked to the buildup of heat in Earth's atmosphere and oceans (Kawachi and Wamala 2007). In 2005, the Atlantic Ocean experienced its worst hurricane season in recorded history. Two years later, in July 2007, a heat wave in the United States killed at least forty-four people. Heightened surface temperatures also result in shifts in the geographic range of certain infectious diseases.

Of the 600,000 deaths worldwide in the 1990s due to weather-related natural disasters, 95 percent occurred in poor countries (WHO 2005). It is a cruel irony that those who contribute least to carbon dioxide output are hardest hit by its adverse effects.

Reproductive Health

Reproductive health may be the most controversial area of healthcare in the world. Contraception, the "morning after" pill, abortion, stem cells, gender-based violence, and sexually transmitted infections are divisive subjects. The World Health Organization states:

> Reproductive health . . . implies that people are able to have a responsible, satisfying and safe sex life and that they have the capability to reproduce and the freedom to decide if, when and how often to do so. Implicit in this are the right of men and women to be informed of and to have access to safe, effective, affordable and acceptable methods of fertility regulation of their choice, and the right of access to appropriate health care services that will enable women to go safely through pregnancy and childbirth and provide couples with the best chance of having a healthy infant. (WHO 2007d)

Though the main focus of reproductive health is on women, its purview also extends to men, through responsible parenting, respect for and equal treatment of women, use of condoms to prevent pregnancy and sexually transmitted infections, including HIV/AIDS, as well as concern about breast and testicular cancers.

Articles 16 and 25 of the Universal Declaration of Human Rights, in particular, speak to reproductive rights:

- *Article 16:* Men and women of full age, without any limitation due to race, nationality or religion, have the right to marry and to found a family. They are entitled to equal rights as to marriage, during marriage and at its dissolution.
- *Article 25:* Motherhood and childhood are entitled to special care and assistance. All children, whether born in or out of wedlock, shall enjoy the same social protection.

Family Planning and Contraception

The word *contraception* can trigger some of the most virulent debates around the world. What is the role of women in a culture? In a family? Who defines what a family is? What do Christianity, Judaism, Islam, and Hinduism say about childbearing? What are the "rules" in Buddhism,

Animist, and other religious traditions? What has the state decided about availability of contraceptives? Should abortion be a method of family planning? When does life begin? Whose life is more important, mother's or baby's? Should human embryonic tissue be used for stem cell research?

Over 100 million women of childbearing age around the world are not able to access contraception (Singh 2003). What might this mean for a young woman? In regions of the world where women marry at very early ages, this will likely mean that she will begin bearing children immediately after marriage; there will be no other option for her. Whether or not her body is developed enough to bear children will not matter, even though complications and death during pregnancy, and particularly during delivery, are significantly higher the younger the woman. It will also likely mean that if the young woman was in school, she will no longer be able to attend, if she was attending at all.

Contraception and education are inextricably linked. Internationally, boys are favored over girls to attend school. Girls often must stay home to watch younger siblings or to work; many families believe that educating girls, who will inevitably "just get married and have children," is a waste of time and resources. Girls in families that have fewer children are more likely to be sent to school, and to stay in school longer, because resources are available. Educating girls has many societal and health impacts, the most significant of which is improving child health and reducing infant mortality; educated women marry later, use contraception, and have fewer and healthier children. Another factor is economic: educated women have improved nutritional levels and better health, including contraceptive and prenatal health, allowing them to provide greater economic contributions to their household (Access to Education 2007).

More than 500 million women in the developing world have access to contraception, preventing 187 million unintended pregnancies each year, including 105 million abortions and 215,000 pregnancy-related deaths. If contraceptives were available to all women in the developing world, an additional 52 million unintended pregnancies and 22 million abortions would be averted (Sonfield 2006).

Reproductive health is often considered a "woman's issue"; however, local reproductive education and intervention programs for boys and men can change this cultural view. Contraception, education, and enlisting men as reproductive health partners provide the foundation for brighter futures for children, and for society as a whole.

Maternal Health

One woman dies every minute of every day from causes relating to pregnancy and giving birth. Table 12.4 reveals the magnitude of this problem. And for every woman who dies, twenty more suffer from injuries including

Table 12.4 Maternal Health, Selected Countries, 2000

	Number of Maternal Deaths	Lifetime Risk of Maternal Death	Contraceptive Prevalence (any method) (percentage)[a]
Brazil	8,700	1 in 140	76.7
China	11,000	1 in 830	83.8
India	136,000	1 in 48	48.2
United States	660	1 in 2,500	76.4
Zambia	3,300	1 in 19	34.2

Source: World Health Organization, "Reproductive Health Indicators Database" (2007), available at http://www.who.int/reproductive_indicators/alldata.asp.
Note: a. Data range from 1995 to 2002.

fistula (a tearing of the tissue between the vagina and urethra, or vagina and large intestine, during delivery), infection, and disability. Pregnancy and giving birth are the leading causes of death around the world for women of childbearing age, resulting from hemorrhage, sepsis, complications of unsafe abortion, prolonged or obstructed labor, hypertensive disorders including edema, as well as delays in the decision to seek care, in transportation to a healthcare facility, and in receiving appropriate care at the healthcare facility (WCRWC 2007).

The Reproductive Health Response in Conflict Consortium, a nongovernmental organization, has developed intervention strategies and policies to decrease maternal mortality in situations of conflict. Through planning and preparation, existing facilities can be enhanced to meet emergency conditions. Training and provision of clean delivery kits to midwives, and providing emergency transport from refugee camps to local hospitals, are two ways of reducing maternal mortality (RHRC 2007a).

Mental Health

Defining Mental Health and Mental Illness

What is mental health? For the past decade or so, mental health has come to be associated with treatment options for people with mental illness; however, this significantly limits the concept of and right to mental health. A person who is mentally healthy can be seen as "having a positive sense of well-being, resources such as self-esteem, optimism, sense of mastery and coherence, satisfying personal relationships and resilience or the ability to cope with adversities" (WHO 2007b: 80). It is important to understand mental health as separate from, and at the same time inextricably linked to, mental illness, each being one side of the same coin.

Most broadly, mental health disorders include several types: mental illness, neurological, behavioral, and substance abuse. Defining mental illness can be tricky, for multiple reasons. Unlike discrete diseases that have specific, limited symptoms and manifestations (e.g., lung cancer), many mental illnesses present overlapping symptoms, making diagnosis difficult. A person can suffer from depression, anxiety, bulimia, and substance abuse all at the same time. Some symptoms may be behavioral, requiring psychological treatment, while others may be physiological, requiring medical treatment.

A second difficulty is that the definition of an "illness" dictates treatment. Western countries tend to focus on the individual when treating mental illness, while other countries and regions tend to focus on the group, or community. This is particularly relevant to intervention. In some cases, one-on-one psychotherapeutic approaches may be antithetical to healing, and traditional healing ceremonies that link the individual with the group may be much more appropriate and successful.

A third difficulty is that "diagnosis" of mental illness can be used to quell dissent in a society. Much has been written on this subject, from Sidney Bloch and Peter Reddaway's *Psychiatric Terror: How Soviet Psychiatry Is Used to Suppress Discontent* (1985), to a recent speech by New York Law School professor Michael Perlin who said:

> Placing dissidents in psychiatric hospitals rather than prisons served three points: it avoided the already limited procedural safeguards of a criminal trial, stigmatized people to subordinate them, and confined dissenters indefinitely. By 1989 conditions had begun to improve in the Soviet Union, but tools of coercive psychiatry still were used in what some call the "criminalization of dissent." And this practice was not limited to Russia; the expression of political opinions was perceived as delusional throughout the Soviet block. (Perlin 2006)

Today, depression is the leading cause of disability in the world, and is the fourth greatest contributor to the global disease burden; by 2020, it will be second. It affects about 154 million people internationally, and is associated with almost 850,000 suicides annually. Depression is readily treatable with appropriate interventions, including traditional healing such as ritual cleansing or purification ceremonies, as well as with medication and therapy; however, less than 25 percent of those affected have access to treatment services (WHO 2007a).

Schizophrenia affects about 25 million people around the world; 90 percent of people with untreated schizophrenia reside in developing countries. Schizophrenia is a chronic and potentially deeply debilitating disease;

people with schizophrenia often don't realize that they have the illness or don't want to take medication (at a cost of about $2 per month) due to significant side effects. A World Health Organization report indicates that people with schizophrenia can be treated providing that the following are in place: appropriate training of primary-care health personnel, provision of essential drugs, strengthening of families for home care, referral support from mental health professionals, and public education to decrease stigma and discrimination (WHO 2007a).

The Burdens of Mental Illness

Mental illness, unlike other illnesses, brings two burdens, an undefined burden and a hidden burden. The undefined burden includes the economic and social impacts of mental illness for families, communities, and countries that result from decreased or lost productivity, as well as from premature death due to suicide.

The hidden burden of mental illness includes both stigma and violations of human rights and freedoms; this burden is borne more by the individual. People with mental illness often experience rejection by friends, family, and society, and tend to suffer from other illnesses as well, including vascular disease, diabetes, cancer, and HIV/AIDS, resulting in shorter lifespans (WHO 2001a). People with mental illness are also often denied basic human rights. The World Health Organization has developed a framework of violations and legislative actions that can be taken to ensure that these violations are rectified, as shown in Table 12.5.

Table 12.5 Mental Health and Human Rights

Human Rights Violation	Legislative Solution
Lack of access to basic mental healthcare and treatment	Mandate and fund mental health services at the community level
Inappropriate forced admission or treatment in mental health facilities; violations within psychiatric institutions	Develop and implement monitoring bodies to ensure that human rights are respected in all mental health facilities
Discrimination against people with mental disorders, largely due to stigma	Develop and enact laws to ensure antidiscrimination in all areas of life, including work, housing, and education
Inappropriate detention in prisons	Mandate that people with mental disorders be diverted to mental health facilities instead of prisons

Source: World Health Organization, "Mental Health, Human Rights and Legislation: WHO's Framework" (2001), available at www.who.int/mediacentre/factsheets/fs218/en.

A Holistic Approach to Mental Illness

People with mental illnesses need to be seen as people first. Saying that someone "is a schizophrenic" is very different from saying that he or she "is a person who has schizophrenia." The first stance equates the person with their illness. The second recognizes the dimensionality of the person, something most everyone wants, resulting in better interventions and legislation regarding mental illness.

Since mental illness and mental health affect overlapping areas of a person's life, interventions must involve entire societies, governments as well as communities. According to the World Health Organization, respect for the individual and his or her needs and values—social, cultural, ethnic, religious, and philosophical—must be maintained. Care and treatment should be provided in the least-restrictive environment, and should seek to promote the individual's self-determination, personal responsibility, and highest attainable level of health and well-being (WHO 2001b).

Conclusion

According to international institutions, health is not merely the absence of disease, but rather a state of well-being. It is a human right, guaranteed around the world, but not yet universally upheld, even in developed countries.

Health is determined by multiple, globally interlinked factors, such as climate change, food security and production, the cost and availability of medication, and nutrition. Therefore, improved health requires the creation and implementation of inclusive, cross-border solutions. This includes work at the policy level as well as research and development of vaccines and medications for all people, not solely the ones who can afford them. On the ground, it mandates the availability and affordability of quality health services. Health as a human right requires immediate work on climate change, with a forward-looking approach to determine the impacts of and solutions for the new health problems that will occur.

Although many battles remain to be fought, the crusade toward global health has achieved many victories. Smallpox, once an uncompromising killer, has been eradicated. Children are receiving vaccines to prevent measles, mumps, and tetanus. Simple, inexpensive, and effective ways to combat malaria have been established. Once-a-day medications for HIV/AIDS have been developed, and generic drugs are being introduced to help make treatment affordable. The Millennium Development Goals represent a significant step toward health as a human right; the world must embrace them and ensure they are met. Then we will need to embrace a new set of goals to ensure health for all.

Discussion Questions

1. What would it mean to truly link health with human rights?
2. How do political and economic will work to help or hinder health?
3. What impact does culture have on health?
4. What would you do, globally, to reduce under- and overnutrition?

Suggested Readings

Bloch, Sidney, and Peter Reddaway (1985) *Psychiatric Terror: How Soviet Psychiatry Is Used to Suppress Discontent.* New York: Basic Books.

Kawachi, I., and S. Wamala, eds. (2007) *Globalization and Health.* New York: Oxford University Press.

Koop, C. E., C. E. Pearson, and M. R. Schwarz, eds. (2002) *Critical Issues in Global Health.* San Francisco: Jossey-Bass.

Koplow, D. (2003) *Smallpox: The Fight to Eradicate a Global Scourge.* Los Angeles: University of California Press.

National Health and Medical Research Council (2007) "The NHMRC Roadmap: A Strategic Framework for Improving Aboriginal and Torres Straight Islander Health Through Research." Available at http://www.nhmrc.gov.au/publications/synopses/_files/r28.pdf.

National Institute of Allergy and Infectious Disease (2007) "Treatment of HIV Infection." Available at http://www.niaid.nih.gov/factsheets/treat-hiv.htm.

Reproductive Health Response in Conflict (2007) "Minimum Initial Services Package (MISP)" (November 15). Available at http://www.rhrc.org/rhr_basics/misp.html.

United Nations Children's Fund (2007) "Facts on Children." Available at http://www.unicef.org/media/media_fastfacts.html.

United Nations Department of Economic and Social Affairs (1998) *Health and Mortality: A Concise Report.* New York.

World Health Organization (2007) "Reproductive Health Indicators Database." Available at http://www.who.int/reproductive_indicators/countrydata.asp.

——— (2007) "World Health Statistics." Available at http://www.who.int/whosis/whostat2007.pdf.

PART 4

The Environment

13

Sustainable Development

Pamela S. Chasek

DEVELOPMENT STRATEGIES IN THE EARLY PART OF THE TWENTY-first century have been shaped by the increasing tensions between economic health and ecological health. Capitalism, the predominant economic system, depends on repetitive expansion. This process of expansion needs and supports industrial processes with insatiable appetites for resources such as oil, coal, wood, and water, and it depends on increasing use of land, sea, and atmosphere as sinks for the deposit of wastes. Clearly, there is a conflict between the economic system's relentless demand and the world's limited and shrinking supplies. Although there is concern that the ravenous economy will gobble up and despoil the natural environment, some scholars and policymakers suggest that this catastrophe can be avoided: with sustainable development, we can have economic growth while protecting the environment (WCED 1987; UN 1992). Sustainable development involves many global actions—from the development of concepts, to the negotiation, monitoring, and financing of action plans. A large number of international organizations are involved in these activities. For over three decades the United Nations and its member states have tried to implement a global program on sustainable development, with limited success (Rogers, Jalal, and Boyd 2006). This chapter examines the evolution of the concept of sustainable development and the contribution of the UN's global conferences toward understanding and implementing development that is economically, socially, and environmentally sound.

Overview of Sustainable Development

Concepts of sustainability and sustainable development appeal to many people because they hold out the promise of reconciling these divergent views about the relationship between economic development and environmental health. Yet, while one can argue that reconciling the tension between ecology and economy is the central goal of sustainable development, there is little agreement on what sustainable development actually means. As a result, sustainable development has a multiplicity of definitions. Generally, it implies that it is possible to achieve sound environmental planning without sacrificing economic and social improvement (Redclift 1987). Some definitions emphasize *sustainability,* and therefore, the focus is on the protection and conservation of living and nonliving resources. Other definitions focus on *development,* targeting changes in technology as a way to enhance growth and development. Still others insist that sustainable development is a contradiction in terms, since development as it is now practiced is essentially unsustainable.

The Brundtland Commission underlined concern for future generations by asserting that sustainable development is development that "meets the needs of the present without compromising the ability of future generations to meet their own needs" (WCED 1987: 8). The commission, formally known as the World Commission on Environment and Development but commonly known by the name of its chair, Norway's Gro Harlem Brundtland, was convened by the United Nations to formulate a long-term agenda for action on the environment and development. Definitions of sustainable development tend to focus on the well-being of humans, with little explicit attention to the well-being of nature. However, the World Conservation Union (IUCN), the United Nations Environment Programme (UNEP), and the World Wide Fund for Nature (WWF) have proposed a definition that includes nature and highlights the constraints of the biosphere: sustainable development is "improving the quality of human life while living within the carrying capacity of supporting ecosystems" (IUCN, UNEP, and WWF 1991: 10). Although this definition observes the traditional hierarchy that places human beings above the natural world, it does emphasize our dependence on the biospheric envelope in which we live.

Over time, the concept of sustainable development has evolved. It has been recognized that efforts to build a truly sustainable way of life require the integration of action in three key areas: economic growth, conserving natural resources and the environment, and social development. Essentially, according to Peter Rogers, Kazi Jalal, and John Boyd, economic criteria cannot be maximized without satisfying environmental and social constraints; environmental benefits cannot necessarily be maximized without satisfying economic and social constraints; and social benefits cannot be maximized without satisfying economic and environmental constraints (2006: 46).

If development is to be genuinely sustainable, policymakers will have to make substantial modifications to their strategies and their assumptions. However, the difficulties of actually delivering on the hopes that many people around the world have attached to the idea of sustainable development have become increasingly evident. In part, these difficulties reflect political problems, grounded in questions of financial resources, equity (justice or fairness), and the competition of other issues, such as terrorism, war, and weapons of mass destruction, for the attention of decisionmakers. In part, they reflect differing views about what should be developed, what should be sustained, and over what period. Additionally, the political impetus that has carried the idea of sustainable development so far since the 1980s in public forums has also increasingly distanced it from the scientific and technological base necessary to make some headway.

The Environmental Pillar of Sustainable Development

Supportive ecosystems are the sole sources of the necessities of life, including air, fresh water, food, and the materials necessary for clothing, housing, cooking, and heating. In addition and equally important, it is only within ecosystems that vital life-supporting processes can take place: these include the regeneration of soil, pollination of plants, and global circulation of carbon, oxygen, and other elements necessary for life (Munro 1995). The vast majority of the world's peoples depend upon ecosystems that are often far from where they live. Trade and technology today bring produce of land and sea to shops from all over the world. Cotton grown in Egyptian soil with water from the Nile River ends up as undergarments in shops in the United States and Europe. Thus, in a sense, the Egyptian ecosystem supports American consumers. Phenomena that disrupt ecological processes have similar extended effects—ozone depletion, acid rain, and climate change affect everyone regardless of where they live on the planet, whether or not their emissions of chlorofluorocarbons, sulfur, or carbon dioxide have contributed to the problem.

The threat to the global environment has been salient enough to make the environment a long-standing item on the international agenda. From 1900 to 1968, conferences on environmental issues, many of them focusing on particular species or regions, resulted in a series of bilateral and multilateral agreements that, for the most part, took the optimistic stance that existing problems could be fixed by the employment of financial and technological resources. But by the time of the Biosphere Conference (formally known as the Intergovernmental Conference of Experts on the Scientific

Basis for Rational Use and Conservation of the Resources of the Biosphere) in 1968, that optimism had dimmed, and conference participants were emphasizing that environmental deterioration had reached a critical threshold (Caldwell 1990). As the international community became more concerned about the state of the environment and the relationship with economic development, this sense of crisis came to be reflected in the global conferences held in 1972, 1992, and 2002.

The 1972 Stockholm Conference

The Stockholm Conference (formally known as the United Nations Conference on the Human Environment), held in 1972, was not the first environmental conference to draw representatives from many of the world's nations. But it was the first large-scale environmental conference to look beyond scientific issues to broader political, social, and economic issues. It put the environment squarely on the official international agenda. Delegates came from 113 nations, 21 United Nations agencies, and 16 intergovernmental organizations (UN 1972). In addition, more than 200 nongovernmental organizations (NGOs) sent observers.

The Stockholm Conference took place at a time when issues of global equity were becoming more prominent in international forums. Developed countries clashed with developing countries over global economic relations and environmental politics. Before the conference, developed countries had identified a particular set of issues to be addressed—such as pollution, population explosion, conservation of resources, and limits to growth. But the developing countries wanted to enlarge the agenda to include issues such as shelter, food, and water. They were able to use their voting power in the UN General Assembly to press developed countries to adopt a more inclusive agenda (McCormick 1989).

The most significant institutional outcome of the Stockholm Conference was the creation of the United Nations Environment Programme. In the decades since the conference, UNEP has played a significant role in shaping the environmental policy agenda and in coordinating environmental policy within the United Nations system. The conference output also included the Declaration on the Human Environment, together with a declaration of principles and recommendations for action. One of the minor resolutions addressed plans for a second UN conference on the human environment.

The Stockholm Conference put the environment firmly on the international political agenda and thereby paved the way for intensified multilateral environmental cooperation and treaty-making. By stressing that environmental issues inherently are political and need to be subject to political negotiations and decisionmaking, the Stockholm Conference rejected the earlier notion that environmental issues were primarily relevant only to scientists and other experts. The conference also identified a theme that has been at the center of international environmental discourse ever since: the possibility of simultaneously achieving economic development and environmental management.

The Brundtland Commission

In the wake of the Stockholm Conference, the United Nations assumed a direct and coordinating role in efforts to raise international environmental awareness through its network of associations with governments, nongovernmental organizations, and the world's business and scientific communities. However, environmental and development issues were often addressed separately and in a fragmented fashion. Stockholm successfully brought international attention to the environmental crisis, but did not resolve any of the inherent tensions in linking environmental protection with social and economic development.

To bring back the focus on the broader issues of environment and development that had been discussed at the Stockholm Conference, in 1983 the UN General Assembly decided to establish an independent commission to formulate a long-term agenda for action. Over the next three years, the newly established Brundtland Commission held public hearings and studied the problem. Its 1987 report, *Our Common Future,* stressed the need for development strategies in all countries that recognized the limits of the ecosytem's ability to regenerate itself and absorb waste products. Recognizing "an accelerating ecological interdependence among nations," the commission emphasized the link between economic development and environmental issues, and identified poverty eradication as a necessary and fundamental requirement for environmentally sustainable development. In addition to contributing to the debate on development and the environment, the Brundtland Commission popularized the term *sustainable development,* describing it as "a process of change in which the exploitation of resources, the direction of investments, the orientation of technological development, and institutional change are all in harmony and enhance both current and future potential to meet human needs and aspiration" (WCED 1987: 46).

The 1992 Earth Summit

On the twentieth anniversary of the Stockholm Conference, governments gathered again, this time in Rio de Janeiro, Brazil, to move the sustainable development agenda forward. The Earth Summit, or Rio Conference, as it was popularly known (formally designated the United Nations Conference on Environment and Development [UNCED]), attracted greater official and unofficial interest than had the Stockholm Conference. More than 170 nations sent delegates, and 108 of these nations were represented by their heads of state or government. Thousands of NGOs sent representatives, and nearly 10,000 members of the media attended (UNDPI 1997).

The summit and the preparatory work that preceded it showed that there were still significant differences between the developed and the developing world. Consequently, each group provided different inputs to the agenda-setting process: developed countries wanted to focus on ozone depletion, global warming, acid rain, and deforestation, while developing countries also wanted to explore the relationship between their sluggish economic growth and the economic policies of the developed countries. The concern was that an "environmentally healthy planet was impossible in a world that contained significant inequities" (Miller 1995: 9).

The major output of the Earth Summit was a nonbinding agreement called "Agenda 21" (referring to the twenty-first century), which set out a global plan of action for sustainable development. In 294 pages, comprising 40 chapters covering 115 separate topics, Agenda 21 demonstrated the emergence of a clear international consensus on the issues affecting the long-term sustainability of human society, including domestic social and economic policies, international economic relations, and cooperation on issues concerning the global commons (see Figure 13.1).

The summit also produced two nonbinding sets of principles—the Rio Declaration of Environment and Development, and the Statement of Forest Principles—that helped create norms and expectations. The United Nations Framework Convention on Climate Change, and the Convention on Biological Diversity, which were negotiated independently of the UNCED process on parallel tracks, were opened for signature at the Earth Summit and are often mentioned as UNCED agreements.

The Earth Summit also marked a watershed in advancing the concept of sustainable development. As Tommy Koh, the diplomat from Singapore who chaired the UNCED preparatory committee, stated:

> It used to be fashionable to argue in the developing countries that their priority should be economic development and that, if necessary, the environment should be sacrificed in order to achieve high economic growth. The sentiment was to get rich first and to clean up the environment later. . . . Today, developing countries understand the need to integrate environment into their development policies. At the same time, developed countries have become increasingly aware of the need to cut down on their wasteful consumption patterns. The new wisdom is that we want economic progress but we also want to live in harmony with nature. (Koh 1997: 242)

Along these lines, one of the issues that delegates focused on was the need for more sustainable patterns of consumption. Sustainable consumption is the use of goods and services that satisfy basic needs and improve quality of life while minimizing the use of irreplaceable natural resources and the byproducts of toxic materials, waste, and pollution (Sierra Club

Figure 13.1 UNCED's Agenda 21

Section I: Social and Economic Dimensions

- International cooperation to accelerate sustainable development in developing countries
- Combating poverty
- Changing consumption patterns
- Demographic dynamics and sustainability
- Protecting and promoting human health conditions
- Promoting sustainable human settlement development
- Integrating environment and development in decisionmaking

Section II: Conservation and Management of Resources for Development

- Protection of the atmosphere
- Integrated approach to the planning and management of land resources
- Combating deforestation
- Managing fragile ecosystems: combating desertification and drought
- Managing fragile ecosystems: sustainable mountain development
- Promoting sustainable agriculture and rural development
- Conservation of biological diversity
- Environmentally sound management of biotechnology
- Protection of the oceans, all kinds of seas, and coastal areas; and the protection, rational use, and development of their living resources
- Protection of the quality and supply of freshwater resources
- Environmentally sound management of toxic chemicals
- Environmentally sound management of hazardous wastes
- Environmentally sound management of solid wastes and sewage-related issues
- Safe and environmentally sound management of radioactive wastes

Section III: Strengthening the Role of Major Groups

- Women
- Children and youth
- Indigenous people
- Nongovernmental organizations
- Local authorities
- Workers and trade unions
- Business and industry
- Scientific and technological community
- Farmers

Section IV: Means of Implementation

- Financial resources and mechanisms
- Transfer of environmentally sound technology; cooperation and capacity building
- Science for sustainable development
- Promoting education, public awareness, and training
- National mechanisms and international cooperation for capacity building in developing countries
- International institutional arrangements
- International legal instruments and mechanisms
- Information for decisionmaking

2007b). While poverty results in environmental stress, the major cause of global environmental deterioration is an unsustainable pattern of consumption and production, particularly in the industrialized countries, which aggravates poverty and imbalances. Inequalities in consumption are stark. The 12 percent of the world's population who live in North America and Western Europe account for 60 percent of private consumption spending, while the one-third living in South Asia and sub-Saharan Africa account for only 3.2 percent. The United States, with less than 5 percent of the global population, uses about a quarter of the world's fossil fuel resources—burning up nearly 25 percent of the world's coal, 26 percent of its oil, and 27 percent of its natural gas.

Between 1970 and 2005, the number of private vehicles on US roads more than doubled. In 2005, the United States had over 31 million more private cars than licensed drivers, and gas-guzzling sport utility vehicles and other "light trucks" made up 41 percent of the vehicle fleet, up from 13 percent in 1970 (USDOT 2007). New houses in the United States were 61 percent bigger in 2005 than in 1970, despite having fewer people per household on average (National Association of Homebuilders 2007). The runaway growth in consumption of the past fifty years is putting strains on the environment never before seen. The world's spending priorities also show the effect that consumption patterns have on sustainable development priorities (see Table 13.1).

According to estimates by the United Nations Development Programme in its 2005 *Human Development Report,* the wealthiest 20 percent of the global population earn 82.7 percent of the total global income, and account for 81.2 percent of world trade, 94.6 percent of commercial lending, 80.6 percent of domestic savings, and 80.5 percent of domestic

Table 13.1 Global Spending on Luxury Items Compared with Funding Needed to Meet Selected Basic Needs, 2004

Item/Need	Global Spending (US$ billions)
Makeup	18
Reproductive healthcare for all women	12
Pet food (Europe and the United States)	17
Elimination of hunger and malnutrition	19
Perfume	15
Universal literacy	5
Ocean cruises	14
Clean drinking water for all	10
Ice cream (Europe)	11
Immunization of every child	1

Source: Worldwatch Institute, *State of the World 2004* (New York: Norton).

investment (UNDP 2005). By contrast, the share of total global income of the poorest 20 percent is a mere 1.4 percent, and their contribution to world trade (1 percent) and commercial lending (0.2 percent) is statistically negligible (Rogers, Jalal, and Boyd 2006).

However, all of this could change as a result of China's economic growth. China has now replaced the United States as the leading consumer for basic commodities. According to Lester Brown (2006), in 2005, China consumed 380 million tons of grain versus 260 million tons in the United States. China's meat consumption, 67 million tons, far exceeded the 38 million tons eaten in the United States. And China burned 960 million tons of coal, easily exceeding the 560 million tons used in the United States. In fact, China is now the world's largest producer of greenhouse gases that contribute to climate change, although on a per capita (per person) basis it still produces less than the United States. Yet this leads to a serious concern: If the Chinese economy continues to grow at 8 percent a year, asks Brown (2006), by 2031, China's per capita income will equal that in the United States in 2004. China's estimated 1.45 billion people in 2031 will consume roughly two-thirds of the global grain harvest, 305 million tons of paper (twice the current world production), and 99 million barrels of oil a day, eclipsing current world oil production of 84 million barrels. It is fair to say that the planet cannot survive that level of consumption.

So while the Earth Summit set out a vision of sustainable development, and established goals and principles for achieving it, turning goals into action has become much more challenging. While the twenty-year period between Stockholm and Rio was characterized by growth in scientific understanding of environmental problems, the emergence of global public awareness and concern, and an increase in both domestic and international environmental law, there was just as much that did not change. Mistrust and suspicion still govern relations between North and South, governments still tenaciously embrace traditional views of national sovereignty, and tension between the long-term vision necessary for ecologically sound planning and the short-term concern for economic growth and political stability still preoccupies most governments (Conca and Dabelko 2004). Global consumption of goods and services has increased, but is still as unequal as ever, despite China's phenomenal economic growth in recent years. These divisions highlight the political challenges that remain, just like the global environmental problems themselves.

The Economic Pillar of Sustainable Development

The challenges faced in implementing Agenda 21 were compounded by changes in the global economy. The years following the Earth Summit saw the rise of globalization as a dominant concern, as rapidly increasing trade and capital flows, coupled with the revolution in information and communication technologies (such as the Internet), made the world more interdependent and interconnected than ever before. Since 1990, the value of world trade has tripled. Exports and imports of goods and services exceeded $26 trillion in 2005, or 58 percent of total global output, up from 44 percent in 1980. Also since 1990, flows of foreign direct investment (companies investing in other countries) into developing countries have increased by ten times (World Bank 2007). Absolute priority has been given to expanding the scope for trade and investment in adherence with neoliberal economics (a political-economic movement, increasingly prominent since 1980, that de-emphasizes or rejects government intervention in the economy, focusing instead on achieving progress and even social justice through freer markets, especially an emphasis on economic growth). For example, neoliberals argue that the best way to protect the environment is by overcoming poverty through increased privatization, foreign direct investment, and free trade. As a result, the institutions governing the global economy have grown stronger, while those promoting social equity, poverty alleviation, and environmental cooperation remain weak. When globalization arrived on the scene, the barriers to trade and investment began to fall, and the belief that poor countries could grow themselves out of poverty by boarding the liberalization express train took on an almost religious force—at least in the rich countries (Halle 2002). In other words, it is believed, if poorer countries support trade liberalization, many of their economic development problems will be solved.

Globalization—the rapid growth and integration of markets, institutions, and cultures—is viewed by some as a threat, by others as an opportunity. Many policymakers and industrialists in the developed world are optimistic about the phenomenon. However, the majority of people living in developing countries, as well as a variety of nongovernmental organizations, worry that globalization is a "race to the bottom" in terms of environmental and labor standards, and is exacerbating social and economic disparities. Many NGOs are worried that developing countries, in order to attract foreign direct investment, will eliminate or ignore environmental and labor standards. In other words, host countries that permit pollution, child labor, and long workdays may attract more corporate investment.

The establishment of the World Trade Organization (WTO) in 1994, at the conclusion of the Uruguay Round of trade negotiations under the General Agreement on Tariffs and Trade (GATT), was widely applauded at the time, but some feared that it would give rise to a new type of "all-powerful" international organization, and that UN treaties and agreements on environment, development, human rights, labor, women, and children would be relegated far down the list of priorities. Without the necessary political commitment, these pressing issues would not receive the attention they deserved.

The goal of the WTO is the harmonization of international standards for trade in goods and services, as well as for intellectual property rights. This includes standards for product testing, health and safety, product liability, paperwork, and the like. Such harmonization means that corporations will find fewer regulatory obstacles to their operations in host countries. Government regulations requiring that imported products meet certain standards—including environmental standards—with regard to content or process, are subject to challenge. For example, a country might impose measures to restrict the exports of forestry resources or fish for conservation purposes, but this could be ruled an unfair trade practice. If a local corporation can buy the resource, then that right also has to be extended to foreign-based corporations. Preferential treatment cannot be given to local interests. If the WTO panel rules against a country, it has to make the recommended change within a prescribed time or face financial penalties or trade sanctions. These changes in trade rules have clear implications for sustainable development, especially as local jurisdictions lose access to some of the regulatory tools that might be used to shape development.

For many environmentalists, trade liberalization raises questions about the potential impact on Earth's ecosystems and on governments' development choices. They see liberalization as driving the demand for greater consumption of natural resources, and as creating pressures to dismantle environmental regulations (Gallagher 2003). Trade liberalization has both direct and indirect effects on sustainable development. As an example of the former, a 2001 study of the increasing levels of transportation due to the North American Free Trade Agreement (NAFTA) found that NAFTA trade increased air pollution in five key transportation corridors that link North American commerce: Vancouver to Seattle, Winnipeg to Fargo, Toronto to Detroit, San Antonio to Monterrey, and Tucson to Hermosillo (NACEC 2001). A second direct effect is the introduction of alien invasive species—species that are not native to a particular country, region, or ecosystem. The trade-based global economy stimulates the spread of economically important species, often with funding from development agencies to establish, in countries far from their place of origin, plantations of rubber, oil palm, pineapples, and coffee, and fields of soybeans, cassava, maize, sugarcane, wheat, and other plants. But the trade-based global economy also stimulates the accidental spread of species by air, land, and water pathways. They arrive as hitchhikers on commodities, stowaways in transportation, or disease in wildlife. Examples of these pathways include recreational boating and international trade in plants, animals, and plant and animal products.

Another pathway is the discharge of ballast water. Ballast water is carried in empty ships to provide stability. It is taken aboard at port before the voyage begins, and tiny stowaways, in the form of marine organisms, are taken aboard with it. The discharge of ballast water, when the ship is loaded again, introduces harmful aquatic organisms, including diseases, bacteria, and viruses, to both marine and freshwater ecosystems, thereby degrading commercially important fisheries and recreational opportunities. For example, the WWF reports that more than sixty invasive species have been found in the Baltic Sea, many of which apparently arrived in ballast water from ships. The zebra mussel was introduced into the Great Lakes through ship ballast water, and the Asian tiger mosquito was introduced through imports of used tires. The rapid decline in frog populations in Queensland is attributed to a virus that is exotic to Australia and that may have been introduced through the thriving international trade in ornamental fish for home aquarium use (McNeely 2001).

Trade liberalization also has indirect effects on sustainable development. Kevin Gallagher (2003) outlines four such indirect mechanisms: scale, composition, technique, and regulatory. Scale effects occur when there is an expansion of economic activity. For example, increased industrial production and transportation of goods around the world have resulted in ever-increasing levels of carbon dioxide emissions, which contribute to climate change. Composition effects occur when increased trade leads nations to specialize in the sectors where they enjoy a comparative advantage. When comparative advantage is derived from differences in the strictness of regulations, then composition effects will exacerbate environmental and social problems in countries with lax regulations, because "dirty" or "socially irresponsible" industries concentrate in these countries. For example, 90 percent of the world's ships were dismantled in Bangladesh, India, and Pakistan in 2006, whereas twenty years earlier, seventy-nine countries engaged in dismantling decommissioned ships (Mohaiemen 2006).

Technique effects, or changes in resource extraction and production technologies, can potentially lead to a decline in pollution. Trade and investment liberalization could encourage the transfer of environmentally sound, cleaner technologies to developing countries (Gallagher 2003). However, in 2005, out of $916 billion in global foreign direct investment flows, only $334 billion went to developing countries, of which 48 percent went to just five countries: Brazil, China, Hong Kong, Mexico, and Singapore (UNCTAD 2006). While foreign direct investment flows to developing countries are increasing, these figures suggest that many of the world's poorer nations will not benefit from the possible transfer of cleaner technologies, especially since there the technologies may be outdated and environmentally harmful.

Regulatory effects of economic integration can crowd out the creation of development-friendly policies and institutions, particularly in developing countries. The World Bank estimates that the cost to implement just three of the WTO agreements that involve restructuring or reforming domestic laws and regulations, comes to $130 million. This is more than the annual development budget for the world's poorest nations (Wise and Gallagher 2005). Thus, developing countries' social and environmental policies may be seriously harmed.

Civil society groups and NGOs, including Oxfam, the World Development Movement, the WWF, and the Third World Network, have been calling for a balance of rights and responsibilities in trade liberalization. They argue that regulations that protect the health and safety of a country's citizens, as well as its natural environment, are necessary complements to liberalization. Developing countries will see the benefits from trade only when economic, environmental, and social development policies work in concert.

The Social Pillar of Sustainable Development

Concerns over globalization and other emerging issues such as the HIV/AIDS epidemic added a new dimension to the sustainable development debate as the millennium approached. There was a growing recognition that the world was failing to achieve most of the goals for a more sustainable society set out in Agenda 21 and elsewhere. Between 1992 and 2000, official development assistance from the industrialized countries plunged, while HIV/AIDS rolled back life expectancies in some countries to pre-1980 levels as the number of people living with the disease approached the 40 million mark. The world's population climbed above 6.1 billion in 2000, up from 5.5 billion in 1992—a significant increase in just eight years. The total number of people living in poverty dropped slightly—from 1.3 to 1.2 billion—but most of the gains were in Southeast Asia and virtually no progress was made in sub-Saharan Africa, where almost half the population lives in poverty. There remained at least 1.1 billion people lacking access to safe drinking water, and 2.4 billion lacking adequate sanitation (UNDPI 2002).

To help prepare the United Nations to meet these challenges, the General Assembly designated the fifty-fifth session, in September 2000, as the "Millennium Assembly," at which world leaders agreed on a far-reaching plan to support global development objectives for the new century. The world's leaders reaffirmed their commitment to work toward a world of peace and security for all, and a world in which sustainable development and poverty eradication would have the highest priority. The Millennium Declaration, which was adopted by all UN member states, set out key challenges facing humanity, and identified eight overarching development goals. The first seven goals are directed toward eradicating poverty in all its forms: halving extreme poverty and hunger, achieving universal primary education and gender equity, reducing the mortality of children under age five by two-thirds and maternal mortality by three-quarters, reversing the spread of HIV/AIDS, halving the proportion of people without access to safe drinking water, and ensuring environmental sustainability. The final goal, that of building a global partnership for development, is viewed by some as developing the sort of North-South pact first envisaged in Rio in 1992 (see Table 8.2, p. 156).

The momentum achieved with the adoption of the Millennium Development Goals was reinforced over the next two years by other major international meetings (see Figure 13.2). In Brussels, governments addressed the needs of the least-developed countries, while in Doha they expressed the need to link sustainable development and trade. Meanwhile, in Monterrey they supported the mobilization of resources to finance development, and in Rome they confirmed the global commitment to eradicate hunger. A series of UN special sessions promoted issues relating to women, social development, human settlements, and children. As delegates turned their attention toward the ten-year review of Agenda 21, there was great concern about the state of the world, the emerging trends, and globalization. It was recognized that rather than providing a place to negotiate or renegotiate another set of principles, the World Summit on Sustainable Development would have to be a forum where various partners from different sectors of society could set out clear programs for future action and agree on projects and initiatives with specific targets and timetables.

Figure 13.2 The Millennium Round of World Conferences

- Women 2000 (gender equality, development, and peace for the twenty-first century), New York, June 5–10, 2000
- World Summit for Social Development and Beyond (achieving social development for all in a globalizing world), Geneva, June 26–July 1, 2000
- Millennium Assembly, New York, September 2000
- Third UN Conference on the Least-Developed Countries, Brussels, Belgium, May 14–20, 2001
- Review of the implementation of the UN Conference on Human Settlements, New York, June 6–8, 2001
- Fourth World Trade Organization Ministerial Conference, Doha, Qatar, November 9–14, 2001
- International Conference on Financing for Development, Monterrey, Mexico, March 18–22, 2002
- Review of the implementation of the World Summit for Children, New York, May 8–10, 2002
- World Food Summit: Five Years Later, Rome, June 10–13, 2002

The 2002 World Summit on Sustainable Development

Ten years after Rio, the United Nations convened the World Summit on Sustainable Development in Johannesburg, South Africa, to map out a detailed course of action for implementation of Agenda 21. The summit sought to overcome the obstacles to achieving sustainable development and to generate initiatives that would deliver results and improve people's lives while protecting the environment. The summit did not aim to renegotiate Agenda 21, but it did attempt to fill some key gaps that have impeded its implementation and the shift to sustainable development.

The World Summit on Sustainable Development opened on August 26, 2002, and brought together government representatives from over 190 countries, including 100 world leaders. An estimated 37,000 people attended either the summit or one of the many other gatherings held alongside the main event (UN 2002). The difficult negotiations that followed focused on an ambitious plan of implementation to eradicate poverty, change unsustainable patterns of consumption and production, and protect and manage natural resources. As in Stockholm and Rio, there were divergent views among governments on how to tackle issues ranging from water and sanitation to desertification, climate change, biodiversity, oceans, health, education, science and technology, and trade and finance. Indeed, there were moments when negotiations came to a standstill and when skeptics questioned the negotiators' commitment to multilateralism and sustainable development.

The North continued to argue that development in the South needed to be environmentally sound, while the South continued to argue that development had to come first and it was the responsibility of the North to help the South in this regard. After all, the North got rich while destroying its environment and was now asking the South not to do the same. This was seen, by some, as condemning the South to poverty. As Indira Gandhi said at the 1972 Stockholm Conference, "Are not poverty and need the greatest polluters?" (Khator 1991: 23). To put it simply, the idea in Rio was that the North should act first, shoulder most of the adjustment burden, offer access to environmental technology, and finally engage in some financial redistribution—then the South would come aboard and eventually share in commitments. Ten years later in Johannesburg, the South argued that the North had not fulfilled its part of the bargain and therefore that it, the South, did not have to fulfill its commitments toward environmentally sound development.

After ten days of hard work in Johannesburg, the negotiations concluded. The summit produced three key outcomes. The first was the Johannesburg Declaration, a pledge by world leaders to commit themselves fully to the goal of sustainable development. The second was an implementation plan, which set out a comprehensive program of action for sustainable development, and included quantifiable goals and targets with fixed deadlines. Finally, the summit produced nearly 300 voluntary partnerships and other initiatives to support sustainable development. Unlike the Johannesburg Declaration and the implementation plan, this major outcome was not the result of multilateral negotiations involving the entire community of nations. Instead, it involved numerous smaller partnerships composed of private-sector and civil society groups, as well as governments, that committed themselves to a wide range of projects and activities.

Many of the commitments and partnerships agreed to in Johannesburg echoed the Millennium Development Goals. For example, countries agreed to commit themselves to halving the proportion of people who lack clean water and proper sanitation by 2015. In energy, countries committed themselves to expanding access to the 2 billion people who do not have access to modern energy services. In addition, while countries did not agree on a target for phasing in renewable energy (e.g., a target of 15 percent of the global energy supply from renewable energy by 2010), which many observers said was a major shortcoming of the summit, they did commit to green energy and the phase-out of subsidies for types of energy that are not consistent with sustainable development.

On health issues, in addition to actions to fight HIV/AIDS, reduce waterborne diseases, and address the health risks of pollution, countries agreed to phase out, by 2020, the use and production of chemicals that harm human health and the environment. There were also many commitments made to protect biodiversity and improve ecosystem management: reduction in biodiversity loss by 2010, restoration of fisheries to their maximum sustainable yields by 2015, establishment of a representative network of marine protected areas by 2012, and improvement of developing countries' access to environmentally sound alternatives to ozone-depleting chemicals by 2010.

Yet among all the targets, timetables, commitments, and partnerships that were agreed upon at Johannesburg, there were no silver-bullet solutions to the fight against poverty and a continually deteriorating natural environment. In fact, as an implementation-focused summit, Johannesburg did not produce a particularly dramatic outcome—there were no agreements that would lead to new treaties, and many of the agreed targets had already been agreed at other meetings, including the Millennium Summit. As then–UN Secretary-General Kofi Annan told the press on the last day of the World Summit on Sustainable Development, "I think we have to be careful not to expect conferences like this to produce miracles. But we do expect conferences like this to generate political commitment, momentum and energy for the attainment of the goals" (UNDESA 2002).

Among the summit's accomplishments was a balancing of the three pillars of sustainable development: social development, economic growth, and protection of the environment. This was a decisive shift from the view during the previous decade that sustainable development was equivalent to protection of the environment. Johannesburg was the first true summit on sustainable development, in the sense that the advocates of all three pillars were under one roof arguing their cases, raising real issues, and confronting those with different interests and perspectives. It was not a social summit, dealing only with poverty, exclusion, and human rights. It was not an economic and globalization summit, addressing only trade and investment, finance for development, and transfer of technology. And it was not an environmental summit, focusing only on natural resource degradation, biodiversity loss, climate change, and pollution. Johannesburg was instead a summit about the intersections of all of these issues (Speth 2003).

However, not everyone was pleased with the outcome of the World Summit on Sustainable Development. To many, the summit fell short of expectations and thus was a lost opportunity, because governments failed to take the sustainable development agenda forward. Some NGOs felt that the summit did not go far enough in setting targets for increasing the use of renewable energies. Jonathan Lash, president of the World Resources Institute, said, "We have missed an opportunity to increase energy production from non-polluting sources like solar, biomass, and wind, and to provide the many companies taking action to reduce emissions with a secure framework for their actions." However, as Lash added, "This Summit will be remembered not for the treaties, the commitments, or the declarations it produced, but for the first stirrings of a new way of governing the global commons—the beginnings of a shift from the stiff formal waltz of traditional diplomacy to the jazzier dance of improvisational solution-oriented partnerships that may include non-government organizations, willing governments and other stakeholders" (UNDESA 2002).

WWF director-general Claude Martin was also critical of the summit: "Negotiations . . . more often resembled a 'race to the bottom' than any real attempt to move forward. While in their speeches world leaders emphasized the importance of sustainable development, in the negotiating rooms many countries worked to protect their own interests by preventing the Summit from reaching new targets and timetables. The compromises and weakening of language in the plan of implementation were to such an extent that in some cases it actually went back on previous commitments" (Martin 2002).

Some took their criticisms one step further. Sunita Narain, from India's Center for Science and the Environment, wrote:

> It now seems to me that this conference was designed to fail and the incompetence of its organizers was not accidental. Why? Simply because

> the multilateral system is now an "unnecessary restraint" for the world's most powerful nation, the US. Weakening this system is a key objective of US foreign policy. The game plan is to shift focus from global responsibility on issues such as climate change, onto national governance, by arguing that poverty and environmental degradation have little to do with global trade or financial systems, but are caused by corrupt and irresponsible governments of the South. This also becomes a convenient argument against aid, which they claim does not work because of corrupt national governments. Instead, they promote funds from the private sector. In this process, UN agencies are emasculated, either by driving them to bankruptcy or by destroying their credibility with failures such as the WSSD. (Narain 2002: 5)

In short, Narain believes that the rich countries, like the United States, control international conferences and the entire UN and Bretton Woods systems, and want to use their control to pursue their own domestic agendas of increased greenhouse gas emissions and consumption of natural resources.

Like the Earth Summit before it, the World Summit on Sustainable Development marked the beginning of a new era of international cooperation on sustainable development—not just environmental protection, not just economic development, not just social development. The summit updated and expanded the activities called for in Agenda 21, to recognize realities of the first few years of the twenty-first century that had not even been envisaged in 1992, and recognized that the challenges of the day were global in nature and required international solutions. Today the sustainable development debate continues. Growing knowledge and awareness, organizational adjustments, and occasional breakthroughs may reveal possibilities for further change and effective international cooperation. At the same time, enduring divisions underscore the depth of the political challenges posed by environmental and development problems (Conca and Dabelko 2004).

Conclusion

Where will the world be in terms of sustainable development by 2050? The global economic system depends on endless growth, which many believe is clearly unsustainable. The governments of industrialized countries support this economic system, because with growing economies they can pacify the less affluent members of their societies with a slice of a larger pie, instead of having to share the existing pie more equitably. Developing-country governments also want their economies to grow; to that end, many have embraced the Western model of development. However, the planet cannot support everyone in such a resource-intensive lifestyle.

The most serious indication of losing our way on the path to a more sustainable future would be an increase in absolute levels of poverty in the world, increasing gaps between rich and poor countries, and increasing gaps between specific countries. The results, so far, are mixed.

According to the United Nations (UN 2007), some progress has been achieved since the Millennium Development Goals were established in 2000:

- The proportion of people living in extreme poverty fell from nearly one-third to less than one-fifth between 1990 and 2004.
- The number of extremely poor people in sub-Saharan Africa has leveled off, and the poverty rate has declined by nearly 60 percent since 2000.

However, health and environmental sustainability challenges still persist. The lack of employment opportunities for young people, gender inequalities, rapid and unplanned urbanization, deforestation, increasing water scarcity, and high HIV prevalence are pervasive obstacles. Moreover, insecurity and instability in conflict and postconflict countries make long-term sustainable development extremely difficult.

Some of the key challenges that need to be addressed (UN 2007):

- Over half a million women still die each year from treatable and preventable complications of pregnancy and childbirth. The odds that a woman will die from these causes in sub-Saharan Africa are 1 in 16 over the course of her lifetime, compared to 1 in 3,800 in the developed world.
- The number of people dying from AIDS worldwide increased to 2.9 million in 2006, and prevention measures are failing to keep pace with the growth of the epidemic. In 2005, more than 15 million children had lost one or both parents to AIDS.
- Half the population of the developing world lacks basic sanitation. In order to meet the Millennium Development Goal target, an additional 1.6 billion people will need access to improved sanitation over the period 2005–2015.
- Widening income inequality is of particular concern in East Asia, where the consumption share of the poorest people declined dramatically between 1990 and 2004.
- Warming of the climate is now unequivocal. Emissions of carbon dioxide, the primary contributor to global climate change, rose from 23 to 29 billion metric tons from 1990 to 2004.
- From 1990 to 2005, the world lost 3 percent of its forests, an aver-

age decrease of 0.2 percent a year. Deforestation results primarily from conversion of forests to agricultural land in developing countries.
- In 2005, only 22 percent of the world's fisheries were sustainable, compared to 40 percent in 1975.

Can the Millennium Development Goals be achieved by 2015? Is the planet on a path to sustainable development, or disaster? Despite the apparent tension between economic health, social health, and ecological health, in the long run economic health is dependent on social and ecological health. The economy cannot thrive in the face of the total devastation of our biospheric envelope, nor can it survive in the face of increasing poverty and disease. If we accept this fact, then there are only two ways to resolve this tension: we can let it continue until the integrity, the ecology, and the economy deteriorate and snap like old rubber bands; or we can make the economic, social, and cultural changes that are needed to support sustainable development at the community, national, and global levels. It is comparatively easy to agree that the shift to sustainable development is necessary. The challenge is for governments to muster the political will to make this shift, and for the people of the world to demand change.

Discussion Questions

1. What are key elements of the North-South debate over sustainable development? Why are they so difficult to reconcile?
2. How has the concept of sustainable development evolved since the 1980s?
3. Why are environmentalists concerned about trade liberalization?
4. What are some of the tensions between the three pillars of sustainable development?
5. Do you think sustainable development is possible? Should it be a priority? Why or why not?

Note

I acknowledge the contribution of the late Marian A. L. Miller, who wrote the chapter on sustainable development for the second edition of this volume.

Suggested Readings

Chasek, Pamela S., David L. Downie, and Janet Welsh Brown (2006) *Global Environmental Politics.* Boulder: Westview.

Chasek, Pamela S., and Richard Sherman (2004) *Ten Days in Johannesburg: A Negotiation of Hope*. Cape Town: Struik.
Conca, Ken, and Geoffrey D. Dabelko (2004) *Green Planet Blues: Environmental Politics from Stockholm to Johannesburg*. 3rd ed. Boulder: Westview.
Redclift, Michael (1987) *Sustainable Development: Exploring the Contradictions*. New York: Methuen.
Rogers, Peter P., Kazi F. Jalal, and John A. Boyd (2006) *An Introduction to Sustainable Development*. Cambridge: Harvard University Press.
Speth, James Gustave (2004) *Red Sky at Morning: America and the Crisis of the Global Environment*. New Haven: Yale University Press.
Worldwatch Institute (2006) *Vital Signs 2006–2007*. New York: Norton.

14

Regulating the Atmospheric Commons

Mark Seis

IN GARRETT HARDIN'S ARTICLE "THE TRAGEDY OF THE COMMONS" (1968), he argued that common property will be destroyed by human greed and overexploitation. Hardin used the hypothetical example of cattle herders on a pasture (not owned by any particular individual). He argued that if an unsustainable number of cattle were added in an effort to maximize profits, then the pasture would eventually be overgrazed and thus destroyed. The problem is that as the herders attempt to maximize their profits and add more cattle, then collectively the herders will work against one another's ability to maximize profit. Individual restraint is unlikely, because each herder knows that if they reduce the number of cattle they have, they will lose profits, at least in the short term, while other herders continue to make profits from unsustainable grazing practices. The end result is that when grazers are free to exploit a pasture with no community control, driven strictly by the ethic of profit maximization, the result is the "tragedy of the commons." This example can be applied to other global commons, such as the atmosphere, space, and the oceans.

Two current approaches to addressing the destruction of the commons by the ethic of profit maximization are complete privatization and government regulation. In the former, a private actor controls the pasture that has been privatized. So, for example, in the case of a pasture that has been transformed from a commons to a private property owned by a single herder family with the intention of owning the land forever, the owner would want to ensure that they could continue to graze cattle on the land. Therefore, theoretically, the family would use it sustainably. If the government took control of the previously community-controlled pasture, then it would need to establish laws to regulate its use and to ensure that it is developed sustainably.

Hardin's concept of the commons, however, has been seriously criticized for mistaking what were actually community-regulated public lands for "'open access' regimes in which anything goes" (Athanasiou and Baer 2002: 145). Historically, socially regulated commons have been maintained by subsistence-based economics guided by cultural practices and a spiritual sense of belonging to the land that "stabilize people's relationships with their ecosystems" (Cronon 1983: 12). Food production, work, recreation, and spiritual fulfillment form a cultural context that affirms and ensures the health of the land and therefore the community. For instance, Aborigines, many Native Americans, and other indigenous peoples around the world have had such cultural practices that affirm their relationship to their ecosystem. Many people feel that the only way to preserve the atmospheric commons is "to convert it from an open-access resource into a commons, a limited socially regulated global commons in which access is appointed to us each in equal measure, by virtue of . . . our common humanity" (Athanasiou and Baer 2002: 145). Maintaining air quality and atmospheric stability as essential components to all life on this planet is an example of a socially regulated commons that affirms our relationship with our biosphere. Conversely, some support the idea of converting the commons into private property. Treating air as private property assumes that rationing air through markets is superior to cultural practices that ensure quality air for all. An example of this privatization approach is the creation of carbon dioxide (CO_2) trading regimes.

This chapter examines specific atmospheric issues—such as global warming, ozone depletion, and acid rain—that relate to the degradation of the global atmospheric commons. Because the atmosphere belongs to all as common property by virtue of our need for it in order to live, nation-states have an assumed responsibility to provide all people equal access and equal shares to a healthy atmosphere. The majority of the atmospheric treaties explored in this chapter are attempts by nation-states to assert their responsibility to preserve the atmospheric commons.

The Threat of Global Warming

Since Charles Keeling set up laboratories in 1958 at the South Pole and Hawaii, he has shown that carbon dioxide levels have been rising (Bates 1990; Leggert 1990; McKibben 1989). Current concentrations of CO_2 are still increasing. In fact, CO_2 levels based on samples of air bubbles trapped in ice cores extracted from the Antarctic show that "the levels of carbon dioxide and methane in the atmosphere (these are the two principal greenhouse gases) are now higher than they have been for 650,000 years" (Monbiot 2007a: 3). There is widespread concern that increased greenhouse gases are leading to an increase in Earth's temperature, known as global warming.

A few critics (Kerr 1989; Ray and Guzzo 1992; Michaels 1992; Lindzen 1993; Bellamy 2004), however, question the actual amount of warming that will occur, and the underlying projections of computer models. Skeptics suggest that Earth's natural atmospheric processes (for example, the interplay of oceans, forests, and sulfate aerosols) will be able to mitigate the greenhouse effect. Other skeptics argue that the current warming trend is a natural fluctuation in global temperatures rather than a result of human activities.

Despite the critics, the overwhelming majority of scientific literature on the subject is mounting a very strong case that warming (frequently referred to as climate change) is occurring. The twentieth century was already 0.60 degrees Celsius (1.08 degrees Fahrenheit) warmer than the nineteenth century (Monbiot 2007a), and the World Meteorological Organization states that "the increase in temperature in the twentieth century is likely to have been the largest in any century during the past 1000 years" (WMO 2003). The most convincing evidence for global warming is the loss of ice sheets in the Northern Hemisphere. The thickness of Arctic ice "has declined by 42 percent since the 1950s, and Norwegian researchers estimate that the Arctic summers may be ice free by 2050" (Dunn 2001b: 87). In Antartica in 2002 the Larsen B ice shelf slid into the sea, and recent measurements in the Arctic show that sea ice "has shrunk to the smallest area ever recorded" (Monbiot 2007a: 5). The accelerated rate of warming has been attributed mostly to the activities of human beings (Bates 1990; Flavin 1996; Dunn 2001b; Athanasiou and Baer 2002; Kolbert 2006; Monbiot 2007a) (see Table 14.1).

Causes and Consequences of Global Warming

Earth is constantly bombarded by solar radiation, some of which is absorbed and some of which is reflected back into space. This process is known as the greenhouse effect. When the natural carbon cycle is altered by fossil fuel burning (automobiles, power plants, industry, and heating are

Table 14.1 World Carbon Emissions from Fossil Fuel Burning, 1950–2005

	1950	1960	1970	1980	1990	2000	2005
Carbon emissions (billion tons)	1.6	2.5	4.0	5.2	5.9	6.3	7.6
CO_2 (parts per million)	n/a	316.9	325.7	338.7	354.2	369.5	379.7

Source: Worldwatch Institute, *Vital Signs 2007–2008* (New York: Norton, 2007).
Note: n/a indicates data not available.

the most common fossil fuel–burning activities), large amounts of CO_2 are released into the atmosphere. The result is that more solar radiation is trapped in Earth's atmosphere and less is reflected, and therefore the planet begins to warm. Gases like carbon dioxide, chlorofluorocarbons (CFCs), methane, tropospheric ozone, and nitrogen oxides trap solar radiation and cause the atmosphere to warm. Carbon dioxide accounts for roughly 50 percent of total greenhouse gases; chlorofluorocarbons, 20 percent; methane, 16 percent; tropospheric ozone, 8 percent; and nitrogen oxide, 6 percent (McKinney and Schoch 1996). As the atmosphere warms, it retains more water vapor due to evaporation. Water vapor is also a powerful greenhouse gas, because it traps long-waved solar radiation. This phenomenon has been recently demonstrated by satellite measurements from the Earth Radiation Budget Experiments, which showed that as ocean surface temperatures increase, more infrared radiation is trapped in the atmosphere (Leggett 1990).

Increasing temperatures set into motion various feedback loops that escalate the problem of global warming. A warming planet means that there is less ice and snow in the mountain and polar regions to reflect back solar radiation. In addition, as Earth warms, large amounts of methane are released from ice, tundra, and mud in the continental shelves. More methane means more greenhouse gases to trap solar radiation, which in turn means hotter temperatures, and hotter temperatures mean more thawing, which means more methane gas—creating a vicious cycle of warming.

With global warming comes increased precipitation in some areas, and drought in others, because of increased evaporation due to the heat and changing wind patterns. Based on a plethora of global climate data, Tom Athanasiou and Paul Baer (2002) report that sea levels are rising, summers are becoming longer and warmer, serious storms are becoming more frequent and severe, and water shortages have become a chronic problem for eighty countries, constituting 40 percent of the world's population. Based on estimates by the World Health Organization, 150,000 people a year are now dying due to climate change (Monbiot 2007a).

It is estimated that increased global warming, which creates rapidly changing habitats, will bring 15–37 percent of the world's species to extinction by 2050 (Mobiot 2007a). The resulting storm and drought damage may also lead to possible food shortages, especially when one considers the expected increase in human population (Athanasiou and Baer 2002).

Major Contributors to Global Warming

Since the Industrial Revolution, human activity has added 271 billion tons of carbon to the atmosphere (Dunn 2001b). Annually, about 7.6 billion tons of carbon are currently being emitted into the atmosphere (Worldwatch Institute 2007). Since 1990, Western industrial nations have increased their carbon emissions by 9.2 percent, and developing nations by 22.8 percent (Dunn 2001a). Some countries emit much more carbon than others, which has created problems in the international community in formulating binding strategies for regulating greenhouse gas emissions. Table 14.2 reveals the major CO_2 contributors for 2005, with the United States in the lead. However, in 2006, China overtook the United States in CO_2 emissions by about 7.5 percent, according to a report by the Netherlands Environmental Assessment Agency (Vidal and Adam 2007). Another way to view CO_2 emissions is by the amount produced per person in a given country. For example, China, which has a population of 1.3 billion people, releases about 4 metric tons of carbon dioxide per person, while the United States releases nearly 20 metric tons per person (Chang 2007). Table 14.3 shows the dramatic difference between carbon dioxide emissions in the North versus the South for 2005. Three of the reasons for such high carbon dioxide emissions in the United States (and other wealthy countries like Australia and Canada) are low energy prices, large houses, and heavy automobile use (Flavin 1996: 31).

The fact that the United States has failed to take mandatory and legally binding steps to reduce its carbon emissions makes it difficult to persuade industrializing nations to slow their rate of carbon emissions and look to alternative technologies. Most US measures to reduce the threat of global warming, such as those promulgated in President Bill Clinton's climate action plan, have been voluntarily based. The Natural Resource Defense Council claims that even aggressive enforcement of Clinton's strategies

Table 14.2 Largest Carbon Emissions Producers, 2005

	CO_2 Emissions (million metric tons)
United States	5,956.98
China	5,322.69
Russia	1,696.00
Japan	1,230.36
India	1,165.72
Germany	844.17
Canada	631.26
United Kingdom	577.17
South Korea	499.63
Italy	466.64

Source: Energy Information Administration, "World per Capita Carbon Dioxide Emissions from the Consumption and Flaring of Fossil Fuels, 1980–2005" (2007), available at http://www.eia.doe.gov/pub/international/iealf/tableh1cco2.xls.

Table 14.3 Per Capita Carbon Dioxide Emissions from Consumption and Flaring of Fossil Fuels, Selected Countries, 2005

	CO_2 Emissions per Capita (metric tons)
United States	20.14
Germany	10.24
Japan	9.65
Poland	7.38
China	4.07
Brazil	1.94
India	1.07
Ethiopia	0.06

Source: Energy Information Administration, "World per Capita Carbon Dioxide Emissions from the Consumption and Flaring of Fossil Fuels, 1980–2005" (2007), available at http://www.eia.doe.gov/pub/international/iealf/tableh1cco2.xls.

would not have met target reductions, because "in 1994 Congress approved only half the funds called for; the 1995 Congress made even more drastic cuts, and weakened appliance and lighting standards that had been enacted by the 1992 Congress" (Flavin 1996: 32). To date, President George W. Bush has shown little concern about reducing carbon emissions in particular, and about global warming in general. The Bush administration rejected emission targets at the 2007 Group of Eight summit in Heiligendamm, Germany, but has stated that the United States will convene a series of meetings with the world's largest greenhouse gas emitters in 2008 to create a long-term reduction strategy.

A recent report by the Central Intelligence Agency suggested that "climate change should be elevated beyond a scientific debate to a US National Security concern" (Townsend and Harris 2004). The Bush administration suppressed the report for four months. Given the suggested threat to world security, some scientists and foreign leaders believed that the report might catalyze President Bush into addressing global warming and climate change. Unfortunately, the Bush administration continued to evade the topic. In a 2006 survey conducted by the Union of Concerned Scientists, it was found that 58 percent of 279 scientists working in US federal agencies had been asked to eliminate terms like *climate change* and *global warming* from their reports. The survey also found that scientific reports were edited to change their meaning, and that scientific findings were deliberately misrepresented. Over a five-year period, the survey reported 435 incidents of political interference (Monbiot 2007b).

Some industrialized nations, such as Germany, Britain, and Japan, have proposed cutting global greenhouse gas emissions by 50 percent by 2050.

Denmark, Switzerland, and the Netherlands have implemented a variety of strategies to reduce carbon emissions. And the European Union continues to promulgate legislation aimed at reducing global warming.

International Climate Control Policies

In 1972, the United Nations Conference on the Human Environment, also known as the Stockholm Conference, was held. This conference brought together 114 governments and was attended by many nongovernmental organizations. It was the first time in history that those nations of the world came together to discuss issues surrounding the destruction of the environment (Switzer 1994; Valente and Valente 1995). The Stockholm Conference did not create any binding obligations, but served more as a catalyst to generate an international discourse on global environmental issues.

It was not until the 1992 United Nations Conference on Environment and Development (UNCED) (also known as the Rio Conference or Earth Summit) that the world seriously discussed reducing carbon dioxide emissions to curtail global warming. The Rio Conference was attended by 178 countries and 110 heads of state (Switzer 1994). One of the five major documents produced at the conference was a framework convention on climate change. Its purpose was "stabilization of greenhouse gas concentrations in the atmosphere at a level that would prevent dangerous anthropogenic interferences in the climate system" (Flavin 1996: 36).

Although the United States, headed by the George H. W. Bush administration, was a major actor at the Rio Conference, it fought the binding of targeted CO_2 reductions to 1990 levels. Despite US reluctance to agree to the targeted reductions, many industrialized European nations signed a separate declaration to reaffirm their commitment to reducing their own CO_2 emissions to 1990 levels (Gore 1992: xiv), following the lead of Germany and Japan. In 1993, President Bill Clinton reversed the Bush administration's position, announcing that the United States would reduce CO_2 emissions to 1990 levels by 2000. But many of the efficiency initiatives enacted by the US Congress in 1992 (regarding appliance and lighting, for example) were weakened in 1994 and 1995, and remained severely underfunded (Flavin 1996).

Persuading the industrialized nations to make serious commitments to CO_2 reductions has become a major concern of the Alliance of Small Island States (AOSIS) and worldwide insurance companies. AOSIS is a small coalition of island nations that are extremely threatened by rising seas. A rise of one meter in sea level could flood their lands and destroy their economies. At the 1995 Berlin Conference, which focused on climate change, AOSIS proposed that industrial nations reduce their CO_2 emissions by 20 percent. This proposal was endorsed by seventy-seven non-AOSIS nations that participated in the conference, but was resisted by the majority of oil-producing states, like Kuwait and Saudi Arabia, and by the larger carbon-consuming countries, like the United States and Australia (Brown 1996; Flavin 1996).

The world's largest insurance companies also obviously have a stake in any losses that may occur due to global warming. With sea levels projected to rise and temperatures projected to increase, insurance companies are reporting that "economic losses related to climate change could top $304 billion a year in the future" (Abramovitz 2001: 117).

The primary objectives of the Berlin Conference were to design measures that would reduce global carbon emissions and to create a series of trial projects aimed at exchanging alternative low-carbon technologies among nations (Flavin 1996). Despite the fact that no legally binding carbon reduction targets were established, the Berlin Conference did provide a sense of renewed hope in formulating a global policy for mitigating climate change. The agreement reached at Berlin, known as the Berlin Mandate, instructs governments to negotiate a treaty "to elaborate policies and measures, as well as to set quantified limitations and reduction objectives within specified time-frames such as [by] 2005, 2010, and 2020" (Flavin 1996: 35).

The meeting to establish this treaty took place at the end of 1997 in Japan and resulted in the Kyoto Protocol, which was adopted on December 11, 1997, and opened for signature on March 16, 1998. The protocol contains legally binding emission targets for key greenhouse gases, especially carbon dioxide, methane, and nitrous oxide. The agreement requires ratification by fifty-five countries; these fifty-five must include developed countries representing 55 percent of the total carbon dioxide emissions of the ratifiers. According to the United Nations: "The overall commitment adopted by developed countries in Kyoto was to reduce their emissions of greenhouse gases by some 5.2 percent below 1990 levels by a budget period of 2008 to 2012. While the percentage did not seem significant, it represented emissions levels that were about 29 percent below what they would have been in the absence of the Protocol" (UN 1998).

One of the problems with Kyoto was the vagueness of negotiators in elaborating how countries could achieve reductions through emission trading (an alternative to socially regulated commons). Through emission trading, a country that reduces its emissions below its allotted level can "trade" its unused emissions to countries that are achieving less success in reducing carbon emissions. Also vaguely elaborated at Kyoto was the degree to which carbon sinks—forests, rangelands, and croplands that absorb carbon—should count toward a country's effort to reduce global warming.

At the Hague Conference in 2000, the question of how to count reductions in carbon emissions remained the major obstacle. The United States, Canada, Japan, and Australia favored using flexible and creative approaches such as carbon sinks and market-based mechanisms, including emission trading. Other European countries, however, interpreted US advocacy of flexible methods as an attempt to avoid reducing emissions from cars, factories, and power plants. Many European countries want to reduce greenhouse gas emissions through the development of non–fossil fuel technology. However, according to a study of five European nations, only the United Kingdom and possibly Germany will meet Kyoto's emission reductions targets; also, it is unlikely that the United States will reach the targets, since it would have to reduce its emissions by 30 percent (Kerr 2000). In contrast to Europe, the United States wants to achieve the Kyoto targets without reducing greenhouse gases, because such cuts would likely involve increased coal and oil prices. Negotiations at the Hague Conference ultimately ended in deadlock over how to measure emission reductions.

In November 2001, negotiators from 160 countries met in Marrakesh, Morocco, and agreed on the details of the Kyoto Protocol. In the treaty, industrialized countries (with the glaring exception of the United States) agreed to reduce CO_2 emissions by an average of 5.2 percent below 1990 levels. Countries have until 2012 to accomplish this task. The United States, under President George W. Bush, withdrew from negotiations ostensibly to protect its "economic best interest" (Athanasiou and Baer 2002: 117). Until the end of 2004, without the United States, negotiators had not been able to enlist the required support of countries responsible for generating 55 percent of carbon dioxide emissions. However, at the end of 2004, Russia, which had been reluctant to sign the treaty, announced its support for Kyoto and ratified it. The protocol officially went into effect early in 2005.

The new treaty combines emission trading and carbon sinks as well as required emission cuts. Critics question that developing nations are exempt from this agreement, due to economic infeasibility and low levels of CO_2 emissions. They also question the significance of the treaty, given the absence of the United States.

The Kyoto Protocol, the world's only international treaty on global warming, is set to expire in 2012. At the 2007 Group of Eight summit in Germany, a major goal was to secure a commitment from the richest nations to prevent an average rise in global temperature of more than 3.6 degrees. The second phase of the Kyoto Protocol talks was also convened in 2007, in Bali, to create a set of mandatory emission goals. A week before the Group of Eight summit, however, President Bush announced that the United States would convene a series of meetings in the fall of 2007 to design a strategy to reduce greenhouse emissions. Bush's plan calls for the ten to fifteen highest-emitting greenhouse gas nations to construct a post-Kyoto agreement that sets no mandatory or binding emission targets and focuses only on technological solutions, with each country free to set its own goals and reduction targets.

Critics suggested that Bush was attempting to undermine existing efforts by the UN and the European Union to secure mandatory emission reductions. In fact, the only agreement that came out of the Group of Eight summit was to make "substantial cuts" in greenhouse gas emissions. At Bali, some progress was made, as the conference delegates agreed to a two-year process to generate a future, post-Kyoto agreement. Countries also agreed in principle that "deep cuts in global emissions will be required" to reduce the effects of climate change. However, key obstacles still remain. The United States has refused to go along with the European Union's proposal for mandatory cuts, and wants key developing countries, like China and India, to be treated the same as developed countries. Since the two-year negotiation timetable will overlap with the 2008 US presidential election, some are optimistic that the new US president will be more concerned about climate change.

The Threat of Ozone Depletion

Although record-low ozone levels were recorded throughout the 1990s, scientists have recently reported that the rate of ozone depletion has slowed in the Arctic, if not so much in Antarctica. Betsy Weatherhead, of the US National Oceanic and Atmospheric Administration, and Signe Bech Anderson, of the Danish Meteorological Institute, have recently reported that ozone loss in the Arctic is beginning to level out due to the efforts of the Montreal Protocol. According to May 2006 measurements over the Northern Hemisphere, the ozone hole is smaller than in previous years at this time (Weatherhead and Anderson 2006). However, the 2006 Antarctic hole exhibited some of the strongest stratospheric ozone depletion seen in recent years. According to Bryan Johnson, "Although ozone destroying halocarbons are decreasing in the atmosphere in response to controls put into effect by the Montreal Protocol, there are still sufficient concentrations of these compounds available to completely remove ozone in a large portion of the Antarctic stratosphere." Johnson notes that these "low values bring home the point that the Antarctic ozone hole will continue to be an annual feature well into this century" (Johnson 2006). International policies are reducing the sources of ozone depletion, but the damaging effects of a thinning ozone layer have yet to be fully realized, and it may take 50 to 100 years for ozone levels to return to concentrations prior to the creation of CFCs.

Causes and Consequences of Ozone Depletion

There are two types of ozone we often hear about. Ozone in the stratosphere is what protects us from harmful ultraviolet radiation. If compressed, the ozone layer would be about a tenth of an inch thick (three millimeters). Ozone in the troposphere, on the other hand, is extremely poisonous to most forms of life. A large portion of the ozone found in our troposphere is generated by human sources of atmospheric pollution that interacts with solar radiation to create a pale blue gas with a strong, pungent odor, which can sometimes be smelled after it rains (Gribbin 1988). The issue of concern here, in contrast, is the depletion of protective, stratospheric ozone resulting from the human production of chemicals like CFCs, halons, and halocarbons. This stratospheric ozone layer protects us from excessive amounts of ultraviolet radiation.

The ozone layer is created from oxygen that escapes from the troposphere. Oxygen is created from living organisms that process carbon dioxide and exhale oxygen. Just as oxygen is vital for the creation of the ozone layer, so is the ozone layer vital for the creation of oxygen, by protecting the living organisms that produce it.

CFCs and halon molecules, which have relatively long atmospheric lives, contain chlorine and bromine atoms. Once these molecules float their way up into the atmosphere, they interact with sunlight, which breaks apart their molecular structure, releasing the chlorine and bromine atoms that destroy ozone. As John Gribbin explains, "A single chlorine atom can scavenge and destroy many thousands of ozone molecules" (1988: 48). "At an altitude of about 11 miles above the ground, more than half the ozone above Antarctica was destroyed in the spring of 1987. And changes in the amount of chlorine oxide present marched precisely in step with changes in the amount of ozone. When chlorine oxide went up, ozone went down, showing clearly that chlorine was destroying the ozone" (1988: xi).

CFCs are solely products of human industry, and are most often used as propellants in aerosol spray cans. They are also used in air conditioners, refrigerators, computer chips, and Styrofoam. Halons are employed mostly in equipment used to suppress fires. There are no natural processes in the troposphere that react with CFCs and halons to break down their molecular structure (McKinney and Schoch 1996).

Another major destroyer of ozone is nitrous oxide (N_2O). Like CFCs and halons, N_2O is not a friend to the stratosphere. When bombarded with high-energy solar radiation, it breaks down into nitric oxide (NO), which reacts with ozone. Nitrous oxide is emitted into the atmosphere by plants, combustion of coal and oil, and aerosol spray cans. Plants naturally emit N_2O, which helps balance ozone levels between too little, which would leave Earth unprotected from harmful radiation, and too much, which would inhibit life as we know it (Gribbin 1988; McKinney and Schoch 1996). The use of chemical fertilizer to increase food production, however, has resulted in a nonnatural increase of N_2O in the troposphere (Gribbin 1988).

Effects of ozone depletion on human beings are several. The most obvious is that ozone depletion causes an increase in Ultraviolet B (UV-B) radiation, which is known to increase our chances of developing cataracts as well as skin cancer, especially malignant melanoma (Meadows, Meadows, and Rander 1992). Increases in UV-B radiation have been most pronounced in countries located in the Southern Hemisphere, such as Australia, New Zealand, and South Africa. Australia has the highest rate of skin cancer in the world, and research suggests that two out of three people who grow up there will develop skin cancer during their lifetimes, and that one in sixty will develop the most deadly type, malignant melanoma (Meadows, Meadows, and Rander 1992: 145). In the Southern Hemisphere, skin cancer and cataracts are becoming increasingly common. In fact, Al Gore reported in 1992 that in "Queensland, in northeastern Australia, for example, more than 75 percent of all its citizens who have reached the age of sixty-five now have some form of skin cancer, and children are required by law to wear large hats and neck scarves to and from school to protect against ultra violet radiation" (1992: 85). In addition to increased rates of cataract skin cancer, UV-B radiation also suppresses the human immune system, making us much more vulnerable to disease and viruses (Gore 1992; Gribbin 1988; Meadows, Meadows, and Rander 1992; McKinney and Schoch 1996).

Increased UV-B radiation on plants such as soybeans, beans, sugar beets, potatoes, lettuce, tomatoes, sorghum, peas, and wheat has been shown to inhibit growth, photosynthesis, and metabolism (McKinney and Schoch 1996). High levels of UV-B radiation also affect freshwater and marine ecosystems, especially ocean plankton, the base of the ocean's food chain (McKinney and Schoch 1996). There have been reports from Chile that sheep are going blind and that rabbits are developing myopia so severe that one can walk into a field and pick them up by the ears (Lamar 1991). Cattle are also known to develop eye cancer and pinkeye when exposed to high levels of UV-B radiation (Gribbin 1988).

Major Producers and Users of Ozone-Depleting Substances

Chlorofluorocarbons, like many other chemicals, became highly popular after World War II. The production of the two common types of CFCs increased from 55,000 tons in 1950 to 800,000 tons in 1975:

> From 1950 to 1975 world production of CFCs grew at 7% to 10% per year—doubling every 10 years or less. By the 1980s the world was manufacturing a million tonnes of CFCs annually. In the United States alone, CFC coolants were at work in 100 million refrigerators, 30 million freezers, 45 million home air conditioners, 90 million car air conditioners, and hundreds of thousands of coolers in restaurants, supermarkets, and refrigerated trucks. (Meadows, Meadows, and Rander 1992: 142)

Approximately 90 percent of the CFCs are immediately released into the atmosphere during their use, with the remainder being released after the product is discarded (for example, refrigerators and air conditioners).

Like most pollution caused by modern technology, there is an immense disparity in the amount of ecological destruction caused by industrialized versus industrializing countries. The countries of the North, such as the United States, the European Union, Japan, New Zealand, and Australia, have used much larger quantities of CFCs than industrializing countries, such as China and India (USCC&AN 1991). Per capita, North Americans and Europeans used an average of two pounds of CFCs a year in 1988, compared to less than an ounce per year for people in developing countries like China and India (Meadows, Meadows, and Rander 1992).

By 1985 there was no longer any doubt that CFCs were destroying the ozone layer. When British scientists measured a 40 percent decrease in ozone over Halley Bay in Antarctica, it was perceived as an error. But after checking their measurements and other monitoring sites, it became apparent that ozone depletion was a reality. In 1987, using the *Nimbus 7* satellite, the National Aeronautics and Space Administration (NASA) confirmed that there was indeed a hole in the ozone layer over Antarctica. This discovery prompted a series of international meetings and ultimately led to targeted reductions and phase-outs of ozone-depleting substances.

International Policies on Ozone Depletion

Mario Molina and Sherwood Rowland's 1974 paper documenting the depletion of ozone led to CFC policies both in the European Community and in the United States. In 1978 the United States banned the use of CFC propellants in aerosol spray cans, and the European Community legislated a 30 percent reduction in such propellants (Gribbin 1988; USSCEPW 1993). Despite the fact there were no international agreements at the time, use of CFC propellants decreased between 1974 and 1982 in most of the developed world, except the Soviet Union (Gribbin 1988). However, use of other types of CFCs did not decrease.

In 1985, the same year the ozone hole over Antarctica was discovered, an international agreement titled the Vienna Convention for the Protection of the Ozone Layer was signed by twenty nation producers of halocarbons. However, the agreement did not entail any binding phase-out or reduction of ozone-depleting substances. Instead, it focused on international research efforts to document the ozone-depleting potential of halocarbons and other CFC and non-CFC ozone-depleting substances (Gribbin 1988; USSCEPW 1993).

After NASA's confirmation of the Antarctic ozone hole in 1987, it took only nine months of negotiations before twenty-seven countries signed the 1987 Montreal Protocol on Substances That Deplete the Ozone Layer. The

Montreal Protocol is by far the strongest piece of international environmental policy to date. The agreement became effective in January 1989 and required a "freeze in world wide production of CFCs and halons (at 1986 levels, for CFCs in 1989 and halons in 1992) and a 50 percent reduction in the production and consumption of CFCs by mid-1988" (USSCEPW 1993: 108). The twenty-seven nations that signed this protocol accounted for 99 percent of the producers and 90 percent of the consumers of ozone-depleting substances. However, many industrializing nations, like India and China, declined to sign, because the protocol made no stipulation for technical and financial assistance for developing nations.

Due to record-low ozone levels over the Northern Hemisphere reported by the scientific community in 1989 and again in 1992, the Montreal Protocol has been amended twice since. The first amendment took place in June 1990 in London, and included an agreement by all the parties "to a complete phase-out of CFCs, halons and carbon tetrachloride by the year 2000" (USSCEPW 1993: 110). Further agreements included a ban in 2005 on the production of methyl chloroform. The London amendment also generated an agreement between the signers of the original Montreal Protocol to assist developing nations convert to the use of ozone-friendly substances.

The second amendment was signed in November 1992 in Copenhagen by 126 countries. It established more rapid phase-out dates for the major ozone-depleting substances. In addition to CFCs, hydrochlorofluorocarbons (HCFCs), the substance used to replace CFCs, were also targeted for phase-out, which began in 2004. While HCFCs are much more environmentally friendly to the ozone layer (95–98 percent less damaging than CFCs), they are destructive nonetheless (McKinney and Schoch 1996). The full effects of HCFCs on the atmosphere and on human health are not yet known, but the Copenhagen amendment's total ban of HCFCs by 2020 ensures that their overall impact on the environment will be mitigated.

A remaining impediment to the Montreal Protocol is the illegal trafficking of CFCs. Chlorofluorocarbons illegally produced in developing countries have been making their way back into the black markets of industrialized countries, including the United States and European nations, where CFCs are still sought for use in refrigerators and automobile air conditioners. According to Hilary French and Lisa Mastny: "By 1995, CFCs were considered the most valuable contraband entering Miami after cocaine. Following a subsequent crackdown on large consignments throughout East Coast ports, much of the illegal trade shifted to the Canadian and Mexican borders. Between April 1998 and March 1999, the US Customs Office in Houston, Texas, reported 619 seizures of Freon, totaling nearly 20 tons" (2001: 182). The good news, however, is that the illegal trade in CFCs in the United States seems to be on the decline (French and Mastny 2001).

The Montreal Protocol is a testimonial that positive policy can be promulgated among nations when environmental degradation is taken seriously. It shows us that serious environmental degradation has a way of smoothing over ideological differences among nations. The Montreal Protocol was passed because a diverse group of international scientists, politicians, and corporations agreed that preservation of the ozone layer—a necessity for life as we know it—outweighed everything else.

The Problem of Acid Rain

Acid rain (often referred to as transboundary air pollution or acid deposition) has become a major problem throughout the world, but it is most publicized in the United States, Canada, and Europe. China and India are now starting to experience a serious acid rain problem as both undergo rapid industrialization. Acid deposition is now recognized both in the United States and in Europe as more chemically diverse than other environmental problems, such as climate change, mercury accumulation, ground-level ozone formation, and overfertilization of the environment.

Acid rain poisons lakes and rivers, damages the soil, and endangers the health of animals and humans. Like most pollution, acid rain does not stop at the political border of one country and ask permission to enter another. Acid rain has been a source of conflict between the United States and Canada, and among many European nations.

Causes and Consequences of Acid Rain

Acid rain is created almost immediately after sulfur dioxide (SO_2) has been emitted into the atmosphere. Almost all fossil fuels contain sulfur, and sulfur content is extremely high in coal. In the United States, coal-fueled utility power plants account for 66 percent of US sulfur dioxide emissions (USEPA 2003: app. A), and in China, coal accounts for 69 percent of all energy production (Larssen et al. 2006). When fossil fuels burn, sulfur combines with oxygen to create SO_2, which is an odorless and colorless gas. Sulfur dioxide is a known lung irritant that, in low concentrations, can trigger asthmatic attacks and make it difficult for people with other respiratory problems to breathe. In the United States, sulfur dioxide by itself has not been a major problem, but the transformation of atmospheric SO_2 into sulfuric acid (H_2SO_4), or acid rain, is a major environmental health problem.

In the atmosphere, SO_2 combines with oxygen to form sulfate (SO_4). Sulfate is a small particle that floats in the air or settles on leaves, buildings, and the ground. When sulfate interacts with mist, fog, or rain, it becomes acid rain (USCC&AN 1991). When inhaled into the lungs, sulfate is also transformed into sulfuric acid.

In the northern United States and in Canada, the death of fish due to acid rain has been well known since the early 1970s. According to a 1984 report, "In the Adirondack mountains, at least 180 former brook trout ponds will no longer support populations" (USCC&AN 1991: 3652). According to a 1975 study of 214 Adirondack lakes, "90 percent . . . were entirely devoid of fish life" (USCC&AN 1991: 3652). The Office of Technology Assessment estimated in 1982 that "in the Eastern United States approximately 3,000 lakes and 23,000 miles of streams have already become acidified or have no acid neutralizing capacity left" (USCC&AN 1991: 3655). In Europe, "fish have disappeared from lakes in Sweden and Norway, as well as Scotland and England" (Switzer 1994: 258). Fish populations in lakes have also been declining, and in some cases dying out altogether, in Russia and Romania (Switzer 1994). Over the past decade, however, in the United States and in parts of Europe, both sulfur dioxide and nitrous oxide levels have noticeably decreased, leading to less acidity in the environment (Boynton 2005; Slanina 2006).

The most serious acid rain problem in the world currently is in China, where over a quarter of the country's land has been affected by sulfur dioxide (Bright 2000; Larssen et al. 2006). China's sulfur dioxide emissions have swollen to nearly 26 million tons per year, a 27 percent increase, since 2000. It is estimated that China will experience $65 billion in economic losses due to its growing acid rain problem, which is affecting crops, forests, and freshwater supplies. The overall impact of acid rain in China in the future is expected to be far worse than it was in the United States and Europe at its peak in the late 1970s and through the 1980s (Staff Writers 2006).

Acid rain has also been linked to forest decline in various parts of the United States. For example, in Appalachia, "the death rate of oaks appears to have doubled and the hickories to have nearly tripled from 1960 to 1990," and acidity in forest soil from Illinois to Ohio has led to "a decline of soil organisms—earth worms, beetles, and so on" (Bright 2000: 34). In Eastern Europe forests are dying rapidly, and in southwestern Poland in 1990 the army was mobilized to fell large tracts of dead forest killed by acid rain (Switzer 1994).

The effects of acid rain on human health have been noted throughout the world. In Santa Catarina, Brazil, for example, "the environment secretary estimates that 80 percent of local hospital patients have respiratory ailments caused by acidic pollutants" (Switzer 1994: 265). High acid levels have been correlated with increased colds, bronchial infections, asthma attacks, and death. Harvard public health research suggested that an approximately 5 percent annual excess of mortality in the United States was due to sulfate and fine particles (USCC&AN 1991).

Major Producers of Acid Rain

As in the cases of ozone depletion and global warming, it is the largest industrialized nations that are the major generators of acid rain–causing pollutants. Coal-fueled power plants and factories account for the highest emissions of sulfur dioxide leading to acid rain worldwide. As noted previously, in the United States, 66 percent of total sulfur emissions come from coal-fueled power plants, and in China, 69 percent. According a 2005 progress report on acid rain, "SO_2 emissions from electric power generation were more than 5.5 million tons below 1990 levels. NO_X [nitrogen oxide] emissions were down by about 3 million tons below 1990 levels. The program's emission cuts have reduced acid deposition and improved water quality in US lakes and streams" (EPA 2006).

The biggest problem with the countries that produce the most sulfur dioxide and nitrogen oxide is that their pollution becomes the problem of other countries. Air pollution goes wherever the wind blows. Most of Canada's acid rain problems have been attributed to coal-fueled power plants operating in the Ohio Valley. Likewise, coal-fueled power plants and factories in eastern Germany, Poland, the Czech Republic, and the Slovak Republic destroy forests to the east. Sulfur emissions from Russia have been implicated in acid rain problems in Finland and Sweden, and emissions from Great Britain have contributed to an acid rain problem in Norway. Sulfur emissions from China create acid rain problems in Japan and South Korea.

In Europe, although there has been a substantial emissions drop due to desulphurization systems and a move away from coal, acidification still remains a major environmental problem (EEA 2004). Recent studies suggest that "the maximum sulfur emissions in western Europe occurred in the period from 1975 to 1980, at a level of 40 million tons of SO_2. Sulfur emissions have gone down since then and are currently [2006] at a level of about 30% of those maximum values" (Slanina 2006: 2).

Given the fact that coal-powered electricity is the major culprit in acid rain generation, reducing our reliance on coal to generate electricity seems to be the obvious answer. No solutions between nations, however, are easy. Coal is a cheap and widely available fossil fuel when compared with more expensive energy sources (for example, hydroelectric power, natural gas, and oil) as well as more dangerous alternatives (for example, nuclear power). Thus, solving what we know to be a simple problem becomes extremely complicated when we figure domestic and international politics and economics into the solution.

International Policies on Acid Rain

There have been no United Nations multilateral agreements regarding the abatement of acid rain. Unlike global warming and ozone depletion, which causes worldwide problems, acid rain is more regional. Accordingly, most acid rain agreements tend to be signed between bordering nations.

Probably the most extensive acid rain discussions have occurred between the United States and Canada. Canadian environmentalists contend that sulfur dioxide emissions from the United States are responsible for damaging as many as 16,000 Canadian lakes (Switzer 1994). Serious talks between Canada and the United States began in 1978 and culminated in a nonbinding treaty titled the Great Lakes Water Quality Agreement. The treaty stipulated emission reductions on the part of both nations. The progress made in 1978 was negated in 1980 when President Jimmy Carter, trying to counteract the Middle East oil crisis, decided to convert 100 oil-fired utility plants to coal (Switzer 1994).

From 1980 to 1990, acid rain talks between the United States and Canada did not make much progress. It was not until 1990, when President George H. W. Bush signed the Clean Air Act, that serious efforts were initiated to reduce US sulfur dioxide and nitrogen oxide emissions in both countries ("Canada–United States Air Quality Agreement" 2006).

With respect to sulfur dioxide emissions, the Clean Air Act promulgates a marketable allowance program. Each electrical utility is granted a set level of emissions, equivalent to roughly half its emissions output prior to the act. If a coal-fired utility company reduces its emissions below the granted level, it can sell its excess allowances to other companies. Likewise, a utility company can exceed its granted level of emissions by buying excess allowances (Seis 1996).

Europe has been somewhat successful with international strategies to curtail acid rain. The European Union and the UN's Economic Commission for Europe have been instrumental in initiating efforts to abate acid rain. The European Union consists of fifteen member nations, and the Economic Commission for Europe consists of all the nations of Europe. Though the European Union has experienced difficulty in attempting to promulgate uniform sulfur dioxide emission reductions across nations, definite progress is being made (Slanina 2006).

The Economic Commission for Europe has made more progress, beginning in 1979 with enactment of the Convention on Long-Range Transboundary Air Pollution. This agreement of intent obliged each nation to develop technology to abate acid rain and, where appropriate, to share the technology. However, the agreement established no emission reduction standards, nor did it require any uniformity in implementation of acid rain reduction technology. It was not until 1985 in Helsinki that thirty European nations agreed to a 30 percent reduction in sulfur dioxide emissions by

1993. Those that signed the protocol are known as members of the "30 percent club." Great Britain, Spain, Ireland, Greece, and Portugal did not sign the protocol (Switzer 1994).

Norway, Sweden, Finland, and Japan have made major inroads into generating sulfur dioxide reduction technology and abating their own sulfur dioxide emissions. Due to the effects of acid rain on their forests, Norway, Sweden, and Finland loaned Russia $1 billion to utilize Finish desulphurization technology (Switzer 1994). These same countries have been extremely influential in convincing most of Europe of the seriousness of acid rain. Japan has also been instrumental in helping China develop desulphurization technology. Unfortunately, many environmentalists contend that good desulphurization may not be enough for China, which has huge coal reserves and is accelerating industrialized development (Brown 1993, 1996; Leggett 1990; Postel 1994; Bright 2000). China is expected to build 500 coal-fueled power plants in the next decade, averaging almost one new plant per week (Lim 2007).

Many nations are recognizing the damaging effects of acid rain and are beginning to establish sulfur dioxide reduction strategies. Still, acid rain is a complicated problem. Coal is the cheapest and most abundant fossil fuel remaining, and for developing nations is the only economically viable option for pursuing development. Even many of the more-developed nations are reluctant to abandon coal as a major fuel. Although some reduction strategies are being implemented, it seems apparent that, given the increasing energy demands of the growing global population, acid rain will remain a major environmental problem for some time to come.

Conclusion

Far more progress has been achieved in addressing ozone-depleting chemicals and the air pollutants responsible for acid rain than in addressing global warming, the effects of which are reported nightly on the news. The lack of serious commitment to global warming is largely due to the inextricable relationship between carbon dioxide emissions and economic growth and development. The growth that is endemic to global corporate capitalism is not, at least in its current form, commensurate with a reduction in carbon dioxide emissions. This has made global warming a less serious policy issue, as judged by the commitment and legal teeth of international agreements, which could prove to be the world's most serious mistake. Radical climate unpredictability could create worldwide food and housing shortages. On a more optimistic note, however, current international efforts toward reducing greenhouse gases are much more serious than they have ever been historically.

International policy regarding ozone depletion is ecologically sound and moving toward a complete worldwide phase-out of major ozone-depleting substances. Unfortunately, it will be decades before international agreements produce major reductions in ozone-depleting substances, and a few decades more before ozone restoration becomes discernible. Nevertheless, the Montreal Protocol epitomizes the type of international environmental agreement that can be achieved when nation-states recognize how they are interconnected ecologically.

Acid rain is also forcing nation-states to look beyond their political boundaries. Some countries are making major efforts to reduce emissions responsible for acid rain, but many nations rich in low-grade coal are in need of expensive, advanced desulphurization technology. Unfortunately, from an ecological standpoint, low-grade coal is the most available fossil fuel reserve for many countries, which means that acid rain will most likely continue to cause bioregional and geopolitical problems. The best hope for acid rain reduction worldwide lies in the use of low-sulfur coal for industrialized nations, and easy availability of high-tech desulphurization equipment for industrializing nations.

Discussion Questions

1. What are some of the ways the largest greenhouse-contributor nations could reduce their emissions?
2. What are some of the major differences between the ozone problem and the global warming problem? Why is it so difficult for the world community to reach a viable solution regarding global warming, like they did with ozone depletion?
3. In what ways is acid rain an international problem? Describe some solutions to the acid rain problem if nations were to work together.
4. Does the North have an obligation to help the South develop in a more environmentally safe way? Does it have an interest in helping?
5. Should the South be expected to ratify the Kyoto Protocol?

Suggested Readings

Brown, Lester R., Christopher Flavin, and Hilary French, eds. (2008) *State of the World 2007.* New York: Norton.

Commoner, Barry (1990) *Making Peace with the Planet.* New York: Pantheon.

Dunn, Seth (2001) "Atmospheric Trends." In Worldwatch Institute, *Vital Signs 2001.* New York: Norton.

Gore, Al (2006) *An Inconvenient Truth.* New York: Rodale.

Kolbert, Elizabeth (2006) *Field Notes from a Catastrophe: Man, Nature, and Climate Change.* New York: Bloomsbury.

Leggett, Jeremy K. (2001) *The Carbon War: Global Warming and the End of the Oil Era.* London: Routledge.
Meadows D. H., D. L. Meadows, and J. Rander (1992) *Beyond the Limits: Confronting Global Collapse, Envisioning a Sustainable Future.* Mills, VT: Chelsea Green.
Monbiot, George (2007) *Heat: How to Stop the Planet from Burning.* Cambridge, MA: South End Press.
Schnaiberg, A., and K. A. Gould (1994) *Environment and Society: The Enduring Conflict.* New York: St. Martin's.

15

Conflict over Natural Resources

Jane A. Winzer and
Deborah S. Davenport

NATURAL RESOURCES ARE THE FOUNDATION OF THE WORLD economy. No wealth can be created without nature's resources. Thus it is not surprising that wars have been fought over natural resources. Since few states are self-sufficient, they cooperate with other countries to obtain the resources they need. If cooperation is not possible, violent conflict may become a viable alternative. The role of natural resources is reflected in the term *geopolitics,* which refers to the importance of geographic factors in international politics, and is also a component of what has been termed *environmental security.*

Environmental security scholars argue that resource depletion or degradation will be a source of increased violent conflict in the future, and attempt to identify characteristics that make violent conflict more or less likely. Most research supports the idea that political factors, such as the capacity of states to manage group demands, are highly significant in determining whether natural resource scarcity leads to violence. This chapter focuses on characteristics of the resources themselves that may exacerbate the potential for conflicts over scarcity.

Types of Natural Resources

Natural resources may be categorized according to whether they are renewable or nonrenewable, and whether they are located within the borders of one state or are transboundary. Such differences may play a role in whether resource conflicts turn violent.

Renewable vs. Nonrenewable Resources

Renewable resources are those that regenerate themselves, such as trees, fish, and animals, unlike a nonrenewable resource such as copper. This distinction is not always simple; most resources are actually renewable, but the question is whether the resource can be renewed over a comparatively short time in terms of human lifespan. For instance, oil regenerates, but over millions of years. So for policymaking purposes, it is more accurate to consider oil and other fossil fuels as nonrenewable resources.

Theoretically, renewable and nonrenewable resources might have different impacts on the international system: because renewable resources regenerate, states should have less need to fight over them. If states are unable to meet their own needs for a renewable resource, they can cooperate to meet those needs through trade agreements and economic integration. Conflict increases, however, when a renewable resource is overconsumed or "mined" at an unsustainable rate. Fishing and whaling conflicts thus seem similar to conflicts over nonrenewable resources in which overconsumption leads to competition for dwindling stocks.

Conflict and violence have historically taken place more often over nonrenewable resources. If a resource is needed but only a finite amount is available, states will sometimes fight to obtain that resource. Most of the time, even vital, nonrenewable resources are peacefully exchanged via international trade. Violent conflict over a nonrenewable resource becomes more likely if that resource, like oil, is vital to a state's existence, if substitutes are not easily found, and if doubt is cast on peaceful and acceptably affordable access. In this context, violent conflict may appear to be the most reasonable course to ensure supply of that resource.

Boundary vs. Transboundary Resources

A resource can be characterized as "boundary" or "transboundary" depending on whether or not it occupies a fixed geographic location with well-defined ownership. A forest is an example of a boundary resource: most forests are located within state borders and ownership is clear. Conflict is less likely to take place over a boundary resource because of the international principle of sovereignty: states have the sovereign right to control their own resources, and the established market mechanisms for "trading" the resource usually prevail.

Most boundary resources are fixed and do not move across boundaries. However, the distinction between boundary and transboundary resources relates fundamentally to the clear ownership of the resource and associated property rights. A fixed geographic location alone does not result in the benefits associated with a boundary resource if that geographic location does not result in clear ownership, such as with "commons" resources (see

Chapter 14). Even natural resources that are not commons but are bounded within a sovereign state may provide transboundary services, such as the benefits that tropical rainforests provide the atmosphere. Clear-cutting a forest may lead to less local rainfall, or even lead to second-order effects of global significance, such as climate instability due to the loss of a forest's carbon absorption ability.

Second-order effects from overharvesting a forest reflect what are often called environmental or ecosystem services—intangible "goods" provided by the natural functions of a resource. These environmental benefits are rarely incorporated into the market value of a natural resource. Even if a natural resource is enclosed in a boundary that determines its physical ownership, it may provide transboundary environmental services. Claims to these benefits are usually trumped by the physical owners' claims to dispose of the natural resource as they see fit, regardless of the second-order impacts.

A fully transboundary resource, such as a river, may be more likely than a boundary resource to cause interstate conflict. A river may define a border between countries or may travel from one country to another; in either case, the river must be shared by two or more states. The possibility of conflict over a transboundary resource hinges on whether the states that share it can cooperate. The more complicated the transboundary resource, the greater the potential for conflict, although such conflict may not necessarily lead to violence.

■ Defining Cooperation and Conflict

Conflict exists in any situation in which parties (states or other actors) disagree over preferred outcomes. For example, a pair of states sharing a river may always have conflicting views over how that river is used. Each country may prefer that the other country draw no water from the river (so as to maximize their own potential access) and dump no pollution into the river (so as to maximize their own potential to use the river as an ecological "sink"). Conflict over shared resources is thus to some extent ever-present, as states would always prefer to have exclusive access. States may also share an interest regarding a natural resource (such as Canada and the United States sharing an interest in the restoration of Pacific salmon stocks), but have a conflict—disagreement—over how that shared goal is to be achieved.

Conflict becomes more salient when it leads to violence. However, just as conflict need not automatically mean violent conflict, cooperation should not be assumed to imply that states are effectively addressing an underlying natural resource scarcity. Cooperation is the conscious choice of parties to change their behavior in order to achieve a joint outcome. States may choose to create a cooperative arrangement that "manages" an underlying natural resource conflict only to the extent that it prevents the broadening or intensification of that conflict.

The two dimensions of natural resources—renewable vs. nonrenewable, and boundary vs. transboundary—suggest that nonrenewable resources that are transboundary in nature generate the greatest potential for conflict, even though they are not necessarily more likely to engender violence. Other factors, such as the existing relationship between parties to a natural resource dispute, the perceived chance that a violent conflict would result in a better outcome, and the availability of alternative resources, are likely to affect whether a conflict over natural resources turns violent.

Case Studies

Figure 15.1 presents four natural resources based on renewability and boundaries. These cases show how some natural resource issues can be dealt with through cooperative action, while others can generate or exacerbate conflicts between the states dependent upon those resources. In addition, the degradation of natural resources may create environmental refugees or migrants and thus spark violent conflict, as exemplified recently in the Darfur region of Sudan.

Renewable Boundary Resource: Forests

A boundary resource such as a forest sits clearly within the domain of the sovereign state in which it is found. Nevertheless, a forest's transboundary environmental benefits generate debate at the international level about its use. Countries in the developed world (the North) argue that the needs of the international community are critical—that issues such as global climate change and species extinctions should play a key role in policy decisions regarding forests. Countries in the developing world (the South) are primarily concerned about their development opportunities. Forests may provide good export income that can fund internal development needs. To some

Figure 15.1 Natural Resource Matrix

	Renewable	Nonrenewable
Boundary	Forests	Oil
Transboundary	Fish	Water

extent, the South sees Northern environmental concern as simply a cover to prevent the South from developing and fully competing with the North, and in any case considers environmental concern somewhat hypocritical, given Northern countries' past exploitation of their own natural resources.

The problem of tropical deforestation revolves around the fact that states with tropical forests generally realize a substantial economic benefit only when the forest is cut, either for timber or for conversion to agricultural land. For instance, the desire for inexpensive beef in the 1990s created a trade environment that encouraged deforestation for cattle-rearing. According to the Center for International Forestry Research, Brazilian beef exports more than quadrupled between 1995 and 2002, and exports to the United States alone increased almost threefold. Deforestation in Brazil was explicitly linked to rapid growth in international demand for Brazilian beef, including the demand for hamburgers from fast food chains (Kaimowitz et al. 2003). In recent years, soy production, much of which itself generates feed for livestock in the North, has overtaken illegal logging and ranching as the main engine of deforestation in the Amazon (Howden 2006). Thus, while Northern countries such as the United States press Southern countries to stop deforestation, they are simultaneously creating a market incentive for that deforestation.

International conflicts over forest management are unlikely to result in interstate violence, as the likelihood that a state would perceive a beneficial outcome through violence is very low. Conflicts over forest management typically concern sustainability. Early cooperative efforts to combat deforestation included onetime "debt for nature" swaps between wealthy Northern states or financial institutions, and Southern debtor states that controlled substantial swaths of tropical forest. Debts were forgiven in exchange for commitments to protect tracts of forest.

Timber certification. A much farther-reaching cooperative effort—to achieve sustainable forest management—is found in the nonstate forest or timber certification schemes that began to proliferate in the 1990s. These certification schemes seek to incorporate sustainable forest management into the market price of timber products. They typically involve the development of criteria that assess progress toward sustainable management, which is verified by an independent third party. Product "ecolabeling" is then used to certify and trace products from that sustainably managed forest, all the way to consumers (Gulbrandsen 2004), in much the same way that organic apples, for instance, carry labels certifying them as organic.

Researchers theorize that certification schemes for sustainable forest management may persuade forest managers to modify their practices in order to earn a label indicating that their timber has been sustainably produced. This label permits consumers to choose between sustainably and nonsustainably produced wood products. Practicing sustainable forest management costs producers more, at least in the short term, than other timber production methods, but if consumers are willing to pay more for products that come from sustainably managed forests, producers will have an incentive to join the scheme (Gulbrandsen 2004).

Certification represents an alternative to the boycott campaigns that Northern environmentalists have used in the past to influence the market, particularly with regard to tropical timber (Cashore, Auld, and Newsom 2004). Certification schemes should encourage forest owners to maintain their forests rather than convert them for other economic uses such as agriculture or cattle-ranching, or otherwise harvest them at an unsustainable rate.

Forest certification has evolved up to the global level, constituting perhaps the most dynamic case of nonstate-driven rule-making for any global environmental issue. Forest certification thus represents a case where cooperation has prevailed over potential conflict. Nevertheless, the question remains whether such schemes can achieve sustainable forest management while also allowing forests to be used as a resource for development. Although forest certification schemes are still too new to be judged on their environmental effectiveness, it is believed that they will raise awareness about sustainable management.

Certification schemes vary along several dimensions, including the number of stakeholders involved in developing standards, the strength of the standards, the quality of auditing, and the level of participation by producers and consumers. The first scheme, the Forest Stewardship Council, was established in 1993 by the World Wide Fund for Nature and other environmental organizations, along with various social and indigenous peoples' groups, timber traders, forest owners, and other stakeholders. In reaction, other schemes were established by forest industry associations, such as, in 1994, the American Forest and Paper Association's Sustainable Forestry Initiative (Gulbrandsen 2004). Schemes initiated by industry tend to be more flexible and give more dominance to industry actors; others tend to give environmental and social groups a greater role in policymaking (Cashore, Auld, and Newsom 2003).

There are limitations on the effectiveness of any certification scheme. First, it depends on developing a market for certified wood, to give producers either greater market access or premium prices for their products. This ultimately depends on the willingness of enough consumers to insist on certified wood.

Another question revolves around the limited authority of certification bodies to ensure participation and compliance, given that they are private rather than governmental entities. Regular audits by third-party certifiers are intended to promote compliance with standards and improvements on performance, but certifiers have no authority to enforce compliance, apart from suspending a forest manager's certificate, nor to stimulate participation by producers who feel unable to afford the costs of certification.

Although there is a market for certified timber products in Europe and increasingly in North America, the overwhelming share of timber traded worldwide is not certified. However, the market for certified forest products is growing and will likely continue to encourage forest certification. On the other hand, the differences between existing certification schemes may confuse customers and ultimately weaken the credibility of certification.

For tropical countries, the situation is even more problematic. There is a general lack of awareness of certification schemes in developing countries, and thus tropical timber producers feel little pressure to certify their products; indeed, a large portion of the tropical timber on the world market comes from illegal sources (Gulbrandsen 2004). In addition, a large portion of certified tropical timber comes from plantations rather than natural forests. This is a problem because tropical forests contain numerous natural tree species that together form a habitat for perhaps thousands of animal species, while monoculture plantations only contain planted, commercially valuable tree species. Finally, there are few financial incentives to practice sustainable forest management, as quick money can be made from harvesting and selling existing timber resources, and converting land to other uses is frequently more economically attractive (Gulbrandsen 2004). Certifying timber, whether tropical or temperate, cannot address all deforestation issues, but it is at least a step in the right direction.

Renewable Transboundary Resource: Fish

Fish are a renewable resource. Yet fish stocks are depleting throughout the world, due in part to technologies developed during the nineteenth and twentieth centuries that increased catch size and processing capabilities. This decrease in fish stocks particularly affects over 40 percent of the world's people who rely on fish as their main source of protein (Combes 2006). And as the world's population increases, the demand for inexpensive food sources such as fish will increase as well.

Many of the fish that are most in demand either live in the ocean, a global commons, or cross state boundaries at some point in their lives. In such cases, their conservation depends on cooperation between two or more states. The 1982 Law of the Sea Treaty was in part intended to help resolve fishing conflicts by establishing a state's rights over resources found in waters up to 200 miles off its coast, an area called its exclusive economic zone. An exclusive economic zone is not part of a state, but a state has the exclusive right either to use all the ocean resources found in that area itself or to sell usage rights to others. People thought that the Law of the Sea Treaty would prevent most conflicts over fishing rights, because worldwide recognition of 200-mile exclusive economic zones brought more than 90 percent of the world's commercially fished stocks under the jurisdiction of individual, coastline states (Alcock 2002). Unfortunately, the Law of the Sea Treaty has been inadequate to resolve a number of fishing disputes, and fish stocks throughout the world continue to be overharvested.

The US-Canada salmon dispute. One example of a transboundary dispute over fishing rights is the dispute over salmon in the Pacific Northwest, ongoing since the nineteenth century. Although occasionally called a "war," it has not usually been characterized by violent conflict. Rather, the relationship between the United States and Canada over Pacific salmon has been one of contentious negotiations alternating with periods of cooperation. However, occasionally, a decision by the government of either Canada or the United States, or actions by parties involved in the salmon fisheries, will spark a cross-national clash.

Technology has fueled this resource conflict. With the development of the canning industry in the early 1900s, salmon became a viable export and a rush to exploit this rich industry began (Ralston and Stacey 1997–1998). The expansion of fisheries within the Pacific Northwest's major river systems precipitated a long decline in the salmon population of the region. Since 1980 the situation has become critical, with a sharp fall in salmon populations in the region, particularly in the US states of Washington and Oregon (Knight 2000; Barringer 2008). Figure 15.2 shows the area of transboundary conflict over salmon.

Salmon are anadromous, meaning that they spawn in freshwater streams, migrate to the sea, where they feed and grow for two to four years, and then return to their stream of origin. This makes it necessary to conserve the salmon-spawning habitat (upstream), which takes place within national jurisdictions, even though the (downstream) salmon fishing industry is located in the open sea. It is also necessary to ensure that harvesting of mature salmon is limited, so that enough salmon can return to rivers and lakes to spawn sustainable populations. Harvests are currently managed through catch quotas, open and closed fishing seasons, minimum size limits, and limits on the numbers of licensed commercial fishers (Huppert 1995).

US-Canadian efforts to cooperate over Pacific salmon began with the Fraser River Convention of 1930. This agreement created an international commission to restore the Fraser River, the spawning habitat of the sockeye salmon, and divided annual harvests equally between the United States and Canada. Even this limited agreement took seventeen years to achieve, and covered only one species and one habitat. While it achieved its limited aims, it did not adequately address how to fairly allocate the costs of maintaining the spawning habitat of sockeye salmon. Nor, of course, could it address the severe depletion of other salmon stocks in the US Northwest that began in the early 1980s (Knight 2000).

Figure 15.2 Species Composition of Commercial Harvests in Pacific Salmon Treaty Region, 1990–1994

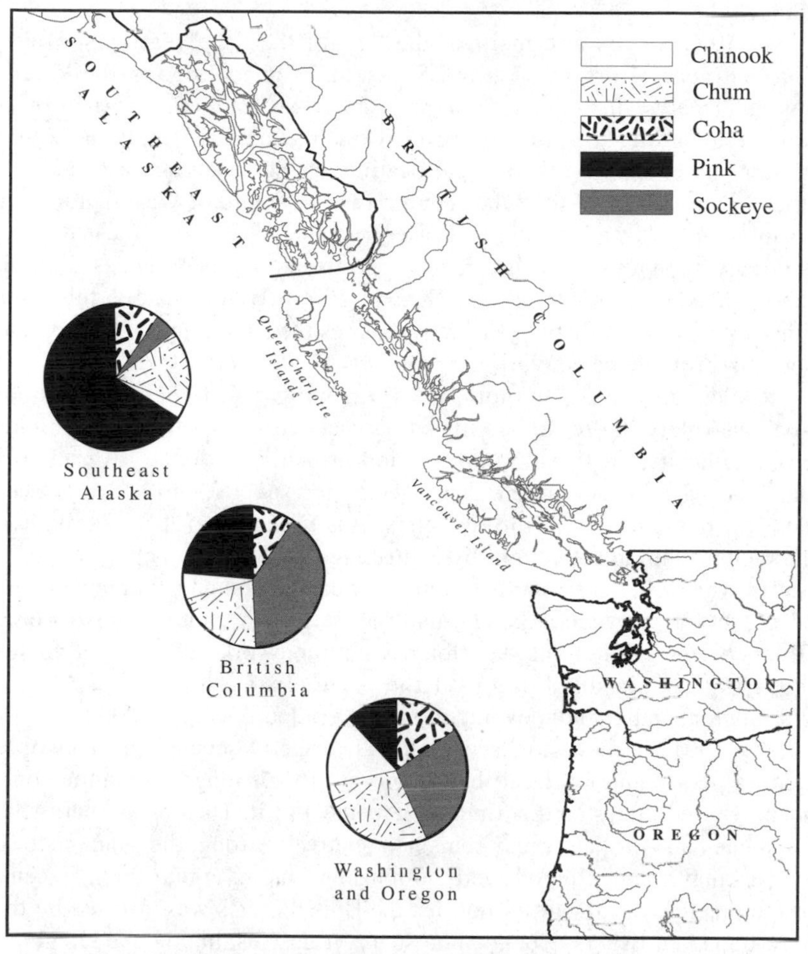

Source: Daniel D. Huppert, "Why the Pacific Salmon Treaty Has Not Brought Peace" (Seattle: University of Washington, School of Marine Affairs, 1995).

The Law of the Sea Treaty changed the international norm of ownership of high-seas resources—from one that applied solely to caught fish, to one based on the "state of origin" of the fish. For salmon, the state in whose freshwater river the salmon spawn is now considered the owner of that resource. However, because salmon cross political boundaries during their migrations, it is impossible to separate them by country of origin. Hence, fishers from both the United States and Canada catch, or "intercept," some of the salmon that originate in each other's rivers. This means that the fishing industries of both countries are mutually dependent, to greater or lesser degrees, on the conservation of habitat within each (Huppert 1995; Knight 2000).

In 1985, in the aftermath of the Law of the Sea Treaty and after a fourteen-year effort, the United States and Canada signed the Pacific Salmon Treaty. This allocated harvests based on equitable distribution of salmon according to state of origin. If one country intercepts more than its fair share, this is offset by allocating an increased share of another stock to the fishers of the other country. For example, if Washington fishers intercept a larger number of Fraser River sockeye than allocated under the treaty, then Canadian fishers to the north can intercept more Columbia River (Alaska) chinook salmon (Knight 2000). Unfortunately, this solution to the political dispute further increases the harvesting of salmon and does not promote conservation.

Under the Pacific Salmon Treaty, the obligation to conserve salmon stocks is solely the responsibility of each country , and the United States gives authority for this implementation primarily to local governments, states, and Native American tribes, which are supposed to arrive at decisions on implementation through consensus (Yanagida 1987). Throughout the region, salmon are caught by competing fleets for competing purposes. Native American tribes harvest salmon for ceremonial and commercial purposes, and recreational fishers also compete with commercial fisheries (Huppert 1995). The weak decisionmaking framework in the United States, requiring consensus among all the various stakeholders in Alaska, Washington, and Oregon, eventually led to gridlock (Knight 2000).

By 1991, salmon stocks, which had formerly seemed inexhaustible, were in poor condition in all but 6 percent of their range within the continental United States, and extinct in 38 percent of it. This meant that, while Canadian and Alaskan stocks remained relatively strong, the salmon stocks in Washington and Oregon were no longer viable (Knight 2000). Because of the practice of offsetting, not just the United States was affected by this loss. Canadian fishers depended on stocks from Washington and Oregon to offset interceptions by US fishing boats of Canadian salmon when they migrated to Alaskan waters.

The decline in health of salmon stocks in Washington led to legal restrictions on their harvesting, which dramatically reduced the number of interceptions of Washington-origin salmon by Canadian fishers. Meanwhile, however, Washington and Alaskan fishers were still able to intercept salmon of Canadian origin. Washington fishers were willing to reduce their own catch of Canadian fish to address this inequity, but Alaskan fishers were not. Their harvests remained large while the rest of the Pacific salmon fishing industry was severely limited.

Canada responded by increasing its fishing off the coast of Vancouver to try to balance Alaskan interceptions of Canadian fish. Cooperation eventually broke down in 1994, when Canada instituted the requirement that US fishing vessels passing through Canadian waters obtain a permit. Then, in 1997, the dispute turned into a real "fish war," as newspapers labeled it, when Canada blockaded an Alaskan ferry and held it for several days to protest Alaskan interceptions of Canadian fish (Nickerson 1997).

Clearly, a new agreement was needed. Negotiations were started in 1995 and took four years to complete. The 1999 agreement replaces one of the annexes, or side agreements, of the 1985 Pacific Salmon Treaty. It mandates the establishment of a new body to oversee salmon originating in the transboundary rivers of Canada and southeastern Alaska, funding to improve resource management and habitat restoration, and scientific cooperation.

The 1999 agreement brought conservation concerns to the fore in allocating harvest allowances, but did not address the problems of consensus-based decisionmaking among US stakeholders that had led to cooperation failure in the 1990s. The Yukon River Salmon Agreement was signed in December 2002, forming another annex to the 1985 treaty and focusing on conservation of salmon stocks originating in the Yukon River in Canada (USDS 2002). Meanwhile, subnational implementation methods have spawned proposals to cut fishing fleets in Alaska ("Report Recommends" 2004) and the rise of salmon farming. Controversy still exists, however, over whether hatchery-spawned salmon pose a threat to wild salmon—such as by infecting them with lice (Froehlich 2004)—or are viable for release into the wild and inclusion in population numbers for allocation purposes ("Coho Salmon" 2004; Wild Salmon Center 2004–2008).

Populations of wild salmon continue to decline, both in the United States and Canada, due to water pollution, loss of habitat, overfishing, dam construction and operation, water use for irrigation and other purposes, predation by other species, diseases and parasites, and climatic and oceanic shifts, as well as genetic and ecological risks posed by hatchery-produced salmon. Increased human consumption compounds all of these other problems. In addition, no one has yet devised a realistic technical solution to reverse the decline of wild Pacific salmon and ensure their survival beyond 2100 ("Experts Embark" 2004). In February 2007, negotiations started up once again to try to address the conflicting interests in salmon harvesting and preservation (Kiffer 2007), but it remains to be seen whether the current regime can survive and lead to full cooperation and recovery of wild salmon in the future. Canada and the United States may be avoiding conflict over Pacific salmon through political cooperation, but without success in managing salmon stocks.

Nonrenewable Transboundary Resource: Water

Water is usually considered renewable, but in fact, more than three-fourths of underground water is nonrenewable, in that it takes centuries to replenish. Water may also be considered nonrenewable when irreparably polluted, for example by a chemical spill. In addition, freshwater sources may be used up during droughts, which can last many years. The water situation around the world is so serious that the United Nations created a new refugee category, "water refugees"—those 25 million people who have faced social and economic devastation as a result of critical shortages of water—and declared 2003 the "International Year of Freshwater." Indeed, water refugees now outnumber war refugees ("Nor Any Drop to Drink" 2001).

Nowhere is the water situation more dire, nor its potential to contribute to conflict greater, than in the Middle East. While the 2003 Iraq War invalidated some experts' predictions that the next protracted war in the Middle East would be fought over water (Gleick 1994; Postel 1993), that region remains one of the most arid in the world. A rarely acknowledged factor in the continuing conflict between Israel and its neighbors is the regional scarcity of water for the growing population.

States in the Middle East depend almost entirely on regional river systems for their water supply, none of which exist solely within one state's borders; 90 percent of all potable (drinkable) water sources are transboundary (Peterson 2000). Generally, upstream countries are in a stronger position, since they can control both the quantity of water (through dams) and the quality of water (by their industrial and agricultural actions). One glaring result of this imbalance in the Middle East is that Palestine and Jordan lack appropriate drinking water, while Israel has a freely flowing supply. This imbalance between upstream and downstream states cannot help but heighten feelings of anger and frustration, although mutual vulnerability to problems such as water pollution and scarcity can also sometimes lead to cooperation (FOEME 2003). There is growing international legal debate over whether states should have absolute sovereignty over water resources found within their borders (the Harmon Doctrine) or, conversely, be obligated to ensure that their use of water resources does not adversely affect other states (the principle of "equitable utilization," as promoted by the United Nations).

The Jordan River basin. The water problems that exist in the Jordan River basin exemplify tangled upstream and downstream relationships. The Jordan River basin is shared by Syria, Lebanon, Jordan, Israel, and Palestine, and comprises the Jordan River as well as its four tributaries, three of which arise in Syria or Lebanon, as seen in Figure 15.3. The Jordan River is modest in terms of length as well as flow. Yet it is the focus of intense attention, given the acute political tension in the area (Dolatyar and Gray 2000).

The Jordan River was able to supply the demand put upon it until the twentieth century. Significant increases in regional population (from under half a million in the late nineteenth century to over ten million by the beginning of the twenty-first century) have noticeably strained the Jordan River basin (Amery and Wolf 2000; Dolatyar and Gray 2000; DOS 2003; and Yaghi 2004). Fortunately, the upstream states, Syria and Lebanon, obtain only 5 percent of their water from the Jordan River basin. Jordan, however, has a serious water deficit and relies heavily on the Jordan River and one of its tributaries, the Yarmouk River, to satisfy its water needs.

Figure 15.3 Jordan River Basin

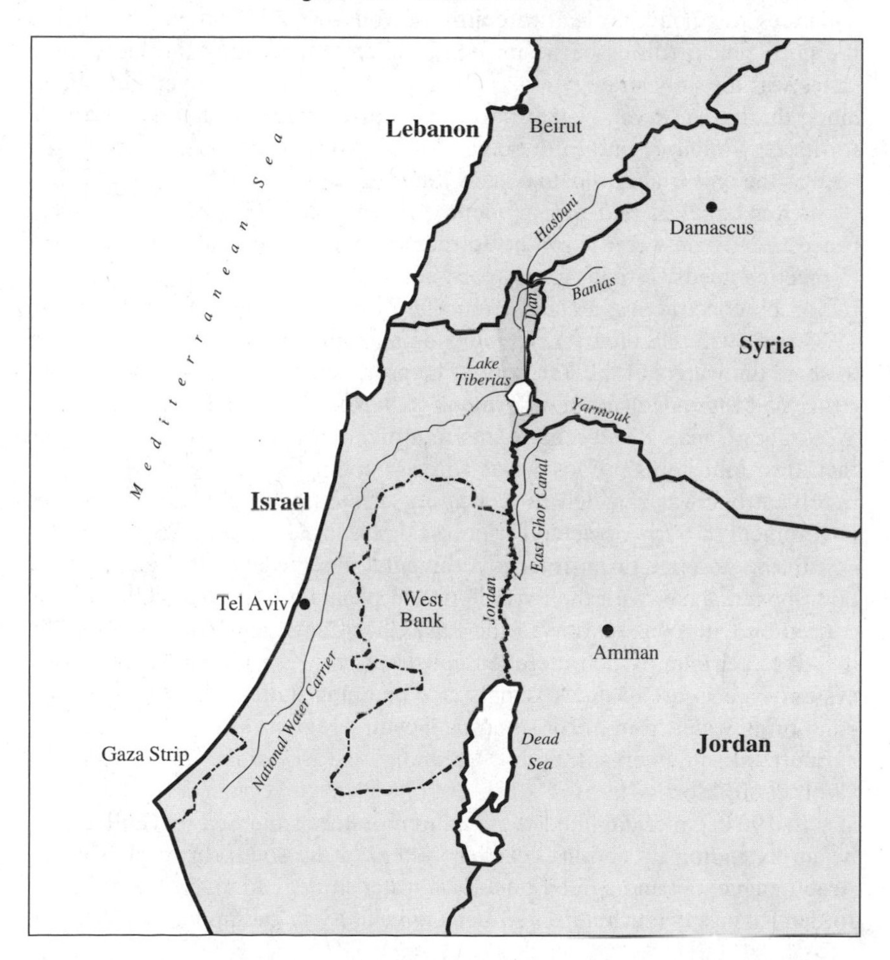

Israel obtains about one-third of its water supply from the Upper Jordan, and has severely restricted Palestinian use of water from the Jordan River directly, and from the aquifers (underground geological structures that can store water) underlying the West Bank of the Jordan (Dolatyar and Gray 2000; Deconinck 2004). Neither Israel nor Jordan nor Palestine have a large enough supply of water to meet the World Health Organization's minimum standard of 500 cubic meters per capita of daily potable water supply (Deconinck 2004).

In the Middle East, water problems are enmeshed with unresolved border issues, massive population increases, diminishing agricultural resources, increasing industrialization, changing living standards, and issues of religion, culture, and tradition. The legacy of British and French control of the region until 1948, and the division of territory after the 1948 war, left the Jordan River divided in such a convoluted way that development of water resources was bound to lead to conflict (Wolf 2000). Unfortunately, due to the political hostilities between them, unilateral action by the individual states was the only strategy available. Each of the riparian (riverside) states along the Jordan River system moved to utilize as much water as it needed for itself, without regard to the others' needs. Military force was frequently seen as the only viable way to secure water rights.

After Israel gained independence, it devised development plans that relied heavily on water from the Jordan, as local aquifers were insufficient to meet its needs. Jordan, for its part, announced a plan to irrigate its Ghor region by constructing a canal from the Yarmouk River in 1951. Between 1953 and 1955, US envoy Eric Johnston attempted to broker an agreement to share the waters of the Yarmouk. The proposed project included the construction of two dams on the Yarmouk, two irrigation canals—the East and West Ghor Canals—and other dams to utilize seasonal flows, and drainage facilities. Johnston's proposal was agreed upon by all the technical experts involved, but was rejected at a meeting of the Arab League because the government of Syria objected for political reasons (Johnston 1958).

In the absence of an intergovernmental agreement, Israel and Jordan both moved ahead with their development plans unilaterally. While Jordan carried on with construction of the East Ghor Canal, Israel carried on with its All Israel Plan, which included construction of a national water carrier system—a network of pipes, canals, and pumping stations that were intended to bring water from the northern and central regions of Israel to southern agricultural settlements, thereby "greening" the Negev desert in the south (Wolf 2000; Morris 1996).

In 1964, Israel dammed the southern outlet of the Sea of Galilee and began extending its national carrier system to the south. In response, the Arab League met and agreed on a joint water strategy to divert the northern Jordan River's two tributaries in Lebanon and Syria and impound the water for use by Jordan and Syria with a dam on the Yarmouk. Israel then launched a series of aircraft and artillery attacks, which halted the diversion project and set off a chain of events that led to the June 1967 Arab-Israeli War (Morris 1996; Wolf 2000).

Israel's victory in that war and resulting occupation of the West Bank and Gaza Strip effectively increased its riparian rights (water rights of riverside landholders) over upstream tributaries to the Jordan by almost 50 percent. The war interrupted the work that had begun on the East Ghor Canal as well, due to Israel's takeover of half the length of the Yarmouk River. Israel then attacked the extant portion of the East Ghor Canal in 1969, in an effort to persuade King Hussein of Jordan to take action against the Palestine Liberation Organization (PLO). King Hussein did expel the PLO from Jordan and the canal was rebuilt, but Johnston's master plan was never implemented and no further development of the East and West Ghor Canals has been possible. Disputes over waters of the Jordan River basin are just one aspect of the conflict over water in the Middle East, a result of the aridity of the region, the growing population, and the desire of all countries to develop and industrialize. Whether water is at the base of conflicts in the Middle East or reflects other underlying tensions, it has clearly exacerbated the political situation and continues to be used as a weapon (Morris 1996).

On the other hand, states' mutual vulnerability to water scarcity could be a source of future cooperation rather than war (Dolatyar and Gray 2000; Coles 2004). The Jordan River is a vital resource for Israel, Palestine, and Jordan, making all three mutually dependent on one another in its administration. The 1994 peace treaty between Israel and Jordan contains promising language on ownership and use of mutually shared water resources (Morris 1996; Izenberg 1997). Despite the upsurge in Israeli-Palestinian conflict in 2000, the governments of Israel and Jordan have generally managed to adhere to the treaty's language on water. In 2002, this cooperative effort was furthered by an agreement between Israel and Jordan to collaborate on a plan to bring water from the Red Sea to replace water that has been diverted from the Jordan River to replenish the receding Dead Sea (Stephens 2002). In the longer term, some scholars have predicted that such collaborative efforts over water resources could increase the chances for regional peace and cooperation more generally (Dolatyar and Gray 2000).

Nonrenewable Boundary Resource: Oil

Technically, oil is a renewable resource; it is created when dead microorganisms accumulate on the ocean floor and eventually become released as hydrocarbon molecules. The key word is *eventually:* fossil fuels are called such because it takes millions of years for supplies to become abundant. Oil reserves are a limited and finite resource, and industrialized states use more of this resource than they can produce themselves. They must therefore import oil; thus, even though it is a boundary resource, oil is the subject of interstate competition and conflict.

The United States is the largest consumer of petroleum, using about 24 percent of global production, or approximately 21 million barrels each day (EIA 2007a). However, the United States produces only about 8 million barrels a day and imports the rest (EIA 2007b, EIA 2007c). Contrary to popular belief, less than 20 percent of US petroleum imports comes from the Persian Gulf region, and Saudi Arabia is only one of five top sources for US petroleum imports. Nonetheless, five of the top fifteen oil-producing countries are in the Persian Gulf, and thus the international market is significantly affected by the stability of Gulf oil production. The importance of Middle Eastern oil is amplified by estimates of oil reserves: with an estimated 711 to 742 billion barrels, the Middle East holds approximately 60 percent of the world's proven reserves (EIA 2007d). Despite any potential efforts to reduce consumption of petroleum, future competition for world production is expected to increase as more countries, particularly large population centers such as China and India, develop and industrialize.

Many powerful developed states depend heavily on petroleum for industrial development, personal consumption, and maintenance of the military. To some extent, power in these states depends on their access to oil, a limited resource that must be imported from other states. Therefore, even though oil-importing and oil-exporting states have a long history of cooperation, competition for oil also gives rise to the potential for conflict.

Since decolonization in the mid-twentieth century, intense conflicts over access to oil have frequently occurred where the ownership or boundary nature of a petroleum deposit is unclear. Much of the history of conflicts in the Persian Gulf relates to competing claims of sovereignty over oil deposits and access routes. Saddam Hussein's attempt to annex Kuwait in 1990 was in part due to his rejection of postcolonial sovereign boundaries.

War in Iraq. On August 2, 1990, Iraqi military forces invaded the sovereign state of Kuwait. While the Iraqi government made many claims to justify the action, one of the major issues was oil. The Rumailla oil fields are on the border between the two states. Iraq claimed that Kuwait was taking more than its rightful share of oil from that reserve (Freedman and Karsh 1993).

This dispute had historical roots. The territories that later became Iraq and Kuwait were under the rule of the Ottoman Empire until Kuwait became a British protectorate in 1899 and Iraq came under British mandate in 1918. Iraq gained independence in 1958 and Kuwait followed three years afterward. It was later discovered that some of the richest oil reserves in the world were within Kuwaiti territory. Kuwait developed rapidly throughout the 1960s, and by the 1970s had one of the highest per capita gross domestic products in the world (Võ 1994). Kuwait's economy was much healthier than Iraq's, especially since Iraq had been paying for a long war with Iran.

The growth of the Kuwaiti economy was a sore point for Iraq, which believed that it had historical rights to the territory and had threatened to fight Britain during Kuwait's protectorate period (Freedman and Karsh 1993). In the 1980s, Iraq's economy was strained due to its eight-year war with Iran. This, coupled with disputes over Iraqi rights to Kuwaiti territory and over the extent of Kuwait's oil-drilling, paved the way for the Iraqi invasion in 1990.

Despite Saddam Hussein's claims regarding Iraq's right to Kuwaiti territory, the international community recognized the sovereignty of Kuwait. Invading another sovereign state was a clear violation of international law, and the United Nations responded. Yet many other clear violations have been ignored by the international community. What made this so different? Many experts suggest that the United States and its coalition allies took swift and decisive military action because of their interest in protecting the oil supply. Were the invasion to stand, Iraq and its leader, Saddam Hussein, would exercise too large an influence in world oil markets (Võ 1994; Pickering and Owen 1994). The North, especially the United States, had many concerns, including apprehension that Iraq would not stop with Kuwait but would attempt to claim the abundant oil reserves in Saudi Arabian territory as well. The North felt that its supply of petroleum was in danger.

The UN Security Council condemned the invasion and demanded that Iraq retreat from the borders of Kuwait. When Iraq refused, sanctions were imposed and a multinational military force was created. A short war ensued, resulting in Iraq's eviction from Kuwait. Although Iraq destroyed some oil fields and set hundreds of oil wells on fire (Warner 1991), the war produced the desired result: protecting access to the oil reserves.

However, the story did not end in 1991. Believing that international pressure would lead to the internal overthrow of Saddam Hussein, and pursuant to cease-fire conditions, economic sanctions were imposed on Iraq, including restrictions on its oil exports. By the early 2000s, having failed to result in the overthrow of Saddam Hussein, the arrangements were breaking down. Concerns about the future of an Iraq still led by Saddam Hussein, in combination with US foreign policy following September 11, 2001, resulted in the decision by US president George W. Bush to remove Saddam Hussein from power. In 2003 the United States led a "coalition of the willing" into Iraq and overthrew Saddam Hussein. Although the United States had pushed for a UN resolution authorizing such a move, to enforce Security Council resolutions from the 1990s, it was unable to persuade the Security Council or the vast majority of the world's states to contribute to the effort. As a result, the United States bore the brunt of the coalition's costs during the invasion and the occupation that followed. Contrary to predictions that the invasion would provide a relatively quick solution to the "Iraq problem" and result in enhanced regional stability, and stability for international petroleum markets, the 2003 invasion and overthrow of Saddam Hussein has resulted in ongoing violent conflict, large refugee flows, and a continuing weakness in Iraq oil exports due to insecurity and sabotage.

Environmental economists suggest that if US prices reflected the true costs of imported oil—including the cost of government subsidies and support, environmental impact, military security, and human lives—Americans would recognize the need to conserve, as the cartoon below suggests. Instead, most of these costs are absorbed in the general budget.

* * *

This survey of four natural resource conflicts demonstrates the impact of ownership and renewability. Clearly defined ownership of boundary resources can reduce conflicts over who has sovereign rights to the exploitation of a natural resource, while the transboundary nature of some resources creates competing sovereign claims. The extent to which a resource is renewable can mitigate the severity of a conflict, with nonrenewable resources posing the highest stakes. According to this typology,

Source: Ed Stein, *Rocky Mountain News,* 1990. Used by permission of Ed Stein.

we should expect a greater number of conflicts, or more complex conflicts, in situations of competition over transboundary, nonrenewable resources.

However, conflicts over natural resources do not necessarily lead to violence. A violent outcome will only result if at least one side anticipates a better outcome from violence than negotiations. Frequent disagreements can also serve as repeated opportunities for states to establish trust, by creating a history of compliance with negotiated cooperative arrangements. Furthermore, there are other important factors to consider, such as the existing relationship between states who are party to a natural resource conflict, and the distribution of power among the parties involved. Research into environmental conflict has suggested that natural resource issues can contribute to interstate war, but are rarely sufficient to cause war. Environmental security studies have suggested that the more likely pathway to violent conflict over natural resources is through intermediary effects on the movement of populations.

Resource Degradation and Violent Civil Conflict

In the 1990s, Thomas Homer-Dixon (1991, 1994) sought to demonstrate that environmental problems would lead to increased violent conflict through competition over depleted or degraded renewable resources. While throughout history interstate conflicts have occurred over nonrenewable resources like oil and minerals, little empirical support was found for the argument that interstate conflicts arise over renewable resources. Freshwater resources have played a role in interstate conflicts, but arguably due to their tendency to resemble nonrenewable resources. Nonetheless, environmental security scholars still emphasize an indirect means by which scarcities of renewable resources can lead to violent conflict. As degradation or depletion of resources occurs, it can trigger large-scale population movements. The creation of ecological migrants or refugees subsequently interacts with existing group-identity conflicts, which can in turn often spark violent subnational conflicts. These subnational conflicts may then spill over into cross-national conflict.

The ongoing conflict in the Darfur region of Sudan has been characterized as an environmental conflict by both UN Secretary-General Ban Ki-moon (Ki-moon 2007; Borger 2007) as well as the United Nations Environment Programme, the latter of which linked the conflict to reduction in crop and grazing land and ultimately to climate change (UNEP 2007).

It is argued that desertification led pastoral Arab Sudanese groups to move south, where they encroached on land farmed by sedentary agricultural African tribes in the Darfur region of Sudan. The conflicts between these two populations, despite their common religion of Islam, ultimately provoked separatist African forces to attack police and Sudanese government forces. The central Sudanese government, preoccupied with other conflicts, chose to arm and encourage the creation of Arab Janjaweed militias. These Janjaweed militias targeted the pastoralist communities in retaliation for allegedly supporting the separatists. Even this oversimplified explanation points to the importance of other factors, beyond ecological resources, underlying the terrible consequences in Darfur.

Nobody disputes the significance of desertification in creating tensions among populations competing for land. Competition for scarce resources can serve as an underlying source of conflict among populations. However, even in cases where ecological scarcities are clear, it is not accurate to say that resource scarcities always cause violent conflict. Additional factors are critical in the case of Darfur, such as the complex interaction of activities by neighboring states (Chad and Libya), long-term economic marginalization of the Darfur region in favor of central Sudan, and, most crucial, actions of the Sudanese government in fostering the violent conflict (de Waal 2007).

Conclusion: The Need for Sustainability

One of the lessons to be drawn from the case studies presented in this chapter is that the potential for both cooperation and conflict exists for all natural resources. Conflict is less likely to occur over renewable, boundary resources—in other words, resources that do not cross boundaries and do not run out. Conflict, particularly violent conflict, is more likely to occur when parties distrust each other and resources are both limited and of critical importance. Competition over both renewable as well as nonrenewable resources can lead to conflict when those resources become depleted. The avoidance of conflict may well require more sustainable use of natural resources. For the South, sustainability requires finding alternatives to many of the development techniques that are currently depleting resources. This might mean finding the inherent value in standing forests rather than deforested land, or developing new ways to share water resources to meet basic human needs. For the North, sustainability means reducing consumption in general, to create the "ecological space" for people in the rest of the world to increase their own consumption to at least the minimum standards necessary to live healthy, safe lives. Both North and South need to cooperate to develop policies that are more efficient and that ensure the provision of basic human needs around the world.

Discussion Questions

1. Do market-based solutions present appropriate responses to tropical rainforest decline? To water shortages?
2. Are salmon a critically important resource?
3. What is the role of technological development in the exploitation and/or conservation of natural resources?
4. Do you think that water will be the cause of the next war in the Middle East?
5. Should the price of gasoline reflect its true costs (government subsidies, environmental impact, military security, human lives, etc.)?
6. If you were president of the United States and faced a threat to Middle Eastern oil reserves, what would you do?
7. Does the United States have a responsibility to reduce its national consumption levels? How?
8. Should countries' security policies address environmental security?

Note

This chapter is based on but significantly revises the work in previous editions by Deborah S. Davenport and Karrin Scapple.

Suggested Readings

Allan, Tony (2002) *The Middle East Water Question: Hydropolitics and the Global Economy.* London: Tauris.

Amery, Hussein A., and Aaron T. Wolf (2000) *Water in the Middle East: A Geography of Peace.* Austin: University of Texas Press.

Cashore, Benjamin, Graeme Auld, and Deanna Newsom (2004) *Governing Through Markets: Forest Certification and the Emergence of Non-State Authority.* New Haven: Yale University Press.

Gulbrandsen, Lars H. (2004) "Overlapping Public and Private Governance: Can Forest Certification Fill the Gaps in the Global Forest Regime?" *Global Environmental Politics* 4, no. 2.

Knight, Sunny (2000) "Salmon Recovery and the Pacific Salmon Treaty." *Ecology Law Quarterly* 27, no. 3.

Petiere, Stephen C. (2004) *America's Oil Wars.* Westport: Praeger.

PART 5

Conclusion

16

Future Prospects

Michael T. Snarr

WHAT WILL THE FUTURE BE LIKE? WILL THINGS GET BETTER OR worse? Will humans organize themselves to more effectively confront global issues? Possible scenarios for the world several decades into the future include world government, regionalism, decentralization, and the status quo.

World Government

Some scholars argue that a world government, consisting of a powerful central actor with significant authority, is the method by which we will organize ourselves in the future. The World Trade Organization's increasing consensus on economic issues, and the emergence of a multitude of free trade agreements, are often cited as evidence that some sort of world government is not out of the realm of possibility. Similarly, for those who argue that the world is moving toward a single global culture, a world government might not seem beyond reach. This possibility is not likely to occur overnight.

In contrast to a slow, evolutionary movement toward consensus on issues like economics, it is conceivable that a world government might be created after a catastrophe. An exchange of nuclear attacks or a destabilizing, worldwide economic crisis might force governments into calling for a central authority, to avoid long, drawn-out negotiations among more than 190 sovereign states.

In theory, a world government would be more effective at addressing salient global issues, since power would be much more centralized. An obvious problem with the world government scenario, and the reason it will not be realized anytime soon, is the unlikeliness that the countries of the world would voluntarily give up their sovereignty. Furthermore, a world government would face many practical problems, such as who would be responsible for enforcing laws. Would a world government have a powerful military? If so, the fear of tyranny would be realistic. If not, its enforcement capabilities would be questionable.

There are other possibilities in addition to a true world government. A *federation* would establish a relatively weaker world government, similar to the model of the United States, where the federal government shares power with the states. Even weaker would be a *confederation,* in which states would be the dominant actors but would give the world government some jurisdiction. Both federate and confederate systems would give a world government more power than the United Nations currently possesses.

Regionalism

In the regionalism scenario, countries would be organized into groups based on geographic proximity, perhaps following the pattern of current economic groupings like the North American Free Trade Agreement (NAFTA), the European Union (EU), and the Asia Pacific Economic Cooperation (APEC) forum. As with a world government, countries would not completely relinquish their sovereignty, but it would likely be significantly reduced. Currently, the European Union is the leader in the movement toward economic and political cooperation. Not only has the EU drastically reduced barriers to economic integration and to the movement of people within its borders, but it has also adopted a single economic currency and made progress toward a common foreign policy. Although NAFTA and APEC are relatively young in comparison to the EU, their formation represents the current popularity of regional arrangements.

Of course, the regionalism scenario also must deal with the reluctance of countries to relinquish their sovereignty, the fear of concentrating too much power in the hands of a central government, and so on. However, these issues may be easier to resolve in smaller groupings of states than in a world government context.

On the positive side, regionalism would facilitate the coordination of regional policymaking on global issues such as the environment, human rights, and trade. Still, the enhanced ability of countries to coordinate policies within their respective regions would not necessarily translate into cooperation between regions. It could be argued that regionalism would simply transform a world in which *countries* compete into one in which *regions* compete, without solving pressing global problems.

Decentralization

At the same time that free trade and environmental agreements are being enacted, creating more centralized authority, there is significant movement toward decentralization of power. One example is the strong separatist movement mounted by Canada's French-speaking province of Quebec. The former Soviet Union and the former Yugoslavia are additional examples of decentralization. Although the various separatist movements have differing motives, many of them do have in common a desire for self-determination—that is, the desire to break away from the dominant culture and govern themselves. Consider that in the 1940s the world had just over 50 countries, but now it has nearly 200. The proliferation of decentralized terrorist cells can also be considered part of this movement toward greater local control.

This trend certainly casts doubt on a future involving "one world" or just a few regions. Each new country would of course be smaller and more culturally homogeneous than today's countries, which ideally would alleviate some of the nationalist tensions discussed in Chapter 3. However, it also would make achieving international consensus on issues like the environment, human rights, and nuclear proliferation more difficult.

There is, however, another type of locally oriented movement, commonly referred to as *civil society,* which has gained momentum in recent years. Civil society comprises nongovernmental, nonprofit groups such as social service providers, foundations, neighborhood watch groups, and religion-based organizations. Jubilee 2000 (Chapter 7) and the Green Belt movement and the Grameen Bank (Chapter 10) are examples of civil society, or grassroots, movements. In recent decades, more and more people have turned to civil society, rather than government, to solve their problems.

Reading this book, you may have noticed the many global nongovernmental organizations (NGOs) mentioned. The number of NGOs has increased dramatically, from about 200 in the early 1900s to nearly 5,000 at the end of the twentieth century. Their ranks include Amnesty International, Greenpeace, CARE, the Mennonite Central Committee, and the World Wildlife Fund. Composed of private citizens in more than one country, they focus on such global issues as the environment, poverty, human rights, and peace.

Those frustrated with government's inability to solve global problems insist that centralized governments are not the most effective way to deal with them. Governments, they argue, are simply too far removed from local communities to understand completely the nature of a particular problem and to offer effective solutions. Advocates of civil society are encouraged by the dramatic increase in NGOs. Critics, however, believe local grassroots efforts will be insufficient to solve global problems like nuclear proliferation, ozone depletion, and global warming. They argue that governments are the only actors with sufficient resources to effectively confront these large-scale issues.

Status Quo

Perhaps the most likely scenario is one in which no dramatic changes occur over the next several decades. This is not to say that change will be absent, but that it will be only a gradual continuation of current trends toward globalization in the areas of economic integration, information flow among countries, the importance of nongovernmental actors (including multinational corporations), and cooperation among countries on environmental and other issues. Citizens will continue to pledge their allegiance to countries, not economic blocs; states, not groups of private citizens, will remain the dominant political actors; and short-term domestic interests will prevent states from surrendering their sovereignty.

Unfortunately, the status quo, while making some significant progress in combating various global issues, has made disappointing progress on areas such as climate change, peacemaking, and poverty.

The Future: Reasons for Hope and Concern

As many of the chapters in this book point out, despite the seemingly overwhelming problems the world faces, positive advancements are being made on many fronts. Smallpox has been eliminated. A cooperative effort is under way to address ozone depletion. Women have been increasingly successful in forming effective grassroots movements and making their voices heard. Fewer wars are being fought. Infant mortality rates are declining. The percentage of people in the world living on less than a dollar per day has declined, as has the percentage of malnourished people.

Despite these positive advancements, the world still faces many daunting issues. One such problem is widespread poverty. Several chapters in this book highlight the issue of poverty. Chapter 12 points out that health is directly related to poverty—that the poorer you are, the more likely you are to suffer from disease or malnutrition. Chapters 9 and 10 discuss how the number of children a woman bears will decrease as poverty is alleviated and women gain more control over their lives. Chapters 13, 14, and 15 show that for those who are desperately poor, issues of immediate survival must take precedence over concerns about the environment. Chapter 2 underscores the vast amounts of money spent on military budgets at the expense of social programs such as health and education. And Chapter 11 presents dramatic statistics on the number of children, primarily the impoverished, who die from preventable diseases.

Central to the issue of poverty is the unequal distribution of wealth, which appears to be getting worse. Chapter 8 highlights this problem. At the same time, as those who live in the North know, poverty is not simply a question of North-South relations: there are many pockets of poverty within the wealthy countries, and evidence suggests that the gap between the rich and poor is increasing within countries as well as between them.

Finding a solution to this problem will be difficult. At the domestic level, a country must achieve economic growth before income can be redistributed, but economic growth does not guarantee better income distribution. Chapters 8 and 11 demonstrate that focusing on taxation, education, healthcare, and other such issues is necessary to foster a more favorable distribution of wealth; however, such an approach typically receives little support from those whose wealth would be transferred. The issue becomes even more complex if we confront the global distribution of wealth. Within a single country, the wealthy are often taxed at higher rates to support social programs. An attempt to tax the wealthy countries in order to pay for social programs in poorer countries would meet a great deal of opposition, not only from the wealthy in the North, but also from the middle- and lower-income populations in the North. Historically, voluntary aid from North to South has helped somewhat, but it has been insufficient to seriously address the poverty gap. Also, as suggested in Chapter 7, foreign aid is a relatively insignificant way to foster development.

What Can I Do?

The good news highlighted above is the result of governments, NGOs, and individuals working together to solve problems and alleviate suffering. It is important to recognize that the future has not yet been written, and that various actions undertaken today can have a critical effect on the issues discussed in this book. Assuming you agree that these issues deserve serious attention, whether on grounds of self-interest, a sense of patriotism, a religious view, or a sense of humanitarianism (see Chapter 1 for an elaboration of these perspectives), the practical question remains: What can I do to make a positive difference?

Common suggestions include: write to your government representatives, vote, buy recycled products, and so on. An increasingly widespread option is to form, join, or support an NGO like those mentioned throughout this book. Take, for example, Equal Exchange, Educating for Justice, and

Social Accountability International. Each of these three NGOs, organized by a small group of people, seeks to relieve the suffering of many. All three share a general concern for poverty and inequality. In particular, they focus on workers, who often are subjected to harsh conditions and receive very low wages.

NGO Case Study: Equal Exchange

Throughout the world millions of people are engaged in growing, harvesting, and processing coffee. Most of the coffee growers are individuals who own small plots of land and sell their coffee to middlemen who export the coffee to markets in the North. As a result of the March 2000 coffee price collapse, the price paid to coffee farmers dropped to as low as four cents per pound in October 2001, the lowest price in a century. As a result, many coffee farmers were forced out of business. Although in subsequent years the prices crept back up to more than a dollar per pound, profits are still meager for coffee farmers.

Equal Exchange is one of a few companies working to improve the lives of coffee farmers throughout the world. Since 1991 it has helped small coffee farms by creating a market for certified "Fair Trade" coffee. In the United States, Fair Trade food and beverage products are certified by TransFair USA and bear its seal. The TransFair designation means, among other things, that farmers will receive a fair price for their coffee. Currently, the minimum price paid to farmers for Fair Trade coffee is $1.26 per pound. This is a significantly better price than they would receive otherwise.

Participating coffee farmers must organize themselves into a cooperative (a business owned by the farmers themselves) and make decisions democratically. The cooperatives are also expected to pursue sustainable development and reinvest much of their profits into improving living conditions in their communities (for example, healthcare, housing, education). Coffee farms are monitored by outside groups to ensure that these practices are followed. Companies committed to fair trading for all their imports use a business model that accepts smaller profits, which enables them to free up more resources for paying the higher coffee prices to farmers.

www.fairtradecertified.org

Due to consumer interest, some of the coffee bought by Starbucks and Dunkin' Donuts is fairly traded. Despite the rapid growth of Fair Trade coffee over the past few years, and interest among large companies like Starbucks, Fair Trade coffee represents only 3 percent of US coffee sales. For this kind of market reform to continue, consumers will have to make a conscious decision to ask for and buy Fair Trade coffee, and similar fairly traded products like chocolate, tea, or bananas. For more information, see the Equal Exchange website at http://www.equalexchange.com.

NGO Case Study: Educating for Justice

This case also deals with fair wages and individuals in the North seeking solidarity with workers in the South. The story of Educating for Justice began with Jim Keady, an assistant soccer coach at St. John's University seeking his master's degree in theology. While researching a paper on Nike labor practices in 1998, Keady discovered that Nike was violating several tenets of Catholic social teaching. Given St. John's Catholic identity, Keady saw a troubling contradiction in the process of the university's negotiating a multimillion-dollar endorsement contract with Nike. After months of trying to seek change through the school's administration, he was essentially forced to resign his position because he could not, in good conscience, follow his boss's order to wear Nike-labeled apparel. In a lawsuit, he charged that because he had been forced to become "a walking billboard" for Nike (evidence indicated he was not given the choice of wearing a uniform without the Nike logo), his freedom of speech had been violated.

Although his lawsuit was unsuccessful, Keady has continued to increase awareness of what he feels is an exploitative wage paid to Nike workers in Southeast Asia. Nike has responded that nobody is forcing these people to work in the factories, and that it does in fact pay a reasonable wage, better than many other Indonesian jobs. In reaction to Nike's claims, Keady offered to work in a Nike factory, manufacturing shoes, in order to judge the fairness of the company's wages. Nike refused the offer, so in 2000, Keady and Leslie Kretzu (cofounder of Educating for Justice) set off for Indonesia to try to live on $1.25 per day, the average wage for those working in Indonesian shoe factories at that time. They spent four weeks living with Indonesian workers, and conducted two subsequent research trips. Since then, Keady and Kretzu have traveled across the United States telling the story of impoverished Indonesian workers.

These two activists have targeted Nike because of its high profile and because it is the industry leader. They point out that Nike chairman Phil Knight has a net worth of nearly $7.4 billion, but refuses to pay third world workers a living wage, which in most countries is three to four times the local minimum wage. Other activists criticize Nike for pulling jobs out of factories that have independent trade unions.

In addition to Educating for Justice's public education program, the organization is also trying an innovative strategy to change Nike's policies. It has raised money in order to buy Nike stock. On the surface this seems like a contradictory policy. However, Nike shareholders are allowed to attend the company's annual shareholder meetings. The hope is that Indonesian factory workers themselves, if permitted to attend these meetings, could influence other shareholders to enact policies more favorable to the workers.

It appears that Educating for Justice is making some progress in its crusade to win better conditions for workers in Asia. Nike has recently admitted some mistakes, and has made some minor changes to its policies. To learn more, see the Educating for Justice website at http://www.educatingforjustice.org.

NGO Case Study: Social Accountability International

Social Accountability International, a multistakeholder initiative, was established in 1997. It seeks to ensure that internationally accepted human rights pertaining to child labor, forced labor, health and safety, freedom of association, discrimination, discipline, working hours, and the like, are upheld in workplaces throughout the world.

Social Accountability International's workplace standards are referred to as "SA8000." A company certified as such demonstrates to consumers and other businesses that it provides a humane workplace. A growing number of companies have suppliers and facilities that are SA8000-certified, including the Gap, Avon Products, Tchibo, Otto, Co-op Italia, Eileen Fisher, and Chiquita Brands International. Government programs in several exporting countries prefer or subsidize the use of SA8000 products.

Social Accountability International provides training programs and works to promote SA8000 certification in many countries around the world, including the United States, France, India, Germany, China, Honduras, Brazil, Italy, and Pakistan. As of 2007, over 650,000 people in over sixty countries, spanning North and South America, Europe, Africa, and Asia, are employed in workplaces that are SA8000-certified.

Consumers play an important role in ensuring the success of such voluntary programs. When individuals buy from certified companies, they help to protect workers' rights. Other businesses get the message that, in order to compete, they too need to respect workers' rights.

In addition to working with companies, Social Accountability International collaborates with a network of other NGOs, such as Amnesty International on the human rights front, as well as with labor organizations and governmental agencies. For more information, see the Social Accountability International website at http://www.sa-intl.org.

* * *

Supporting the work of one of these NGOs, or buying products with social justice in mind, is an important action that concerned individuals can take.

There are many, many choices—thousands of small groups, usually NGOs, working to make the world a better place. In fact, dozens, if not hundreds, of NGOs have organized around each issue discussed in this book. For instance, Fair Trade coffee is sold by several other groups, including Pura Vida Coffee, Just Coffee, and Green Mountain Coffee. Fair trade extends beyond coffee to food and clothing as well. Similarly, there are many NGOs, like Educating for Justice and Social Accountability International, that are involved in ensuring that laborers throughout the world work in decent conditions.

Conclusion

Although national and local (and perhaps regional) governments will continue to play important roles, we cannot depend solely on them to solve all of the problems discussed in this book. It is up to each individual to work to create the world he or she prefers. Dramatic results at the global level can be realized if the world's citizens act. After all, if "citizens 'leave it to the experts,' they simply ensure that the expert's values and interests become policy" (Thompson 2003: 2).

It is important to remember that seemingly insurmountable obstacles have been overcome. Many thought that the scourge of smallpox, apartheid in South Africa, and racism in the US South would never be defeated. However, each of these problems was overcome when groups of individuals organized themselves into social movements, small at first, in the name of a better world. We hope readers will seek to learn more about the issues discussed in this book, educate others, and become active members for positive change in their community and the world.

Discussion Questions

1. Of the four future world scenarios discussed in this chapter, which do you think is most likely to emerge? Which do you think is most desirable?
2. Can you think of another possible world scenario?
3. Would a strengthened United Nations be desirable?
4. What do you think are the most serious challenges confronting humanity?
5. What can you do as an individual to make the world a better place?
6. What do you think of the efforts of Equal Exchange, Educating for Justice, and Social Accountability International?

Suggested Readings

Cohen, David, Rosa de la Vega, and Gabrielle Watson (2001) *Advocacy for Social Justice: A Global Action and Reflection Guide.* Bloomfield, CT: Kumarian.

Commission on Global Governance (1995) *Our Global Neighborhood.* New York: Oxford University Press.

Diehl, Paul F. (2001) *The Politics of Global Governance: International Organizations in an Interdependent World.* Boulder: Lynne Rienner.

Karnes, Margaret P., and Karen A. Mingst (2004) *International Organizations: The Politics and Processes of Global Governance.* Boulder: Lynne Rienner.

Kennedy, Paul, Dirk Messner, and Franz Nuscheler, eds. (2001) *Global Trends and Global Governance.* Sterling, VA: Pluto.

Naidoo, Kumi (1999) *Civil Society at the Millennium.* Bloomfield, CT: Kumarian.

United Nations Development Programme (annual) *Human Development Report.* New York: Oxford University Press.

Worldwatch Institute (annual) *State of the World.* New York: Norton.

——— (2006) *Vital Signs 2006–2007: The Trends That Are Shaping Our Future.* New York: Norton.

——— (2007) *Vital Signs 2007–2008.* New York: Norton.

Acronyms

AHA	American Heart Association
AIDS	acquired immunodeficiency syndrome
AOSIS	Alliance of Small Island States
APEC	Asia Pacific Economic Cooperation
B.C.E.	before the common era
BFW	Bread for the World
CFC	chlorofluorocarbon
CGG	Commission on Global Governance
CGS	Citizens for Global Solutions
CIOSC	China Information Office of the State Council
CNN	Cable News Network
DAC	Development Assistance Committee
DAW	Division for the Advancement of Women (United Nations)
DOS	Hashimite Kingdom of Jordan
ECOSOC	Economic and Social Council (United Nations)
ECSC	European Coal and Steel Community
EEA	European Environment Agency
EIA	Environmental Information Administration
EPA	Environmental Protection Agency
EU	European Union
FAO	Food and Agriculture Organization
FCNL	Friends Committee on National Legislation
FDI	foreign direct investment
FOEME	Friends of the Earth Middle East
FPI	foreign portfolio investment

GATT	General Agreement on Tariffs and Trade
GDP	gross domestic product
GNI	gross national income
GNP	gross national product
HCFC	hydrochlorofluorocarbon
HDI	Human Development Index
HIPC	highly indebted poor country
HIV	human immunodeficiency virus
HRW	Human Rights Watch
HSC	Human Security Centre
IAEA	International Atomic Energy Association
IBRD	International Bank for Reconstruction and Development
ICBM	intercontinental ballistic missile
ICC	International Criminal Court
IFAD	International Fund for Agricultural Development
IGO	international governmental organization
IISS	International Institute for Strategic Studies
ILO	International Labour Organization
IMF	International Monetary Fund
INGO	international nongovernmental organization
INSTRAW	International Research and Training Institute for the Advancement of Women (United Nations)
IPU	Inter-Parliamentary Union
IRIN	Integrated Regional Information Networks
IUCN	World Conservation Union
LDC	less-developed countries
LTTE	Liberation Tigers of Tamil Eelam
MDC	more-developed countries
NACEC	North American Commission for Environmental Cooperation
NAFTA	North American Free Trade Agreement
NASA	National Aeronautics and Space Administration
NATO	North Atlantic Treaty Organization
NGO	nongovernmental organization
NIAID	National Institute of Allergy and Infectious Disease
NTB	nontariff barrier
ODA	official development assistance
OPEC	Organization of Petroleum Exporting Countries
PLO	Palestine Liberation Organization
PPP	purchasing power parity
PRB	Population Reference Bureau
RHRC	Reproductive Health Response in Conflict
SIPRI	Stockholm International Peace Research Institute

TNC	transnational corporation
UDHR	Universal Declaration of Human Rights
UN	United Nations
UNACC-SCN	UN Administrative Committee on Coordination–Subcommittee on Nutrition
UNAIDS	Joint UN Programme on HIV/AIDS
UNCED	UN Conference on the Environment and Development
UNCTAD	UN Conference on Trade and Development
UNDESA	UN Department of Economic and Social Affairs
UNDESIPA	UN Department for Economic and Social Information and Policy Analysis
UNDP	UN Development Programme
UNDPI	UN Department of Public Information
UNEP	UN Environment Programme
UNFPA	UN Fund for Population Activities
UNGA	UN General Assembly
UNHCR	UN High Commissioner for Refugees
UNICEF	UN Children's Fund
UNIFEM	UN Development Fund for Women
UNPD	UN Population Division
UNRWA	UN Relief and Works Agency for Palestine Refugees in the Near East
UNSD	UN Statistical Division
UCS	Union of Concerned Scientists
USCC&AN	US Code Congressional and Administrative News
USDA	US Department of Agriculture
USDHS	US Department of Homeland Security
USDOL	US Department of Labor
USDS	US Department of State
USEPA	US Environmental Protection Agency
USGAO	US General Accounting Office
USSCEPW	US Senate Committee on Environmental and Public Works
UV-B	Ultraviolet B
VER	voluntary export restraint
WCED	World Commission on Environment and Development
WCRWC	Women's Commission for Refugee Women and Children
WHO	World Health Organization
WMD	weapons of mass destruction
WMO	World Meteorological Organization
WTO	World Trade Organization
WWF	World Wide Fund for Nature (also known as the World Wildlife Fund)

Bibliography

Abramovitz, Janet N. (2001) "Averting Unnatural Disasters." In Lester R. Brown, Christopher Flavin, and Hilary French, eds., *State of the World 2001.* New York: Norton.

Access to Education (2007) "Barriers to Education." Available at http://www.netaid.org/global_poverty/education/#barriers.

"African Deluge Brings Misery to 1.5m People" (2007) *The Guardian* (September 20). Available at http://www.guardian.co.uk/naturaldisasters/story/0,,2172943,00.html.

AHA (American Heart Association) (2007) *Obesity and Overweight.* Available at http://www.americanheart.org/presenter.jhtml?identifier=4639.

Alcock, Frank (2002) "Bargaining, Uncertainty, and Property Rights in Fisheries." *World Politics* 54, no. 3.

Allison, Graham (2004) *Nuclear Terrorism.* New York: Times Books.

Alterman, Jon B. (2004) "The Information Revolution and the Middle East." In Nora Bensahel and Daniel L. Byman, eds., *The Future Security Environment in the Middle East.* Santa Monica: RAND.

America Federation of Teachers (2007) "In Our Own Backyard: The Hidden Problem of Farmworkers in America." Available at http://www.ourownbackyard.org.

Amery, Hussein A., and Aaron T. Wolf (2000) *Water in the Middle East: A Geography of Peace.* Austin: University of Texas Press.

Amnesty International (2004a) *Amnesty International Report 2004.* London.

——— (2004b) "Sudan: Arming the Perpetrators of Grave Abuses in Darfur" (November). Available at http://www.amnesty.org/en/alfresco_asset/d2b0481e-a3ad-11dc-9d08-f145a8145d2b/afr541392004en.html.

——— (2006) *Amnesty International Report 2006.* London.

——— (2007a) *Amnesty International Report 2007: The State of the World's Human Rights.* New York.

——— (2007b) "Press Release Report 2007: Politics of Fear Creating a Dangerously Divided World." In *Amnesty International Report 2007.* Available at http://thereport.amnesty.org/page/1670/eng.

——— (2007c) "Sudan: Arms Continuing to Fuel Serious Human Rights Violations in Darfur" (May). Available at http://www.amnesty.org/en/alfresco_asset/8e5995f6-a2b8-11dc-8d74-6f45f39984e5/afr540192007en.html.

Angell, Norman (1909) *Europe's Optical Illusion.* London: Simpkin, Marshall, Hamilton, and Kent.

Arga, Adhityani (2007) "Half of Papuans Unaware of AIDS-Indonesian Report." *Reuters* (June 19). Available at http://www.alertnet.org/thenews/newsdesk/jak69156.htm.

Assadourian, Erik (2004) *State of the World 2004.* New York: Norton.

Athanasiou, Tom, and Paul Baer (2002) *Dead Heat: Global Justice and Global Warming.* New York: Seven Stories.

Bailey, Adrian (2005) *Making Population Geography.* New York: Oxford University Press.

Bane, Mary Jo, and Rene Zenteno (2005) "Poverty and Place in North America." Luxembourg Income Study Working Paper no. 418. Available at http://www.lisproject.org/publications/newsletter/2005nov.pdf.

Barber, Benjamin R. (1992) "Jihad vs. McWorld." *Atlantic Monthly* (March).

Barringer, Felicity (2008) "Collapse of Salmon Stocks Endangers Pacific Fishery." *New York Times* (March 13).

Bates, A. K. (1990) *Climate in Crisis.* Summertown, TN: Book Publishing Company.

Bellamy, David (2004) "Global Warming? What a Load of Poppycock!" *Daily Mail* (July 9). Available at http://www.junkscience.com/july04/daily_mail-bellamy.htm.

Betancourt, Antonio (2004) "Mexico: Report on Juárez Killings." *New York Times* (June 4).

BFW (Bread for the World) (1994) *Hunger 1995: Causes of Hunger.* Washington, DC.

——— (1995) "At the Crossroads: The Future of Foreign Aid." Occasional Paper no. 4. Washington, DC.

——— (1997) *Hunger 1998: Hunger in a Global Economy.* Washington, DC.

Bilmes, Linda, and Joseph E. Stiglitz (2006) "The Economic Costs of the Iraq War: An Appraisal Three Years After the Beginning of the Conflict" (January). Paper presented at the annual meeting of the Allied Social Science Association, Boston.

Birdsall, Nancy, Thomas Pinckney, and Richard Sabot (1996) "Why Low Inequality Spurs Growth: Savings and Investment by the Poor." Working Paper no. 327. Washington, DC: Inter-American Development Bank.

Birdsall, Nancy, David Ross, and Richard Sabot (1995) "Inequality and Growth Reconsidered: Lessons from East Asia." *World Bank Economic Review* 9, no. 3.

Bloch, Sidney, and Peter Reddaway (1985) *Psychiatric Terror: How Soviet Psychiatry Is Used to Suppress Discontent.* New York: Basic Books.

Borger, Julian (2007) "Darfur Conflict Heralds Era of Wars Triggered by Climate Change, UN Report Warns." *The Guardian* (June 23).

Boserup, Ester (1970) *Women's Role in Economic Development.* New York: St. Martin's.

Boulding, Elise (1992) *The Underside of History: A View of Women Through Time.* Rev. ed. Newbury Park, CA: Sage.

Boynton, Gary (2005) "Acid Rain Monitoring in the Adirondacks and Catskills." New York: State Department of Environmental Conservation. Available at

http://www.nyserda.org/programs/environment/emep/conference_2005/boynton.pdf.
"Brazil Says Amazon Deforestation Down" (2007) *Associated Press* (August 13). Available at http://www.enn.com/top_stories/article/21778.
Bright, Chris (2000) "Anticipating Environmental Surprise." In Lester R. Brown, Christopher Flavin, and Hilary French, eds., *State of the World 2000.* New York: Norton.
Broad, William J., David E. Sanger, and Thom Shanker (2007) "US Selecting Hybrid Design for Warheads." *New York Times* (January 7).
Brown, Lester R. (1993) "A New Era Unfolds." In Lester R. Brown, ed., *State of the World.* New York: Norton.
——— (1996) "The Acceleration of History." In Lester R. Brown, ed., *State of the World.* New York: Norton.
——— (2006) *Plan B 2.0: Rescuing a Planet Under Stress and Civilization in Trouble.* New York: Norton.
Brysk, Alison (2003) "Globalization and Human Rights: It's a Small World After All." *Phi Kappa Phi Forum* 83, no. 4.
Brzezinski, Matthew (2004) *Fortress America.* New York: Bantam.
Caldwell, Lynton K. (1990) *International Environmental Policy: Emergence and Dimensions.* 2nd ed. Durham, NC: Duke University Press.
"Canada–United States Air Quality Agreement: Progress Report 2006" (2006) Environment Canada. Available at http://www.ec.gc.ca/cleanair-airpur/caol/canus/report/2006canus/toc_e.cfm.
Carmichael, M. (2007) "Diabetes: A 'Disease of Poverty'?" *Newsweek* (July 9).
Carson, Rachel (1962) *Silent Spring.* Boston: Houghton Mifflin.
Cashore, Benjamin, Graeme Auld, and Deanna Newsom (2004) *Governing Through Markets: Forest Certification and the Emergence of Non-State Authority.* New Haven: Yale University Press.
Castles, Stephen, and Mark J. Miller (2003) *The Age of Migration: International Population in the Modern World.* 3rd ed. New York: Guilford.
Cavanaugh, John, et al., eds. (1992) *Trading Freedom: How Free Trade Affects Our Lives, Work, and Environment.* San Francisco: Institute for Food and Development Policy.
CGG (Commission on Global Governance) (1995) *Our Global Neighborhood.* New York: Oxford University Press.
CGS (Citizens for Global Solutions) (2004) "US Policy on the ICC" (December 3). Available at http://www.globalsolutions.org/programs/law_justice/icc/resources/uspolicy.html.
Chang, Anita (2007) "China Defends Record on the Environment." *Durango Herald* (June 28).
Chen, Shaohua, and Martin Ravallion (2004) "How Have the World's Poorest Fared Since the Early 1980s?" Policy research working paper. Washington, DC: World Bank.
CIOSC (China Information Office of the State Council) (2005) "Human Rights Record of the US in 2004" (March 21). Available at http://news.xinhuanet.com/english/2005-03/03/content_2642607.htm.
Cirincione, Joseph, Jon B. Wolfstahl, and Miriam Rajkumar (2002) *Deadly Arsenals: Tracking Weapons of Mass Destruction.* Washington, DC: Carnegie Endowment for International Peace.
CNN (Cable News Network) (2008) "CNN News at a Glance" (March 29). Available at http://www.cnnasiapacific.com/factsheets/?catid=9.

CNN.com/asia (2007) "Report: Myanmar Recruiting Child Soldiers." October 31. Available at http://edition.cnn.com/2007/world/asiapcf/10/30/myanmar.child-soldiers.ap/index.html.

"Coho Salmon Will Stay on Endangered Species List" (2004) *Douglas County News-Review* (October 31). Available at http://www.newsreview.info/article.

Coles, Clifton (2004) "Water Without War." *The Futurist* 38, no. 2.

Coll, Steve, and Susan B. Glasser (2005) "Jihadists Turn the Web into Base of Operations." *Washington Post* (August 7).

Collier, Paul (2007) *The Bottom Billion: Why the Poorest Countries Are Failing and What Can Be Done About It.* New York: Oxford University Press.

Combes, Stacy (2006) "Are We Putting Our Fish in Hot Water?" Gland, Switzerland: World Wide Fund for Nature. Available at http://www.wwf.org.uk/filelibrary/pdf/int_hotfish_ma.pdf.

Conca, Ken, and Geoffrey D. Dabelko (2004) *Green Planet Blues: Environmental Politics from Stockholm to Johannesburg.* 3rd ed. Boulder: Westview.

Consultative Group for the Reconstruction and Transformation of Central America (1999) "Central America After Hurricane Mitch: The Challenge of Turning a Disaster into an Opportunity." Washington, DC: Inter-American Development Bank. Available at http://www.iadb.org/regions/re2/consultative_group/backgrounder6.htm.

Cronon, William (1983) *Changes in the Land: Indians, Colonists, and the Ecology of New England.* New York: Hill and Wang.

Crossette, Barbara (2001) "U.N. Effort to Cut Arms Traffic Meets a US Rebuff." *New York Times* (July 10).

——— (2002) "Washington Is Criticized for Growing Reluctance to Sign Treaties." *New York Times* (April 4).

Davis, Zachary S. (1991) *Non-Proliferation Regimes: A Comparative Analysis of Policies to Control the Spread of Nuclear, Chemical, and Biological Weapons and Missiles.* Washington, DC: Congressional Research Service.

de Waal, Alex (2007) "Is Climate Change the Culprit for Darfur?" Available at http://www.ssrc.org/blog/2007/06/25/is-climate-change-the-culprit-for-darfur.

Deconinck, Stefan (2004) "Israeli Water Policy in a Regional Context of Conflict: Prospects for Sustainable Development for Israelis and Palestinians?" Available at http://waternet.ugent.be/waterpolicy.htm.

Demeny, Paul, and Geoffrey McNicoll (2003) *Encyclopedia of Population.* Vols. 1–2. New York: Macmillan Reference.

"Dimensions of Health" (2007) *Health Science.* Available at http://www.msjc.edu/hs/www/health_overview.htm.

Dobbs, Lou (2003) "President Bush Has Shown Good Faith in Working Through the UN." *CNN Moneyline* (February 5).

Dodge, Robert (1994) "Grappling with GATT." *Dallas Morning News* (August 8).

Dolatyar, Mostafa, and Tim S. Gray (2000) "The Politics of Water Security in the Middle East." *Environmental Politics* 9, no. 3.

DOS (Department of Statistics [Hashimite Kingdom of Jordan]) (2003) *Population Projection 1998–2002.* Available at http://www.dos.gov.jo.

Dugger, Celia W. (2007) "CARE Turns Down Federal Funds for Food Aid" (August 16). Available at http://www.nytimes.com/2007/08/16/world/africa/16food.html?_r=1&oref=slogin

Dunn, Seth (2001a) "Atmospheric Trends." In Worldwatch Institute, *Vital Signs 2001.* New York: Norton.

——— (2001b) "Decarbonizing the Energy Economy." In Lester R. Brown,

Christopher Flavin, and Hilary French, eds., *State of the World 2000.* New York: Norton.
Easterly, William (1999) "The Lost Decades: Explaining Developing Countries' Stagnation, 1980–1998." Policy research working paper. Washington, DC: World Bank.
——— (2006) *The White Man's Burden: Why the West's Efforts to Aid the Rest Have Done So Much Ill and So Little Good.* New York: Penguin.
Eberstadt, Nicholas (2002) "The Future of AIDS." *Foreign Affairs* 81, no. 6 (November–December).
EEA (European Environment Agency) (2004) "Acidification Introduction" (November 13). Available at www.eea.europa.eu/themes/acidification.
EIA (Energy Information Administration) (2007a) "Table 2.1: World Oil Balance, 2003–2007" (December 12). Available at http://www.eia.doe.gov/emeu/ipsr/t21.xls.
——— (2007b) "Top World Oil Producers and Consumers." Available at http://www.eia.doe.gov/emeu/cabs/topworldtables1_2.htm.
——— (2007c) "US Imports by Country of Origin" (November 26). Available at http://tonto.eia.doe.gov/dnav/pet/pet_move_impcus_a2_nus_ep00_im0_mbbl_m.htm.
——— (2007d) "World Proved Reserves of Oil and Natural Gas, Most Recent Estimates" (January 9). Available at http://www.eia.doe.gov/emeu/international/reserves.html.
EPA (Environmental Protection Agency) (2006) "Acid Rain Program Shows Continued Success and High Compliance, EPA Reports" (October 16). Available at http://yosemite.epa.gov/opa/admpress.nsf/6424ac1caa800aab85257359003f5337/eacba68ebb35276685257209 00630205!opendocument.
"Experts Embark on Salmon Project" (2004) *Salem Statesman Journal* (October 27). Available at http://www.energytoday.it/pages_203443.html.
Fallows, James (1993) "How the World Works." *Atlantic Monthly* (December).
FAO (Food and Agriculture Organization) (2006a) "The State of Food Insecurity in the World, 2006." Available at http://www.fao.org/docrep.
——— (2006b) "World Hunger Increasing, FAO Head Calls on World Leaders to Honour Pledges" (October 30). Available at http://www.fao.org/newsroom/en/news/2006/1000433/index.html.
Farer, Tom (2002) "The United Nations and Human Rights: More Than a Whimper, Less Than a Roar." In Richard Pierre Claude and Burns H. Weston, eds., *Human Rights in the World Community.* Philadelphia: University of Pennsylvania Press.
FCNL (Friends Committee on National Legislation) (2005) "'Extraordinary Rendition': Outsourcing Torture" (March 10). Available at http://www.fcnl.org/issues/item_print.php?item_id=1249&issue_id=70.
——— (2007) "President's Bill on Military Tribunals and the War Crimes Act Heads to Congress" (December 18). Available at http://www.fcnl.org/issues/item.php?item_id=2047&issue_id=70.
Filkins, Dexter (2001) "In Fallen Taliban City, a Busy, Busy Barber." *New York Times* (November 13).
Finkel, M. (2007) "Bedlam in the Blood: Malaria." *National Geographic* (July).
Fiske, Edward B. (1993) *Basic Education: Building Block for Global Development.* Washington, DC: Academy for Educational Development.
Flavin, Christopher (1996) "Facing Up to the Risks of Climate Change." In Lester

R. Brown, ed., *State of the World 1996.* New York: Norton.

Flynn, Stephen (2004) "The Neglected Homefront." *Foreign Affairs* 83, no. 5 (September–October).

FOEME (Friends of the Earth Middle East) (2003) "Good Water Makes Good Neighbors." *Middle East Environment Watch* 5, no. 1 (Spring).

Frantz, Douglas, et al. (2004) "The New Face of Al Qaeda: Al Qaeda Seen As Wider Threat." *Los Angeles Times* (September 26).

Fraser, Arvonne S. (2001) "Becoming Human: The Origins and Development of Women's Human Rights." In Marjorie Agosin, ed., *Women, Gender, and Human Rights.* New Brunswick, NJ: Rutgers University Press.

Freedman, Lawrence, and Efraim Karsh (1993) *The Gulf Conflict, 1990–1991: Diplomacy and War in the New World Order.* Princeton: Princeton University Press.

French, Hilary, and Lisa Mastny (2001) "Controlling International Environmental Crime." In Lester R. Brown, Christopher Flavin, and Hilary French, eds., *State of the World 2000.* New York: Norton.

French, Howard W. (2007) "Child Slave Labor Revelations Sweeping China." *International Herald Tribune* (June 15). Available at http://www.iht.com /articles/2007/06/15/news/china.php.

Friedman, Thomas L. (2005) *The World Is Flat: A Brief History of the Twenty-First Century.* New York: Farrar, Straus, and Giroux.

Froehlich, George (2004) "Suzuki Salmon Crusade Draws Critic's Ire." *Business Edge* 1, no. 22 (October 28). Available at http://www.businessedge.ca/ viewnews.html.

Gallagher, Kevin P. (2003) "The Economics of Globalization and Sustainable Development." *Policy Matters: Trade, Environment, and Investment—Cancun and Beyond* no. 11 (September).

Gansler, Jacques S. (2002) "Transforming America's Military: Protecting Cyberspace." Washington, DC: National Defense University. Available at http://www.ndu.edu/inss/books/books_2002/transforming%20americas%20Mil %20-%20ctnsp%20-%20aug%202002/01_toc.htm.

Gellman, Barton (2000) "AIDS Declared Threat to US Security." *Washington Post* (April 30).

Gleick, Peter H. (1994) "Water, War, and Peace in the Middle East." *Environment* 36, no. 3.

Goklany, I. (2006) "Death and Death Rates Due to Extreme Weather Events: Global and US Trends, 1900–2004." Available at http://member.cox/net/igoklany.

Gore, Al (1992) *Earth in the Balance: Ecology and the Human Spirit.* Boston: Houghton Mifflin.

Gribbin, John (1988) *The Hole in the Sky: Man's Threat to the Ozone Layer.* New York: Bantam.

Grimmett, Richard F. (2006) "Conventional Arms Transfers to Developing Nations, 1998–2005" (October 23). Washington, DC: Congressional Research Service. Available at http://www.fas.org/sgp/crs/weapons/rl33696.pdf.

——— (2007) "Conventional Arms Transfers to Developing Nations, 1999–2006" (September 26). Washington, DC: Congressional Research Service.

Gulbrandsen, Lars H. (2004) "Overlapping Public and Private Governance: Can Forest Certification Fill the Gaps in the Global Forest Regime?" *Global Environmental Politics* 4, no. 2.

Hall, K. M. (2007) "Southern Heat Wave Death Toll Reaches 44." *ABC News* (August 19). Available at http://abcnews.go.com/us/wirestory?id=3497524.

Halle, Mark (2002) "Sustainable Development Cools Off: Globalization Demands Summit Take New Approach to Meeting Ecological, Social Goals." *Winnipeg Free Press* (July 29).

Hardin, Garrett (1968) "The Tragedy of the Commons." *Science* 162 (December 13).

Harper, Charles L. (1995) *Environment and Society: Human Perspectives on Environmental Issues.* Upper Saddle River, NJ: Prentice Hall.

Hellman, Christopher (2001) "Military Spending: US vs. the World" (July 24). Washington, DC: Center for Defense Information. Available at http://www.cdi.org/issues/wme/spendersfy03.html.

Hersh, Seymour M. (1994) "The Wild East." *Atlantic Monthly* (July).

Hertz, Noreena (2004) *The Debt Threat: How Debt Is Destroying the Developing World . . . and Threatening Us All.* New York: HarperCollins.

Hijah, Nadia (2003) *The Situation of Children in Iraq: An Assessment Based on the United Nations Convention on the Rights of the Child.* New York: United Nations Children's Fund.

Hochschild, Adam (1999) *King Leopold's Ghost: A Story of Greed, Terror, and Heroism in Colonial Africa.* Boston: Houghton Mifflin.

Hoffman, Bruce (1997) "Viewpoint: Terrorism and WMD: Some Preliminary Hypotheses." *Non-Proliferation Review* (Spring–Summer).

Hogg, Chris (2004) "Storm Across the Taiwan Strait." *BBC News* (June 22). Available at http://news.bbc.co.uk.

Holm, Hans-Henrik, and Georg Sørensen, eds. (1995) *Whose World Order? Uneven Globalization and the End of the Cold War.* Boulder: Westview.

Homer-Dixon, Thomas (1991) "On the Threshold: Environmental Changes as Causes of Acute Conflict." *International Security* 16, no. 2 (Fall). Available at http://www.library.utoronto.ca/pcs/thresh/thresh1.htm.

——— (1994) "Environmental Scarcities and Violent Conflict: Evidence from Cases." *International Security* 19, no. 1 (Summer). Available at http://www.library.utoronto.ca/pcs/evidence/evid1.htm.

Howden, Daniel (2006) "Moratorium on New Soya Crops Wins Reprieve for Rainforest." *The Independent* (July 26). Available at http://news.independent.co.uk/world/americas/article1197277.ece.

HRW (Human Rights Watch) (2004a) "Trafficking" (October 28). Available at http://www.hrw.org/women/trafficking.html.

——— (2004b) "Women's Rights" (November 21). Available at http://hrw.org/women.

——— (2007) "Child Labor." Available at http://www.hrw.org/about/projects/crd/child-labor.htm.

HSC (Human Security Centre) (2006) *Human Security Report 2005: War and Peace in the 21st Century.* New York: Oxford University Press.

Huntington, Samuel P. (1996) "The West: Unique, Not Universal." *Foreign Affairs* 75, no. 6.

——— (1998) *The Clash of Civilizations and the Remaking of World Order.* New York: Simon and Schuster.

Huppert, Daniel D. (1995) "Understanding the US-Canada Salmon Wars: Why the Pacific Salmon Treaty Has Not Brought Peace." Seattle: University of Washington, School of Marine Affairs. Available at http://www.wsg.washington.edu/salmon/huppertreport.html.

ICC (International Criminal Court) (2004) "Historical Introduction" (December 1). Available at http://www.icc-cpi.int/about/ataglance/history.html.

IFAD (International Fund for Agricultural Development) (2007) "Sending Money Home." Available at http://www.ifad.org/events/remittances/maps/brochure.pdf.

IISS (International Institute for Strategic Studies) (2005) *The Military Balance, 2005–2006.* Oxford: Oxford University Press, 2005.

ILO (International Labour Organization) (1993) *World Labour Report.* Geneva.

——— (2006) "The End of Child Labour: Within Reach." Available at http://www.ilo.org/dyn/declaris/declarationweb.download_blob?var_document id=6176.

India Committee of the Netherlands (2007) "Child Bondage Continues in Indian Cotton Supply Chain" (September 25). Available at http://www.indianet.nl/pdf/childbondagecotton.pdf.

IPU (Inter-Parliamentary Union) (2007) "Women in Parliament: World Classification" (April). Available at http://www.ipu.org/wmn-e/world.htm.

IRIN (Integrated Regional Information Networks) (2007) "Kenya: Climate Change and Malaria in Nairobi" (July 31). Available at http://www.irinnews.org/report.aspx?reportid=73501.

IUCN (World Conservation Union), UNEP (United Nations Environment Programme), and WWF (World Wide Fund for Nature) (1991) *Caring for the Earth.* Gland, Switzerland.

Izenberg, Dan (1997) "An Insider's View of the Jordan Rift." *Jerusalem Post* (May 9).

Jackson, Lisa R., and Jeanette E. Ward (1999) "Aboriginal Health: Why Is Reconciliation Necessary?" *Medical Journal of Australia* (May 3). Available at http://www.mja.com.au/public/issues/may3/jackson/jackson.html#refbody12.

Johnson, Bryan (2006) "Antarctic Ozone Hole Continues to Intensify Late into 2006 Season." Washington, DC: US Department of Commerce, National Oceanic and Atmospheric Administration. Available at http://hotitems.oar.noaa.gov/storyprint_org.php?sid=3752.

Johnston, Eric (1958) "A Key to the Future of the Mideast." *New York Times Magazine* (October 19).

Jubilee USA (2007) "Country Reports." Available at http://www.jubileeusa.org/?id=112.

Kaimowitz, David, Benoit Mertens, Sven Wunder, and Pablo Pacheco (2003) "Hamburger Connection Fuels Amazon Destruction: Cattle Ranching and Deforestation in Brazil's Amazon." Bogor Barat, Indonesia: Center for International Forestry Research. Available at http://www.cifor.cgiar.org/publications.

Kapsos, Steven (2007) "World and Regional Trends in Labour Force Participation: Methodologies and Key Results." Geneva: International Labour Office, Economic and Labour Market Paper no. 2007/1. Available at http://www.ilo.org/public/english/employment/strat/download/kps02.pdf.

Kawachi, Ichiro, and Sarah Wamala (2007) *Globalization and Health.* New York: Oxford University Press.

Kay, Sean (2006) *Global Security in the Twenty-First Century: The Quest for Power and the Search for Peace.* Lanham: Rowman and Littlefield.

Kent, George (1995) *Children in the International Political Economy.* New York: St. Martin's.

——— (2005) *Freedom from Want: The Human Right to Adequate Food.* Washington, DC: Georgetown University Press.

——— (2008) *Global Obligations for the Right to Food.* Lanham: Rowman and Littlefield.

Keohane, Robert O., and Joseph S. Nye (2001) *Power and Interdependence.* 3rd ed. New York: Longman.

Kerr, Richard A. (1989) "Greenhouse Skeptic Out in the Cold." *Science* (December).

——— (2000) "Can the Kyoto Climate Treaty Be Saved from Itself?" *Science* (November).

Khator, Renu (1991) *Environment, Development, and Politics in India.* Lanham: Rowman and Littlefield.

Ki-moon, Ban (2007) "A Climate Culprit in Darfur." *Washington Post* (June 16). Available at http://www.washingtonpost.com/wpdyn/content/article/2007/06/15/ar2007061501857.html.

Kiffer, Dave (2007) "Alaska/Canada Salmon 'War' Was 10 Years Ago" (July 19). Available at http://www.sitnews.us/kiffer/salmonwars/071907_salmonwars.html.

Kirby, Alex. (2003) "Why World's Taps Are Running Dry." *BBC News.* Available at http://news.bbc.co.uk.

Kirby, Peadar (2006) *Vulnerability and Violence: The Impact of Globalization.* Ann Arbor, MI: Pluto.

Klare, Michael (1999) "The Kalashnikov Age." *Bulletin of Atomic Scientists* (January–February).

Knight, Sunny (2000) "Salmon Recovery and the Pacific Salmon Treaty." *Ecology Law Quarterly* 27, no. 3.

Koh, Tommy B. B. (1997) "Five Years After Rio and Fifteen Years After Montego Bay: Some Personal Reflections." *Environmental Policy and Law* 27, no. 4.

Kolbert, Elizabeth (2006) *Field Notes from a Catastrophe: Man, Nature, and Climate Change.* New York: Bloomsbury.

Koop, C. E., C. E. Pearson, and M. R. Schwarz, eds. (2002) *Critical Issues in Global Health.* San Francisco: Jossey-Bass.

Koplow, D. (2003) *Smallpox: The Fight to Eradicate a Global Scourge.* Berkeley: University of California Press.

Lagoutte, Stephanie, Hans-Otto Sano, and Peter Scharff Smith, eds. (2007) *Human Rights in Turmoil: Facing Threats, Consolidating Achievements.* Leiden: Koninklijke Brill NV.

Lamar, B. (1991) "Life Under the Ozone Hole: In Chile, the Mystery of the Bug-Eyed Bunnies." *Newsweek* (December 9).

Larssen, Thorjørn, et al. (2007) "Acid Rain in China: Rapid Industrialization Has Put Citizens and Ecosystems at Risk." *Environmental Science & Technology.* Available at http://pubs.acs.org/subscribe/journals/esthag/40/i02/html/011506feature_larssen.html.

Laurance, Edward J. (1992) *The International Trade in Arms.* New York: Lexington Books.

Leggett, Jeremy. (1990) "The Nature of the Greenhouse Threat." In Jeremy Leggett, ed., *Global Warming: The Greenpeace Report.* New York: Oxford University Press.

Levy, Gideon (2007) "Twilight Zone: The Children of 5767." *Haaretz.* Available at http://www.haaretz.com/hasen/spages/907708.html.

Lieber, Keir A., and Daryl G. Press (2006) "The End of MAD? The Nuclear Dimension of US Primacy." *International Security* 30, no. 4 (Spring).

Lim, Lousia (2007) "China's Coal-Fueled Boom Has Costs" (May 2). Available at http://www.npr.org/templates/story/story.php?storyid=9947668.

Lindzen, R. (1993) "Absence of Scientific Basis." *Research and Exploration* (Spring).

Malthus, T. R. (1826) *An Essay on the Principle of Population.* 6th ed. London: Reeves and Turner.

Mamdani, Mahmood (1996) *Citizen and Subject: Contemporary Africa and the Legacy of Late Colonialism.* Princeton: Princeton University Press.

Martin, Claude (2002) "The Future of Multilateralism." Gland, Switzerland: World Wide Fund for Nature International. Available at http://www.panda.org/news_facts/newsroom/opinions/news.cfm?unewsid=2671.

McCormick, John (1989) *Reclaiming Paradise: The Global Environmental Movement.* Bloomington: Indiana University Press.

McGwire, Michael (1994) "Is There a Future for Nuclear Weapons?" *International Affairs* 70, no. 2.

McKibben, Bill (1989) *The End of Nature.* New York: Random House.

McKinney, M. L., and R. M. Schoch (1996) *Environmental Science: Systems and Solutions.* Minneapolis: West Publishing.

McNaugher, Thomas L. (1990) "Ballistic Missiles and Chemical Weapons." *International Security* 15, no. 2.

McNeely, Jeffrey A. (2001) "The Great Reshuffling: How Alien Species Help Feed the Global Economy." In O. T. Sandlund, P. J. Schei, and A. Viken, eds., *Invasive Species and Biodiversity Management.* Dordrecht: Kluwer Academic.

McNeil, Donald G., Jr. (2007a) "Child Mortality at Record Low; Further Drop Seen." *New York Times* (December 4). Available at http://www.nytimes.com/2007/09/13/world/13child.html?ex=1347336000&en=1f0d2cd7f97947ce&ei=5088&partner=rssnyt&emc=rss.

——— (2007b) "U.N. to Say It Overstated H.I.V. Cases by Millions." *New York Times* (November 20). Available at http://www.nytimes.com/2007/11/20/world/20aids.html?ex=11962260000&en=0b7acd7539a53bef&ei=5070&emc=etal.

Meadows, D. H., D. L. Meadows, and J. Rander (1992) *Beyond the Limits: Confronting Global Collapse: Envisioning a Sustainable Future.* Mills, VT: Chelsea Green.

Michaels, P. (1992) *Sound and Fury: Science and Politics of Global Warming.* Washington, DC: Cato Institute.

Milanovic, Branko (2006) "Global Income Inequality: What It Is and Why It Matters?" DESA Working Paper no. 26. Washington, DC: World Bank.

Miller, Marian A. L. (1995) *The Third World in Global Environmental Politics.* Boulder: Lynne Rienner.

Mitchell, Anthony (2004) "U.N.: HIV/AIDS Fuels Tuberculosis Crisis." *Associated Press* (September 21).

Mitchell, Kathleen T. (2005) "Children Born from Rape: Overlooked Victims of Human Rights Violations in Conflict Settings." Available at http://www.jhsph.edu/academics/degreeprograms/mph/_pdf/capstone_symposium_program_2005.pdf.

Mohaiemen, Naeem (2006) "Between the Devil and the Deep Blue." *Himal Southasian.* Available at http://www.himalmag.com/2006/july/photofeature.htm.

Molina, Mario J., and F. S. Rowland (1974) "The Stratospheric Sink for Chlorofluoromethanes: Chlorine Atom-Catalyzed Destruction of Ozone." *Nature* 249 (June 28): 810–812.

Monbiot, George (2007a) *Heat: How to Stop the Planet from Burning.* Cambridge, MA: South End Press.

——— (2007b) "There Is Climate Change Censorship and It's the Deniers Who

Dish It Out." *The Guardian* (April 10). Available at http://www.truthout.org/docs_2006/printer_041007P.shtml.

Moon, Bruce E. (1998) "Exports, Outward-Oriented Development, and Economic Growth." *Political Research Quarterly* (March).

——— (2000) *Dilemmas of International Trade.* 2nd ed. Boulder: Westview.

Morris, Mary E. (1996) "Water and Conflict in the Middle East: Threats and Opportunities." *Studies in Conflict & Terrorism* 20, nos. 1–13.

Mowlana, Hamid (1995) "The Communications Paradox." *Bulletin of Atomic Scientists* 51, no. 4.

Mueller, John (2002) "Harbinger or Aberration? A 9/11 Provocation." *National Interest* no. 68.

Munro, David A. (1995) "Sustainability: Rhetoric or Reality?" In T. C. Trzyna, ed., *A Sustainable World.* London: Earthscan.

Murray, Christopher J. L., Thomas Laakso, Kenji Shibuya, Kenneth Hill, and Alan D. Lopez (2007) "Can We Achieve Millennium Development Goal 4? New Analysis of Country Trends and Forecasts of Under-5 Mortality to 2015." *Lancet* 370 (September 22). Available at http://www.thelancet.com/journals/lancet/article/piis0140673607614780/abstract.

NACEC (North American Commission for Environmental Cooperation) (2001) "North American Trade and Transportation Corridors: Environmental Impacts and Mitigation Strategies." Montreal.

Narain, Sunita (2002) "The World After." *Down to Earth* 11, no. 9 (September 30).

National Association of Homebuilders (2007) *Housing Facts, Figures, and Trends, May 2007.* Washington, DC. Available at http://www.nahb.org/publication_details.aspx?publicationid=2028.

National Research Council (1986) *Population Growth and Economic Development: Policy Questions.* Washington, DC: National Academy Press.

Newbold, K. B. (2002) *6 Billion Plus: Population Issues for the 21st Century.* Boulder: Rowman and Littlefield.

NIAID (National Institute of Allergy and Infectious Disease) (2007) "Treatment of HIV Infection" (November). Available at http://www.niaid.nih.gov/factsheets/treat-hiv.htm.

Nickerson, Colin (1997) "Canadian 'Fish War' Catches US Tourists on an Alaskan Ferry." *Boston Globe* (July 22).

Nincic, Miroslav (1982) *The Arms Race: The Political Economy of Military Growth.* New York: Praeger.

Nonow, Bob (2004) "Global Networks: Emerging Constraints on Strategy." *Defense Horizons*, no. 43.

"Nor Any Drop to Drink" (2001) *Lancet* 358 (September 29).

Nye, Joseph S. (2002) *The Paradox of American Power.* Oxford: Oxford University Press.

O'Byrne, Darren J. (2003) *Human Rights: An Introduction.* London: Pearson Education.

"Palestine Fact Sheet" (2004) *Palestine Monitor* (October 22). Available at http://www.palestinemonitor.org/factsheet/israeli_settlements_on_occupied.htm.

Pape, Robert A. (2003)"The Strategic Logic of Suicide Terrorism." *American Political Science Review* 97, no. 3.

——— (2005) "Soft Balancing Against the United States." *International Security* 30, no. 1.

Parker, Richard (2002) "From Conquistadors to Corporations." *Sojourners Magazine* (May–June).

Paul, T. V. (2000) *Power Versus Prudence: Why Nations Forgo Nuclear Weapons.* Montreal: McGill-Queen's University Press.

Perlin, Michael (2006) "Human Rights Abuses in Mental Institutions Common Worldwide." *Virginia Law* (February 27). Available at http://www.law.virginia.edu/html/news/2006_spr/perlin.htm.

Peterson, Scott (2000) "Turkey's Plan for Mideast Peace." *Christian Science Monitor* 92, no. 102 (April 18).

Pew Research Center for the People and the Press (2003) "Views of a Changing World 2003" (June 3). Available at http://people-press.org/reports/display.php3?reportid=185.

——— (2004) "A Year After Iraq War" (March 16). Available at http://people-press.org/reports/display.php3?reportid=206.

Pickering, Kevin T., and Lewis A. Owen (1994) *An Introduction to Global Environmental Issues.* London: Routledge.

Pincus, Walter (2001) "Panel Urges $30 Billion to Secure Russian Nuclear Arms." *Washington Post* (January 11).

Polanyi, Karl (1944) *The Great Transformation.* New York: Farrar and Reinhart.

Pomfret, John (1999) "China Ponders New Rules of 'Unrestricted War.'" *Washington Post* (August 8).

Pomfret, Richard (1988) *Unequal Trade.* Oxford: Basil Blackwell.

Postel, Sandra (1993) "The Politics of Water." *Worldwatch* 6, no. 4.

——— (1994) "Carrying Capacity: Earth's Bottom Line." In Lester R. Brown, ed., *State of the World 1994.* New York: Norton.

PRB (Population Reference Bureau) (2004) "World Population Data Sheet" (September 28). Available at http://www.prb.org.

——— (2007) "Data by Geography: United States—Summary." Available at http://www.prb.org/datafinder/geography/summary.aspx?region=70®ion_type=3.

Ralston, Keith, and Duncan A. Stacey (1997–1998) "Salmon Wars and the Crises of the Nineties." *Beaver* 77, no. 6 (December–January).

Ravallion, Martin (2001) "Growth, Inequality, and Poverty: Looking Beyond Averages." Working Paper no. 2558. Washington DC: World Bank.

Ravallion, Martin, and Shaohua Chen (1997) "What Can New Survey Data Tell Us About Recent Changes in Distribution and Poverty?" *World Bank Economic Review* 11, no. 2 (May).

Ray, D. L., and L. Guzzo (1992) *Trashing the Planet.* New York: Harper Perennial.

Redclift, Michael (1987) *Sustainable Development: Exploring the Contradictions.* New York: Methuen.

"Report Recommends Cutting Fishing Fleet" (2004) *Juneau Empire* (October 29). Available at http://www.juneauempire.com/stories.

RHRC (Reproductive Health Response in Conflict) (2007a) "Increasing Access to Quality Reproductive Health for Refugee and Internally Displaced People Worldwide."

Robertson, Geoffrey (2000) *Crimes Against Humanity: The Struggle for Global Justice.* New York: Norton.

Rodney, Walter (1983) *How Europe Underdeveloped Africa.* Washington, DC: Howard University Press.

Rogers, Peter P., Kazi F. Jalal, and John A. Boyd (2006) *An Introduction to Sustainable Development.* Cambridge: Harvard University Press.

Rohde, David, and David E. Sanger (2004) "Key Pakistani Is Said to Admit Atom Transfers." *New York Times* (February 4).

Sachs, Jeffrey (2005) *The End of Poverty: Economic Possibilities for Our Time.* New York: Penguin.

Sagan, Scott D. (1986) "1914 Revisited: Allies, Offense, and Instability." *International Security* 11, no. 2.

Schumacher, E. F. (1993) *Small Is Beautiful: Economics As If People Mattered.* San Francisco: Harper and Row.

Seis, Mark (1996) "An Eco-Critical Criminological Analysis of the 1990 Clean Air Act." PhD diss., Indiana University of Pennsylvania.

Shanker, Thom (2004) "US and Russia Still Dominate Arms Market, but World Total Falls." *New York Times* (August 30).

Sharder, Katherine Pfleger (2005) "Spy Imagery Agency Watching Inside US." *Associated Press* (September 27).

Sheehan, Molly O. (2003) "Atmosphere Trends." In Worldwatch Institute, *Vital Signs 2003.* New York: Norton.

Sierra Club (2007a) "The Millennium Development Goals Report: 2007." Available at http://www.un.org/millenniumgoals/pdf/mdg2007.pdf.

——— (2007b) "Sustainable Consumption" Available at http://www.sierraclub.org/sustainable_consumption.

Singh, Susheela, ed. (2003) *Adding It Up: The Benefits of Investing in Sexual and Reproductive Health.* New York: Allen Guttmacher Institute and UNFPA.

SIPRI (Stockholm International Peace Research Institute) (2006) *SIPRI Yearbook 2006: Armaments, Disarmament, and International Security.* London: Oxford University Press.

Sivard, Ruth Leger (1991) *World Military and Social Expenditures 1991.* Washington, DC: World Priorities.

——— (1996) *World Military and Social Expenditures 1996.* Washington, DC: World Priorities.

Sklar, Holly, ed. (2004) *Putting Dignity and Rights at the Heart of the Global Economy.* Philadelphia: American Friends Service Committee, Working Party on Global Economics.

Slanina, Sjaak (2006) "Impact and Abatement of Acid Deposition and Eutrophication." *Encyclopedia of Earth* (December 21). Available at http://www.ecoearth.org/article/impact_and_abatement_of_acid_deposition_and_eutrophication.

Small, Melvin, and J. David Singer (1982) *Resort to Arms: International and Civil Wars, 1816–1980.* Beverly Hills: Sage.

Smeeding, Timothy (2005) "Poor People in Rich Nations: The United States in Comparative Perspective." Luxembourg Income Study Working Paper no. 419. Available at http://www.lisproject.org/php/wp/wp.php#wp.

Smith, Adam (1910) *An Inquiry into the Nature and Causes of the Wealth of Nations.* London: Dutton.

"Sold As Slaves, Children Are Cheaper Than Animals" (2007) *Asia News* (April 6). Available at http://www.asianews.it/index.php?l=en&art=8946&size=a.

Sonfield, Adam (2006) "Working to Eliminate the World's Unmet Need for Contraception." *Guttmacher Policy Review* 9, no. 1 (Winter).

Speth, James Gustave (2003) "Perspectives on the Johannesburg Summit." *Environment* 45, no. 1.

Staff Writers (2006) "Acid Rain in China Threatening Food Chain." *TerraDaily* (April 6). Available at http://www.terradaily.com/reports/a060806032030.24x2b5ht.html.

Steinberg, Gerald M. (1994) "US Non-Proliferation Policy: Global Regimes and Regional Realities." *Contemporary Security Policy* 15, no. 3.

Stephens, Bret (2002) "Israel, Jordan Announce Project to Save Dead Sea." *Jerusalem Post* (September 2).

Stern, Jessica (1999) *The Ultimate Terrorists.* Cambridge: Harvard University Press.

Stevenson, Richard W., and Janet Elder (2004) "Support for War Is Down Sharply, Poll Concludes." *New York Times* (April 29).

Stiglitz, Joseph E. (2002) *Globalization and Its Discontents.* New York: Norton.

Stipp, David (2004) "Climate Change a National Security Threat." *Fortune* (January 26).

Switzer, Jacqueline Vaughn (1994) *Environmental Politics: Domestic and Global Dimensions.* New York: St. Martin's.

Tannenwald, Nina (2005) "Stigmatizing the Bomb: Origins of the Nuclear Taboo." *International Security* 29, no. 4 (Spring).

Thompson, J. Milburn (2003) *Justice and Peace: A Christian Primer.* 2nd ed. Maryknoll, NY: Orbis.

Townsend, Mark, and Paul Harris (2004) "Now the Pentagon Tells Bush: Climate Change Will Destroy US." Available at http://www.commondreams.org/cgi-bin/print.cgi?file=/headlines04/0222-01.

Tucker, Jonathan B. (2000) "Chemical and Biological Terrorism: How Real a Threat?" *Current History* (April).

Tujan, Antonio, Audrey Gaughran, and Howard Mollett (2004) "Development and the 'Global War on Terror.'" *Race and Class* 46, no. 1.

Tumulty, Brian (1994) "US Industry Confronts Cost of Implementing GATT." *Gannett News Service* (July 18).

"£20k Reward Aims to Stop Female Circumcision" (2007) *The Independent* (July 11). Available at http://news.independent.co.uk/uk/crime/article2753969.ece.

UN (United Nations) (1972) *Report of the United Nations Conference on the Human Environment* (June 5–16). UN Doc.A/Conf. 48/Inf. 5.

——— (1988) *Human Rights: Questions and Answers.* New York.

——— (1992) *Agenda 21: Programme of Action for Sustainable Development.* New York.

——— (1998) "United Nations Press Briefing on Kyoto Protocol" (March 16). Available at the United Nations Framework Convention on Climate Change website: http://www.unfccc.int/resource/kpstats.pdf.

——— (2002) "Report of the World Summit on Sustainable Development; /CONF.199/20." Available at http://www.johannesburgsummit.org/html /documents/summit_docs/131302_wssd_report_reissued.pdf.

——— (2007) *The Millennium Development Goals Report, 2007.* New York. Available at http://www.un.org/millenniumgoals/pdf/mdg2007.pdf

——— (2008) *The State of the World's Children, 2008* (March 25). Available at http://www.unicef.org/sowc08/docs/figure-1.4.pdf.

"UN High Commissioner for Human Rights Louise Arbour: The US War on Terror Is Constantly Being Used by Other Countries As Justification for Torture and Other Violations of International Human Rights" (2007) *Democracy Now* (September 7). Available at http://www.democracynow.org/article.pl?sid =07/09/07/1349246.

UN Millennium Project (2005) "Investing in Development: A Practical Plan to Achieve the Millennium Development Goals" Available at http://www.un millenniumproject.org/reports/index_overview.htm.

UNACC-SCN (United Nations Administrative Committee on Coordination–Subcommittee on Nutrition) (1997) "Update on the Nutrition Situation 1996: Summary of Results for the Third Report on the World Nutrition Situation." Geneva.

UNAIDS (Joint United Nations Programme on HIV/AIDS) (2004) "2004 Report on the Global AIDS Epidemic." New York.

——— (2007) "UNAIDS Policy and Practice." Available at http://www.unaids.org/en/policyandpractice/default.asp.

UNCTAD (United Nations Conference on Trade and Development) (2006) *World Investment Report 2006*. Available at http://www.unctad.org/en/docs/wir2006_en.pdf.

——— (2007a) "Foreign Direct Investment Surged Again in 2006." Available at www.unctad.org/en/docs/iteiiamisc20072_en.pdf.

——— (2007b) "Country Fact Sheets." Available at http://www.unctad.org/templates/page.asp?intitemid=3198&lang=1.

UNDESA (United Nations Department of Economic and Social Affairs) (2002) "The Johannesburg Summit Test: What Will Change" (September 25). Available at http://www.un.org/summit/html/whats_new/feature?story41.html.

UNDESIPA (United Nations Department for Economic and Social Information and Policy Analysis) (1995) "Global Population Policy Data Base." New York.

UNDP (United Nations Development Programme) (1998–2006) *Human Development Report* (annual). New York: Oxford University Press.

——— (2007) "The Millennium Development Goals Report 2007" (July 2). Available at http://www.un.org/millenniumgoals/pdf/mdg2007-globalpr.pdf.

UNDPI (United Nations Department of Public Information) (1995) *World Urbanization Prospects: The 1994 Revision*. New York.

——— (1997) "Earth Summit Review Ends with Few Commitments" (July). Press release.

——— (2002) "Press Summary of the Secretary-General's Report on Implementing Agenda 21" (January). Press release.

UNEP (United Nations Environment Programme) (2007) "Environmental Degradation Triggering Tensions and Conflict in Sudan" (June 22). Press release. Available at http://www.unep.org/documents.multilingual/default.asp?documentid=512&articleid=5621&l=en.

UNFPA (United Nations Fund for Population Assistance) (2004) "Country Profiles." Available at http://www.unfpa.org/profile.

UNGA (United Nations General Assembly) (2001) *We the Children: End-Decade Review of the Follow-Up to the World Summit for Children—Report of the Secretary General*. UN Doc. A/S-27/3. Available at http://www.unicef.org/specialsession/documentation/index.html.

UNHCR (United Nations High Commissioner for Refugees) (1995) *The State of the World's Refugees 1995*. Oxford: Oxford University Press.

——— (2006) *2006 UNHCR Population Statistics (Provisional)*. Population Data Unit. Available at http://www.unhcr.ch.

UNICEF (UN Children's Fund) (1996) "Secretary-General Reports Big Progress for Children." Press release. UN Doc. CF/Doc/PR/1996-24. New York.

——— (2005) "Water, Environment, and Sanitation" (March 23). Available at http://www.unicef.org/wes/index_25637.html.

——— (2007a) "Facts on Children." Available at http://www.unicef.org/media/media_fastfacts.html.

——— (2007b) *The State of the World's Children 2007*. Available at http://www.unicef.org/sowc07/docs/sowc07.pdf.

UNPD (United Nations Population Division) (2001) *World Population Prospects: The 2000 Revision*. New York.

——— (2003) *World Population Prospects: The 2002 Revision—Highlights*. New York.

——— (2004) *World Population Policies 2003*. New York.

——— (2007) *World Population Prospects: The 2006 Revision—Highlights*. New York.

——— (2008) *World Urbanization Prospects: The 2007 Revision—Highlights*. New York.

US Census Bureau (2004) "Poverty Thresholds 2003." Available at http://www.census.gov/hhes/poverty/threshld/thresh03.html.

——— (2008a) "Factfinder" (March 31). Available at http://factfinder.census.gov/servlet/safffacts?_event=search&geo_id=&_geocontext=&_street=&_county=wilmington&_citytown=wilmington&_state=04000US39&_zip=&_lang=en&_sse=on&pctxt=fph&pgsl=010&show_2003_tab=&redirect=y.

——— (2008b) "World Pop Clock Projection" (March 29). Available at http://www.census.gov/ipc/www/popclockworld.html.

UCS (Union of Concerned Scientists) (2004) "Global Warming." Available at http://www.uscusa.org/global_warming.

USCC&AN (US Code Congressional and Administrative News) (1991) 101st Congress, 2nd sess. Legislative History Clean Air Act Amendments, January, no. 10D. Minneapolis: West Publishing.

USDA (US Department of Agriculture) (2007) "Household Food Security in the United States, 2006" (November). Available at http://www.ers.usda.gov/publications/err49.

USDHS (US Department of Homeland Security) (2003) *2002 Yearbook of Immigration Statistics*. Washington, DC: US Government Printing Office.

——— (2007) *2006 Yearbook of Immigration Statistics*. Washington, DC: US Government Printing Office.

USDOL (US Department of Labor) (2007) "Youth Rules." Available at http://www.youthrules.dol.gov.

USDOT (US Department of Transportation) (2007) "Automobile Profile." In *National Transportation Statistics*. Washington, DC: Bureau of Transportation Statistics. Available at http://www.bts.gov/publications/national_transportation_statistics/#appendix_a

USDS (US Department of State) (2002) "US-Canada Yukon River Salmon Agreement Signed" (December 4). Press statement.

USEPA (US Environmental Protection Agency) (2003) "National Air Quality and Emissions Trends Report: 2003 Special Studies Edition" (March 30). Available at http://www.epa.gov/air/airtrends/aqtrnd03.

USSCEPW (US Senate Committee on Environmental and Public Works) (1993) "Three Years Later: Report Card on the 1990 Clean Air Act Amendments, November 15." Washington, DC: US Government Printing Office.

Valente, C. M., and W. D. Valente (1995) *Introduction to Environmental Law and Policy: Protecting the Environment Through Law*. Minneapolis: West Publishing.

Vergano, Dan (2003) "Water Shortages Will Leave World in Dire Straits" (January 1). Available at http://www.usatoday.com/news/nation/2003-01-26-water-usat_x.htm.

Vidal, John, and David Adam (2007) "China Overtakes US as World's Biggest CO_2

Emitter" (June 19). Available at http://www.guardian.co.uk/environment/2007/jun/19/china.usnews.

Võ, X. H. (1994) *Oil, the Persian Gulf States, and the United States.* Westport: Praeger.

Warner, Sir Frederick (1991) "The Environmental Consequences of the Gulf War." *Environment* 33, no. 5.

WCED (World Commission on Environment and Development) (1987) *Our Common Future.* Oxford: Oxford University Press.

WCRWC (Women's Commission for Refugee Women and Children) (2007) "The Minimum Initial Service Package (MISP) for Reproductive Health in Crisis Situations: A Distance Learning Module." Available at http://www.rhrc.org/misp/english/index.html.

Weatherhead, Betsy, and Signe Bech Anderson (2006) "Ozone Layer Slowly Healing Itself." *Times of India* (May 4). Available at http://timesofindia.indiatimes.com/articleshow/msid-1516464-1.cms.

Weeks, John R. (2005) *Population: Introduction to Concepts and Issues.* 9th ed. Belmont, CA: Wadsworth.

White House (2002a) "National Strategy to Combat Weapons of Mass Destruction" (December). Available at http://www.whitehouse.gov/news/releases/2002/12/wmdstrategy.pdf.

——— (2002b) "President Bush Delivers Graduation Speech at West Point." Remarks by the president at the 2002 graduation exercise of the US Military Academy, West Point, NY. Available at http://www.whitehouse.gov/news/releases/2002/06/print/20020601-3.html.

WHO (World Health Organization) (2001a) "Fact Sheet 218." Available at http://www.who.int/mental_health/policy/fact_sheet_mnh_hr_leg_2105.pdf.

——— (2001b) "Mental Health, Human Rights, and Legislation: WHO's Framework." Available at http://www.who.int/mediacentre/factsheets/fs218/en.

——— (2005) "Climate and Health." Available at http://www.who.int/globalchange/news/fsclimandhealth/en/index.html.

——— (2006a) "Nutrition." Available at http://www.who.int/topics/nutrition/en.

——— (2006b) "Obesity." Available at http://www.who.int/topics/obesity/en.

——— (2006c) "Poliomyelitis." Available at http://www.who.int/topics/poliomyelitis/en.

——— (2007a) "Mental Health: WHO Urges More Investments, Services for Mental Health." Available at http://www.who.int/mental_health/en/index.html.

——— (2007b) "Promoting Mental Health." Available at http://www.who.int/entity/mental_health/evidence/mh_promotion_book.pdf.

——— (2007c) "Re-defining 'Health.'" Available at http://www.who.int/bulletin/bulletin_board/83/ustun11051/en.

——— (2007d) "Reproductive Health." Available at http://www.who.int/topics/reproductive_health/en.

——— (2007e) "WHO Releases New Guidance on Insecticide-Treated Mosquito Nets." Available at http://www.who.int/mediacentre/news/releases/2007/en/index.html.

——— (2007f) "World Health Statistics." Available at http://www.who.int/whosis/whostat2007.pdf.

——— (2008) "HIV Surveillance, Estimations, and Monitoring and Evaluation." Available at http://www.who.int/hiv/topics/me/en/index.html.

Wild Salmon Center (2004–2008) "Factsheet: Hatchery Salmon Interactions with Wild Salmon Populations." Available at http://www.wildsalmoncenter.org/hatchery_factsheet.php.

Williams, Phil (2003) "Eurasia and the Transnational Terrorist Threats to Atlantic Security." In James Sperling, Sean Kay, and S. Victor Papacosma, eds., *Limiting Institutions: The Challenge of Eurasian Security Governance.* Manchester: Manchester University Press.

Wise, Timothy A., and Kevin P Gallagher (2005). "Doha Round's Development Impacts: Shrinking Gains and Real Costs." *RIS Policy Briefs* no. 19 (November). Available at http://www.ifg.org/documents/wtohongkong/risdoha benefits.pdf.

Wiseberg, Laurie S. (1992) "Human Rights Nongovernmental Organizations." In Richard Pierre Claude and Burns H. Weston, eds., *Human Rights in the World Community.* Philadelphia: University of Pennsylvania Press.

WMO (World Meteorological Organization) (2003) "Extreme Weather Events Might Increase" (July 2). Available at http://www.wmo.ch/web/press/ press695.doc.

Wolf, Aaron T. (2000) "'Hydrostrategic' Territory in the Jordan Basin: Water, War, and Arab-Israeli Peace Negotiations." In Hussein A. Amery and Aaron T. Wolf, eds., *Water in the Middle East: A Geography of Peace.* Austin: University of Texas Press.

World Bank (2001) *Attacking Poverty: World Development Report 2000/2001.* New York: Oxford University Press.

——— (2002) *World Development Indicators 2002.* Washington, DC.

——— (2007) *World Development Report 2007.* Washington, DC.

——— (2008a) "Kenya Profile" (April 3). Available at http://devdata.worldbank. org/external/cpprofile.asp?ptype=cp&ccode=usa.

——— (2008b) "United States Profile" (April 3). Available at http://devdata.world bank.org/external/cpprofile.asp?ptype=cp&ccode=usa.

——— (2004) *State of the World 2004.* New York: Norton.

——— (2006) *Vital Signs 2006–2007.* New York: Norton.

Yaghi, Abdulfattah (2004) "Water Public Policy: A Study on the Near East Countries." Unpublished paper.

Yanagida, Joy A. (1987) "The Pacific Salmon Treaty." *American Journal of International Law* 81, no. 3 (July).

The Contributors

Mary Ellen Batiuk is professor of social and political studies at Wilmington College, a Quaker school in southwestern Ohio. She currently teaches courses on global issues and the political economy of globalization.

Elise Boulding is professor emerita of sociology at Dartmouth College, and former secretary-general of the International Peace Research Association. She has undertaken numerous transnational and comparative cross-national studies on conflict and peace, development, and women in society. A scholar-activist, she was international chair of the Women's International League for Peace and Freedom in the late 1960s. Among her many publications are *Women in the Twentieth Century World* (1997) and *Cultures of Peace: The Hidden Side of History* (2000).

Pamela S. Chasek is associate professor of government and director of international studies at Manhattan College in New York City. She is also cofounder and executive editor of the International Institute for Sustainable Development's *Earth Negotiations Bulletin,* a reporting service on UN environment and development negotiations. She has been a consultant to the UN Environment Programme, the UN Development Programme, and the UN Department for Economic and Social Affairs. She is the author of numerous articles and publications on international environmental politics and negotiation, including *Global Environmental Politics* (2006), *Ten Days in Johannesburg: A Negotiation of Hope* (2004), and *Earth Negotiations: Analyzing Thirty Years of Environmental Diplomacy* (2001).

Deborah S. Davenport is lecturer in international political economy and director of the master's program in global affairs at the University of Buckingham, England. She is the author of *Global Environmental Negotiations and US Interests* (2006) and numerous other articles on international environmental politics, particularly international forest policy. Formerly director of environmental work of the Carter Presidential Center in Atlanta, she is also a consultant writer and editor for the *Earth Negotiations Bulletin* and other publications of the International Institute for Sustainable Development.

Jashinta D'Costa has led field research teams for the Food and Agriculture Organization of the United Nations on assessing hunger and food security after wars in the former Yugoslavia (2001–2002) and Afghanistan (2003), and for the UN World Food Programme in Azerbaijan (2004). She was formerly a consultant to the UN Development Programme's *Human Development Report* and Save the Children–USA.

Lori Heninger is executive director of HiTOPS, a youth health and wellness center in Princeton, New Jersey. She was formerly director of the Children and Adolescents Program of the Women's Commission for Refugee Women and Children, headquartered in New York City.

Lina M. Kassem is assistant professor of international studies at Zayed University in the United Arab Emirates (UAE). Her teaching and research interests include nationalism and identity, with a special focus on issues of subnational identity. She is currently researching Druze identity in the Middle East and the development of an Emirati national identity in the UAE. She teaches courses on global issues, global politics, and Middle Eastern politics.

Sean Kay is chair of the International Studies Program and professor of politics and government at Ohio Wesleyan University. He is also Mershon Associate at the Mershon Center for International Security Studies at the Ohio State University, and a fellow at the Eisenhower Institute in Washington, DC. He is coeditor of *NATO After 50 Years* (2001) and *Limiting Institutions: The Challenge of Eurasian Security Governance* (2003), and author of *NATO and the Future of European Security* (1998) and *Global Security in the Twenty-First Century* (2006).

George Kent is a professor in the Department of Political Science at the University of Hawaii. His teaching and research interests include human rights, international relations, peace, development, and environmental issues, with a special focus on nutrition and children. He is also co-convener of the Commission on International Human Rights of the International Peace Research Association, and a member of the Working Group on Nutrition, Ethics, and Human Rights of the UN's Standing Committee on Nutrition. His publications include *Freedom from Want: The Human Right to Adequate Food* (2005) and *Global Obligations for the Right to Food* (2008).

Ellen Percy Kraly is William R. Kenan Jr. Professor in the Department of Geography at Colgate University. She is author of numerous articles on international migration to and from the United States, US immigration policy and environmental issues, trends in socioeconomic mobility among immigrant groups, and ethical dimensions of population data systems. She has conducted research for the UN Statistical Commission, the US Immigration and Naturalization Service, the National Academy of Sciences, and the US Commission on Immigration Reform.

Jeffrey S. Lantis is associate professor of political science at The College of Wooster and a 2007 Fulbright Senior Scholar to Australia. His teaching and research interests include foreign policy analysis, international treaties, and international security. He is author of *The Life and Death of International Treaties* (2008) as well as numerous journal articles and book chapters, and coeditor of *The New International Studies Classroom* (2000) and *Foreign Policy in Comparative Perspective* (2002).

Bruce E. Moon is a professor in the Department of International Relations at Lehigh University. He is author of *The Political Economy of Basic Human Needs* (1991) and *Dilemmas of International Trade* (1996, 2000). His articles have appeared in *International Studies Quarterly, International Organization,* the *American Journal of Political Science, Comparative Political Studies, Studies in Comparative International Development, Political Research Quarterly,* and the *Journal of Conflict Resolution,* as well as in several edited volumes. He is currently completing a book on the dangers of trade deficits.

Fiona Mulligan graduated from Colgate University with honors in environmental studies, with thesis research focused on health and migration aspects of environmental justice. She served as a leader of Students for Environmental Action, and also as a fellow of the Upstate Institute at Colgate University.

Heather Parker is a recent graduate of the University of Cincinnati's College of Law and the Women's Studies Graduate Program. She served as Amnesty International's state campaign coordinator for the Stop Violence

Against Women Campaign in Ohio and has authored several publications on violence against women. Her primary interest lies with women's human rights; her current research focuses on the impact of gender bias and sex-stereotyping in protection order cases.

Don Reeves is president of the Center for Rural Affairs in Nebraska, which focuses on practical and policy needs of small rural communities and small and moderate-sized family farms. He also sits on the boards of the local development corporation and the local ethanol plant. He was formerly interim general secretary for the American Friends Service Committee; economic policy analyst for the Bread for the World Institute; legislative secretary for the Friends Committee on National Legislation, a Quaker lobby group in Washington, DC; and founding chair of the Nebraska Farm Crisis Response Program as well as Nebraskans for Peace.

Mark Seis is associate professor of sociology at Fort Lewis College in Durango, Colorado. His primary research interests are sustainable communities and environmental law and policy, with publication topics including economic globalization and the environment, alternative perspectives on sustainable living and environmental crime, and epistemological problems with environmental policy. He is coauthor of *A Primer in the Psychology of Crime* (1993).

D. Neil Snarr is professor of social and political studies at Wilmington College in Ohio. He teaches freshman and senior courses on global issues. *Introducing Global Issues* was partially developed to provide course material for the freshman global issues course. He has published in several sociology and disaster-oriented journals, and has edited several books on a variety of topics.

Michael T. Snarr is associate professor of social and political studies at Wilmington College in Ohio, where he teaches courses on global issues, US-Mexico relations, and nonviolence. His research focuses on Latin American foreign policy, and he is a member of the steering committee of the Christian Peacemaker Teams. He is coeditor of *Foreign Policy in Comparative Perspective* (2002), and coauthor of "Assessing Current Conceptual and Empirical Approaches" in *The Foreign Policies of the Global South* (2002).

Kelsey M. Swindler is a student at Wittenberg University, majoring in English and international relations. She qualified for Phi Delta Sigma honors each year of her high school career, and was recognized as a National Merit Commended Scholar. A citizen of Buckeye Girls State 2007, she was elected to the Ohio State Board of Education.

Anthony N. Talbott is a lecturer in political science at the University of Dayton. He also teaches courses on global issues and Southeast Asian studies at Wilmington College, Ohio. His research and publications focus on popular religion and politics in the Philippines. A ten-year veteran of the US Navy, he has traveled extensively in Asia, Australia, Africa, and elsewhere. In 2000–2001 he conducted political and ethnographic research in the Philippines as a Fulbright Research Fellow.

Jane A. Winzer is a visiting professor of political science at Georgia State University. Her research focuses on the impact of globalization and international governance on labor and environmental issues. She is currently researching the evolving tactics of nongovernmental organizations and activists in addressing environmental and labor concerns.

Index

Abbas, Mahmoud, 51
Abortion, 178, 214, 235
Abu Ghraib prison, 88
Acid rain, 279–284
Advanced Weapons Concept Initiative, 34
Afghanistan: war in, 4, 6, 17
Africa: colonization, 40, 62, 118–120, 123–124; HIV/AIDS, 84–85, 90, 229–232; poverty, 142–144, 156, 261
African Charter on Human and Peoples' Rights, 71
African Development Bank, 134
Agenda 21, 248–251, 255–257, 260
Agreed Framework, 32
Agriculture, 186–189, 198, 204, 292; industrialization, 128–129; sustainable, 152–153
Ahmadinejad, Mahmoud, 31–32
AIDS. *See* HIV/AIDS
Alliance of Small Island States (AOSIS), 271
Al-Qaida, 26, 30, 87–89
American Forest and Paper Association (AF&PA), 292
American Revolution (1776), 39
Amin, Idi, 68
Amnesty International, 71–72, 129, 313, 318
Anderson, Signe Bech, 274
Annan, Kofi, 69, 258
Anthrax, 21, 26, 88
Anti–Ballistic Missile (ABM) Treaty, 28, 33, 126
Anticolonization, 40–41
Arab-Israeli War, 301
Arafat, Yasir, 46, 49, 51
Arbour, Louise, 72
Arms race, 20–21
Arms trade, 29–30, 72
Asian Development Bank, 134
Asia Pacific Economic Cooperation (APEC), 312
Athansiou, Tom, 268
Atmospheric commons. *See* Global commons
Atoms for Peace, 33
Average income ratios, 141–143

Baer, Paul, 268
Balfour Declaration (1917), 45–46
Bangladesh, 140, 143
Barber, Benjamin, 2, 3, 6
Berlin Conference (1884), 118, 271–272
Berlin Mandate, 272
Bin Laden, Osama, 26, 43
Biodiversity, 257–259
Biological weapons, 21, 26–28, 88
Biological Weapons Convention, 28
Biosphere Conference, 245–246
Birth control. *See* Family planning
Body Mass Index (BMI), 227

Bonaparte, Napoleon, 39
Boserup, Ester, 188–190
Boundary resources, 288, 290–292, 301–304
Brain drain, 181
Bread for the World (BFW), 151
Bretton Woods Agreements, 107–108, 121–123, 126–127
British East India Company, 118
Brown, Lester, 251
Brundtland, Gro Harlem, 244, 247
Brundtland Commission, 151, 244, 247
Buchanan, Pat, 110
Bush, George H. W., 271, 282
Bush, George W., 17, 69, 75, 102, 134, 178; global warming, 270, 273–274, 303; tariffs, 107–108; weapons proliferation, 30–35

Camp Crame, 37
Canada: acid rain, 2, 279–282; carbon emissions, 269, 272; civic nationalism, 42; Fisheries Act, 111; fisheries dispute with the United States, 289, 294–297; healthcare, 146; immigration to, 176, 180; military spending, 18; Quebec separatism, 41, 313
Cancer, 89, 110, 226, 234; colon, 222–223; lung, 237–238; skin, 276
Capital flight, 129
Capital flows, 99, 116–127, 133
Capitalism, 62, 69, 123, 243, 283
Carbon sinks, 272–273
CARE, 132, 313
Carson, Rachel, 195
Carter, James Earl, 33, 73, 282
Catholic Church, 38
Center for International Forestry Research (CIFOR), 291
Center for Science and the Environment, 259
Central Intelligence Agency, 124
CFCs. *See* Chlorofluorocarbons
Chemical weapons, 20–21, 26, 28
Chemical Weapons Convention (CWC), 28
Cheney, Richard, 31
Child mortality, 18, 155–156, 209–211, 216, 233, 235, 314; due to HIV/AIDS, 222, 229, 256
Child prostitution, 204–205, 211, 214
Children, 19, 37, 51; armed conflicts, 205–208, 214; education, 153, 155; health, 84–85, 90, 150, 152, 201, 208–217, 229; homelessness, 205, 211; laborers, 147, 174, 189, 202–204, 215; poverty, 138, 201, 211; refugees, 176; rights of, 59, 65, 73–74, 201, 211–215; undernutrition, 225–226; violence, 118, 206–211
Children Bearing Military Arms, 206
Child soldiers, 206
China: acid rain, 283; carbon emissions, 269, 274; children, 202, 211; development, 251, 254; distribution of power, 78–80, 82–85, 90; economy, 121, 124, 126; free trade, 99, 103; globalization, 3; health, 224, 227; human rights, 59, 67–68, 75; nationalism, 44; oil consumption, 302; ozone depletion, 277–281; population, 166–167, 173, 179–180; poverty, 139, 142–145, 147, 149; weapons, 19, 22–25, 29
Chlorofluorocarbons, 274–278. *See also* Ozone depletion
Christianity, 6, 43–44, 53, 119, 234
Churchill, Winston, 124
Civil rights, 59, 120. *See also* Human rights
Civil society, 313
Civil wars, 19, 30, 124, 207
Clean Air Act (CAA), 282
Climate change. *See* Global warming
Clinton, Hillary, 213
Clinton, William Jefferson, 32, 69, 178, 213, 269, 271
Coffee, 316, 319
Cold War, 17; economics, 123–129, 133, 228; nuclear proliferation, 21–25, 28, 33–34; power, 80, 90; state sovereignty, 68
Collier, Paul, 134
Colonialism, 44–45, 62, 116–129, 133, 193, 197
Commission on the Status of Women, 190
Comparative advantage, 97, 128, 134
Comprehensive Nuclear Test Ban Treaty (CTBT), 28
Consumption: CFCs, 278; development, 248–253, 257, 260; natural resources, 270, 288, 297, 302, 306; nutrition, 226; poverty, 139–140, 144, 155; trade, 96–99, 101, 105, 111

Contraception. *See* Family planning
Convention for the Protection of Human Rights and Fundamental Freedoms, 70
Convention on Biological Diversity, 248
Convention on Long Range Transboundary Air Pollution, 282
Convention on the Elimination of All Forms of Discrimination Against Women, 72, 192, 214
Convention on the Rights of the Child, 212–214
Cornucopianism, 169
Crimes against humanity, 68–69. *See also* Human rights, abuses and violations
Cuba: defense spending, 17; health, 223; human rights, 72; nuclear weapons, 29, 33; poverty, 145
Cultural imperialism, 7
Cultural relativism, 65–66
Cultural Survival, 7–8, 71
Currency: stabilization of, 121–122, 126

DAC. *See* Development Assistance Committee
Danish Meteorological Institute, 274
Darfur, 129, 290, 305–306
Debt: global structure of, 133–134
Decentralization, 313
Declaration of Principles, 246
Declaration of the Rights of the Child, 212, 214
Declaration on Human Rights, 190
Declaration on the Human Environment, 246
Declaration on the Right to Development, 62
Decolonization, 40–41, 302
Defense, 17, 79, 109
Deforestation, 196, 248, 261–262, 291
Demographics, 84–85, 161–162, 166, 169, 174; demographic transition model, 169–173; trends, 176–178, 181
Development Assistance Committee (DAC), 131–132
Direct export subsidy, 101
Dixon, Thomas Horner, 305
Dunkin' Donuts: Fair Trade, 317

Earth Radiation Budget Experiments (ERBE), 268
Earth Summit (1992), 247–248, 251, 260, 271
East Asia: poverty, 149, 156; trade, 96
Easterly, William, 133
Ecofeminists, 195
Ecological migrants, 305
Economic dualism, 188, 193
Economic growth: environment, 243; poverty reduction, 146–150, 156
ECSC. *See* European Coal and Steel Community
Educating for Justice (EFJ), 316–319
Education, 152, 181; challenges to, 130–131; gender, 256; global security, 77, 90–91; health, 226, 231, 235, 238; as human right, 62, 74, 127–129, 185, 211; policies, 179; poverty, 137, 144–147, 150; women and girls, 150–153, 170, 178, 191–194, 235; universal, 155
EFJ. *See* Educating for Justice
Eisenhower, Dwight D., 33
Emigration, 45, 173–174, 180
Environmental Protection Agency, 111
Equal Exchange, 316–317
Equal rights movement, 194
Ethnic cleansing, 69, 83. *See also* Genocide
EU. *See* European Union
European Coal and Steel Community (ECSC), 108–109
European Commission of Human Rights, 70
European Union (EU), 32, 51; acid rain, 282; CFC use, 277; distribution of power, 78; emissions, 271, 274; power, 78; regionalism, 312; tariffs, 64, 107–108
Exclusive economic zone (EEZ), 293–294

Fair Trade, 316–319
Family planning, 178–179, 234–235
Fayyad, Salam, 51
FDI. *See* Foreign direct investment
Female circumcision, 65–66, 76
Fertility, 162–163, 167–169, 171–172, 176–183
Feudalism, 38–39
Fish, 110–111, 253–254, 258, 262, 280, 288, 293–297
Fisheries Act, 111

Food and Agriculture Organization (FAO), 216
Force Publique, 118
Foreign direct investment (FDI), 115–116, 127–130, 252, 254
Forest certification, 291–293
Forests, 290
Forest Stewardship Council (FSC), 292
Fourth World Conference on Women, 192
Framework Convention on Climate Change, 271
France: colonialism, 145; human rights, 59, 62, 67; nationalism, 39–40, 46; nuclear technology, 22–24, 79, 81
Fraser River Convention, 294
Free trade, 101, 103, 111; peace, 108
French, Hilary, 278
French Revolution (1789): beginning of nationalism, 39
Friends of the Earth, 110

G8. *See* Group of Eight
Gandhi, Indira, 257
Gates, Bill, 232
GATT. *See* General Agreement on Tariffs and Trade
GDP. *See* Gross domestic product
Gender roles, 105, 155, 185–190, 193–198
General Agreement on Tariffs and Trade (GATT), 107–110, 123, 127, 131, 252
General Council of the Union for Child Welfare, 212
Genocide, 58–59, 66–67, 73, 83, 85, 205, 210
Germany: acid rain, 281; carbon emissions, 269–273; economy, 116, 120, 122, 124, 134; free trade, 109, 111
Girls, 174; education of, 74, 153, 170, 192, 235; human rights, 65–66, 205, 207; living standards of, 151; poverty, 138
Global Campaign for Education, 91
Global commons, 248–250, 259; atmospheric, 265–285; international policies, 283–284
Global economy: poverty in, 146, 156; sustainable development, 251–253; trade, 95–96, 100
Global issues, 2, 9–10, 71, 161, 174; future challenges, 311–314; population, 182
Globalization, 2; advantages and criticism, 5–8; effects of, 87, 148; migration, 169, 174; poverty, 148–157; trade, 96
Global security, 34, 77–78, 80–82, 84–86, 89–91
Global warming, 261, 266–268, 279, 281–283, 290, 305; consequences of, 267; global security, 85–86; greenhouse gases, 251, 254, 257; health, 228, 233, 239; major contributors to, 268–274; skepticism, 267; treaty, 273
GNI. *See* Gross national income
GNP. *See* Gross national product
Gold standard, 122, 126
Gore, Al, 276
Grameen Bank, 197–199, 313
Grant, James, 213
Great Britain: acid rain, 281, 283; nuclear weapons, 22; nationalism, 40. *See also* United Kingdom
Great Depression, 96, 100, 107
Great Lakes Water Quality Agreement, 282
Green Belt movement, 196–197, 313
Greenhouse effect. *See* Global warming
Greenhouse gases, 251, 266–273; reduction of, 283
Greenpeace, 110, 313
Gribbin, John, 275
Gross domestic product (GDP), 7, 96, 104, 131, 139–143, 146, 153
Gross national income (GNI), 131–132, 140, 155
Gross national product (GNP), 131, 133–134, 139–141, 149, 188, 193
Group of Eight (G8), 134; summit, 271–274
Group of 77, 188
Group of Ten (G10), 122
Gulf War (1991), 19, 21, 29, 68, 207

Hague Conference, 69, 272–273
Halon, 275, 278
Hamas, 49–53
Hardin, Garrett, 265
Harmon Doctrine, 298
HDI. *See* Human Development Index
Health, 221; acid rain, 279–284; climate change, 233–234; drug resistance, 231–232; as human right, 239; infectious disease, 227–233; maternal,

234–236; mental, 236–239; nutrition, 224–227
Healthcare: availability, 222–223; funding for, 15, 130, 146, 227; as human right, 62, 213, 221; nutrition, 152–153
Hegemonic stability theory, 109
Helms, Jesse, 110
Hertz, Noreena, 134
Herzl, Theodor, 45
Highly Active Antiretroviral Therapy (HAART), 231
Highly Indebted Poor Countries, 134
Hitler, Adolf, 16
HIV/AIDS, 78, 84, 85, 90, 152, 192, 255–256, 258, 261; effect on population, 171, 179, 182; prevention and treatment, 226–232
Hull, Cordell, 107
Human Development Index (HDI), 144–146
Human Development Report: 2000, 64; 2005, 250. *See also* United Nations Development Programme
Human rights, 57; abuses and violations, 59, 64–69, 72, 74–76, 106; implementation of, 66, 70; NGOs, 71; origin of, 57–59; regional organizations, 70–71; sustained obligation, 215–217
Human Rights Watch, 71–73, 204, 207
Hunger: growth of, 129; in less-developed countries, 144; mortality, 152; reduction of, 84, 151, 155–156, 255–256
Huntington, Samuel, 7, 8
Hussein, Saddam, 19, 22, 30–31, 80–81, 302–304
Hussein-McMahon Correspondence (1915), 45
Hydrochlorofluorocarbons (HCFCs), 278

ICC. *See* International Criminal Court
IFAD. *See* International Fund for Agricultural Development
ILO. *See* International Labour Organization
IMF. *See* International Monetary Fund
Immigration, 173–176, 180–181
Imperialism, 7, 62, 69, 109, 133
Income distribution, 102, 141, 143, 148, 154, 261, 315
India: children, 202–205; development, 118, 254, 259; distribution of power, 78–79, 83–85, 90; environment, 274, 277–279, 302; health, 224, 232; nationalism, 43, 45; population, 166–167, 173, 179–180; poverty, 142–143, 152; weapons, 19, 21–25, 28–29, 34
Indonesia, 100, 133, 149, 231, 317-318
Industrialization, 119, 122, 169, 186, 279, 300
Industrial policy, 100–101, 105
Infanticide, 179
Infectious disease, 152, 182, 222–224, 227-233
INGOs. *See* International nongovernmental organizations
Integrated Child Development Services, 152
Inter-American Commission of Human Rights, 70
Inter-American Court of Human Rights, 70
Inter-American Development Bank, 133
Intercontinental ballistic missiles (ICBMs), 20
Intergovernmental Conference of Experts on the Scientific Basis for Rational Use and Conservation of the Resources of the Biosphere, 245–246
Intergovernmental organizations (IGOs), 1, 9, 216–217, 246
Intermediate-Range Nuclear Forces Treaty, 28
International Atomic Energy Association (IAEA), 32
International Bank for Reconstruction and Development, 122
International Commission of Jurists, 71
International Committee of the Red Cross, 71
International Conference on Population and Development, 178
International Covenant on Civil and Political Rights, 212
International Covenant on Economic, Social, and Cultural Rights, 212
International Criminal Court (ICC), 69–70
International Feminist Network, 193
International Fund for Agricultural Development (IFAD), 133, 216
International Labour Organization (ILO), 147, 202–204, 215

International League for Human Rights, 71
International Monetary Fund, 107, 122–134
International nongovernmental organizations (INGOs), 9, 216
International Population Conference, 178
International Programme on the Elimination of Child Labour, 204
International trade, 3, 10, 85, 95, 97; effects, 102–107
International Trade Commission, 103
International Tribunal on Crimes Against Women, 193
International Women's Decade, 191
International women's movement, 198
International Women's Year, 190, 193
International Year of Freshwater, 298
Internet: as terrorist tool, 21, 83, 86–87
Intifada: first (1987–1993), 46, 47; second (Al-Aqsa), 49, 51
Iran: human rights, 75; nuclear program, 22–26, 30–32, 79–80; oil, 45, 130
Iran-Iraq War (1980–1988), 20, 303
Iraq: children, 207–208, 213; nationalism, 41, 44–45, 54; nuclear program, 22–24, 29; oil, 91; state sovereignty, 68, 80–81, 83; WMD development, 22, 30–31, 80, 83
Iraq War: effect on US economy, 81; 1991 Gulf War, 15, 17, 19, 21–22, 29, 68, 207, 302–304; 2003, 19, 26, 30–31, 298
Islam, 6–7, 31, 49; health, 234; human rights, 53; natural resources, 306; terrorism, 43
Israel: conflict with Arabs, 44–54, 298–301; nuclear weapons development, 21, 23

Japan: acid rain, 281, 283; colonialism, 116, 121; development, 134, 148, 155; distribution of power, 78; environment, 270–271, 277; human rights, 69, 73–74, 78; nuclear bombing of, 20, 27; reconstruction, 122; trade, 98–101, 105, 126–127; weapons, 19–20
Jewish National Fund, 46
Job Corps, 153
Job creation, 7, 95, 102, 115, 130, 147–148, 156
Johannesburg, 257–259
Johannesburg Declaration, 257–258
Johannesburg Summit, 260. *See also* World Summit on Sustainable Development
Johnson, Bryan, 274
Johnston, Eric, 300, 301
Jong-Il, Kim, 32
Jordan River Basin, 298–301
Jubilee 2000, 134
Jubilee USA, 133

Keady, Jim, 317
Keeling, Charles, 266
Kennedy, John F., 33
Kenya: development, 196–197; economy, 115, 133; Green Belt movement, 313; health, 230, 233
Khameini, Ayatollah, 31
Khan, Irene, 72
Kipling, Rudyard, 119
Kiva, 197–198
Knight, Phil, 317
Koh, Tommy, 248
Kretzu, Leslie, 317
Kurds, 41, 68
Kuwait, 272, 302–303
Kyoto Protocol, 272–273

Labor: child, 106, 111, 202–204, 215; forced, 73; migration of, 173, 175, 180; unwaged, 190
Larsen B ice shelf, 267
Lash, Jonathan, 259
Law of the Sea (LOS) Treaty, 293–296
LDCs. *See* Less-developed countries
League of Nations, 40, 212
Lenin, Vladimir, 123
Less-developed countries (LDCs), 4; acid rain, 283; agriculture, 127–128; children, 203–210; colonization of, 40; development, 115–116, 132, 246; economics, 99, 129–130, 133; environmental concerns, 248; malnutrition, 208; population, 96, 161, 166–167, 172–176, 181; poverty, 129; women's roles, 185–188, 193. *See also* Human rights
Liberal economics, 121, 252

Liberalism, 97, 101–104, 109–112
Liberation Tigers of Tamil Eelam (LTTE), 87
Locke, John, 39

Maathai, Wangari, 196–197
Malaria, 84, 222, 226–227, 232–233, 239
Malnutrition, 84, 201, 208–209, 216–217, 222, 225–226, 314. *See also* Undernutrition
Malthus, Thomas, 167
Mandate Period (1919–1947), 46
Manhattan Project, 20, 27
Marcos, Ferdinand, 37
Marshall Plan, 109, 122–123
Martin, Claude, 259
Mastny, Lisa, 278
MDCs. *See* More-developed countries
MDGs. *See* Millennium Development Goals
Mennonite Central Committee, 313
Mental health, 212–223, 236–239
Mercantilism, 97–101, 112, 117, 167
Mexico, 2–3; development, 154; economy, 118, 130; human rights, 74; migration, 173–180; poverty, 155; trade, 100, 103–105; women, 190–192
Microcredit Summit Campaign, 198
Migration, 69, 74; international, 161–163; population change, 167, 169, 173–176; undocumented, 180–182
Milanovic, Branko, 142
Millennium Assembly, 255
Millennium Challenge Accounts, 134
Millennium Challenge Corporation, 134
Millennium Declaration, 192, 255
Millennium Development Goals (MDGs), 131, 155–157, 192, 222, 226, 230, 239, 256–258, 261–262
Millennium Summit, 258
Milosevic, Slobodan, 68
Missile Technology Control Regime, 29
Molina, Mario, 277
Monterey Institute's Center for Nonproliferation Studies, 26
Montreal Protocol on Substances That Deplete the Ozone Layer, 274, 278–279, 284
Moon, Ban Ki, 305
More-developed countries (MDCs), 4; development, 132–133, 246; economy, 116, 127–128; environmental concerns, 248; jobs, 7; population, 161, 166–167, 174–176, 179; trade, 96, 115–116, 179, 185, 188; women's roles, 185, 188
Mortality, 152, 155–156, 162, 166–167, 171, 176, 178, 182; rates, 18, 144, 169. *See also* Children; Women
Mowlana, Hamid, 4
Multinational corporations, 7, 314; development, 188, 195; economy, 130–133
Multinational states, 41
Muslims, 41, 43, 176
"Mutual assured destruction," 33, 79

Nader, Ralph, 110
NAFTA. *See* North American Free Trade Agreement
Narain, Sunita, 259, 260
National Council of Women of Kenya, 196
Nationalism: definitions of, 37–38, 40–41; evolution of, 39–41; exclusivity of, 42; as political tool, 43–44; positive aspects of, 40; role of, 53–54; trade, 108
National missile defense (NMD) system, 33
National security, 78, 106, 270
Nation states, 41, 169
NATO. *See* North Atlantic Treaty Organization
Natural Resource Defense Council, 269
Natural resources, 287–290, 304–306
Neo-Malthusianism, 169
NGOs. *See* Nongovernmental organizations
Nike, 317
Nixon, Richard M., 21, 33, 126
Nongovernmental organizations (NGOs), 9, 313, 315; debt, 134; environment, 246, 271; HIV/AIDS, 230; human rights, 71–74; population, 176; social justice, 319
Nonproliferation of weapons, 15, 26–30
Nonproliferation Treaty, 27, 32–35
Nontariff barriers, 101

North. *See* More-developed countries
North American Free Trade Agreement (NAFTA), 103–105, 111, 253, 312
North Atlantic Treaty Organization (NATO), 78, 81, 124
North Korea, 22, 33; nuclear program, 15, 22–25, 29–32
NPT. *See* Nonproliferation Treaty
Nuclear weapons: clandestine programs, 34; development of, 19–26; hard power, 79–80; nonproliferation, 26–30; proliferation, 26–27; reduction of, 35; terrorism, 88
Nutrition, 145, 224; healthcare, 152, 226; rights, 216–217. *See also* Malnutrition; Overnutrition; Undernutrition

Obesity, 226–227. *See also* Overnutrition
ODA. *See* Official Development Assistance
Office of the UN High Commissioner for Refugees (UNHCR), 175
Official Development Assistance (ODA), 115, 127, 131–132, 255
Oil, 32, 45, 120, 128–130; conflict over, 304; necessity for industrial development, 302; nonrenewable resource, 301; reserves, 91, 302
O'Neill, Paul, 103
Organization of Petroleum Exporting Countries (OPEC), 128
Our Common Future, 247
Overnutrition, 226–227
Oxfam, 255
Ozone depletion, 258, 274–278, 283

Pacific Salmon Treaty (PST), 295–296
Pakistan: human rights, 73; nuclear program, 22–25, 28–29, 79; SAI, 318; weapons, 19, 21, 88
Palestine Liberation Organization (PLO), 46, 86, 301
Palestinian Legislative Council, 51
Partial Nuclear Test Ban Treaty, 27, 33
Patriarchal society, 190–193, 196, 198; rise of, 185–187
Perlin, Michael, 237
Perot, Ross, 110
Philippine People Power Revolution, 37, 43
Physicians for Human Rights, 71
Pinochet, Augusto, 68
Plan of Implementation of the World Summit on Sustainable Development, 258
Polio, 227–229
Pol Pot, 68
Population: analysis, 161, 170; policies, 176, 179
Population growth, 179; policies, 178–181; rate, 161–169, 171–172, 182. *See also* Emigration; Immigration; Migration
Poverty, 137, 314–315; antipoverty policies, 151; children, 201–211; development, 128–129, 134; eradication of, 255; gap, 143, 315; measures of, 139–143, 146; Millennium Development Goals, 255–262; rates, 144, 155; reduction, 149–152, 155, 169, 178, 182, 192, 247, 252; in the United States, 143; women, 192, 196–197; worldwide, 138
Powell, Colin, 30
Power, 77–78, 80–84, 87; distribution of, 78; types, 78–86
PPP. *See* Purchasing power parity
Privatization, 127, 130–131, 252, 265–266
Prostitution, 73, 140, 190, 204, 211, 214
Protectionism, 95, 100–101, 110, 121
PST. *See* Pacific Salmon Treaty
Purchasing power parity (PPP): used in measuring poverty, 140–141, 149
Putin, Vladimir, 35

Qaddafi, Muammar, 26
Quaker United Nations Office, 206
Quebec, 41, 54, 313

Rabin, Yitzhak, 49
Randall, Doug, 270
Reagan, Ronald, 17, 178
Reconstruction, 121–123
Refugees, 52, 59, 67–68, 174–177; environmental, 290, 298; resettlement of, 180–181
Regionalism, 311–312
Renewable energy, 258–259
Replacement fertility, 162, 179
Reproductive Health Response in Conflict (RHRC), 236

Resource degradation, 305–306
Revolutionary United Front, 207
Ricardo, David, 97, 98
Rio conference, 257
Rio Declaration of Environment and Development and the Statement of Forest Principles, 248
Rio Summit (1992), 247, 251, 256–257, 271. *See also* Earth Summit
Robust Nuclear Earth Penetrator, 34
Roosevelt, Franklin D., 20
Rousseau, Jean-Jacques, 39
Rowland, Sherwood, 277
Rubber, 118–119, 253
Russia, 8; acid rain, 280–281, 283; arms deals, 19; carbon emissions, 273; children, 205; distribution of power, 78–79, 85, 88; economy, 123, 134, 147; health, 237; human rights, 67, 78–79; nationalism, 40, 45; nuclear weapons, 22–28, 31–33; ozone depletion, 277; trade, 100
Rwanda, 67–69, 73, 83

Sachs, Jeffrey, 134
SAI. *See* Social Accountability International
Salmon, 111, 289, 294–298
SALT I, 28, 33, 126
Sanitation, 85, 152, 154, 163, 255, 257–258, 261
SAPs. *See* Structural adjustment programs
Sarin, 20
Schumacher, E. F., 195
Schuman, Robert, 109
Schwartz, Peter, 270
Security: and power, 78; effects of globalization on, 77; global, 77–78, 80–82, 84–86
September 11 attacks, 21, 30–31, 35, 75, 86–91, 175, 181, 303; 9/11 Commission, 87
Sharon, Ariel, 49
Sierra Club, 110, 248
Sipila, Helvi, 190
Slave trade, 57, 117–119
Smallpox, 227–228
Smith, Adam, 97
Smoot-Hawley Act, 100
Social Accountability International (SAI), 316, 318–319
Social and economic rights, 59, 64
South. *See* Less-developed countries
Sovereignty, 6–7, 38; free trade, 109–112; natural resources, 288–290, 298, 302–305, 311–314; world government, 217, 251
Soviet Union, 6; decentralization, 313; distribution of power, 80, 90–91; economy, 123; health, 228, 237; human rights, 69, 75; migration, 181; nationalism, 40; nuclear weapons, 20–28, 33, 35; ozone depletion, 277; poverty, 147
Spain: acid rain, 283; colonialism, 118, 145; population, 163–164
Special Supplemental Food Program for Women, Infants, and Children, 152
Stalin, Josef, 123
Starbucks Corporation, 317
State of Food Insecurity in the World 2006, 144
State sovereignty, 6–7, 38, 54; human rights, 65, 67–70; origin and evolution, 38–39
Stiglitz, George, 133
Stockholm Conference. *See* United Nations Conference on the Human Environment
Strategic Arms Limitation Treaty (SALT I), 28, 33, 126
Strategic Arms Reduction treaties, 28, 35
Structural adjustment programs (SAPs), 130–131, 188
Sudan, 129
Suffrage movement, 187, 194
Sulfur dioxide, 279–283
Sulfuric acid, 279–280
Sustainable development, 151–152, 154, 169, 243–244, 248, 306; challenges to, 245, 257; definitions, 151, 244; energy, 258; environment, 244–245; future concerns, 260–262; globalization, 251–253; issues of, 259; objectives of, 151; origin of term, 247; population policies, 178; social development, 255; three pillars of, 259; world population, 255
Sustainable forest management (SFM), 291
Sustainable Forestry Initiative (SFI), 292
Sutherland, Peter, 110

Swallow, Ellen, 195
Sweden: acid rain, 280–281, 283
Swedish International Development Agency (SIDA), 216
Sykes-Picot Agreement, 45

Taliban, 6, 30, 206
Tamil Tigers (Liberation Tigers of Tamil Eelam), 87
Tariffs, 64, 100–103, 107–108, 121, 128–129
Terrorism, 53, 81, 86–90; asymmetric power tactic, 82, 86; war on, 35, 86, 91; WMD, 15, 20, 26–27, 88–90
Third World Network, 255
30 Percent Club, 283
Thirty Years' War (1618–1648), 38
Torture, 59, 66, 72
Trade: consequences of, 96, 101–105; international, 95–97
Trade deficits, 98–101, 104–106
Trade liberalization, 127, 130–131, 134; effects of, 253–255
Trade policies, 100–107, 112
Trafficking: human, 73–74, 82, 203–205; chemical, 278
Transboundary resources, 288–289, 293–301
TransFair USA, 316
Treaty for the Prohibition of Nuclear Weapons, 27
Treaty of Westphalia (1648), 38, 39
Triangular Trade, 117–118
Truman, Harry, 27, 33

UDHR. *See* Universal Declaration of Human Rights
Ultraviolet radiation, 274–275
Undernutrition, 225–226
UNDP. *See* United Nations Development Programme
UNICEF. *See* United Nations Children's Fund
Union of Concerned Scientists, 270
United Kingdom: arms deals, 19; carbon emissions, 273; income inequality, 143
United Nations: human rights, 58, 66–70; oversight of nuclear materials, 27, 30–31; purpose, 40
United Nations Children's Fund (UNICEF), 85, 152, 201, 216
United Nations Commission on Human Rights, 212
United Nations Conference on the Environment and Development (UNCED). *See* Earth Summit
United Nations Conference on the Human Environment, 190, 246, 257, 271
United Nations Declaration of Human Rights, 180
United Nations Development Fund for Women (UNIFEM), 191–192
United Nations Development Programme (UNDP), 19, 64, 140, 144, 188, 191, 198
United Nations Division for the Advancement of Women (DAW), 192
United Nations Economic and Social Council, 74, 190
United Nations Economic Commission for Europe (ECE), 282
United Nations Environment Programme (UNEP), 244, 246, 305
United Nations Financing for Development Conference, 155
United Nations Framework Convention on Climate Change, 248
United Nations Fund for Population Activities (UNFPA), 179
United Nations International Decade for Women, 191
United Nations International Research and Training Institute for the Advancement of Women (INSTRAW), 191
United Nations Millennium Development Goals. *See* Millennium Development Goals
United Nations Millennium Summit, 155
United Nations Monetary and Financial Conference, 121
United Nations Population Division, 167
United Nations Security Council, 126
United Nations World Commission on Environment and Development, 151, 244, 247
United States: acid rain, 279–284; arms sales, 18–19; carbon emissions, 267–272; child labor, 104–106, 109–111, 204, 214–217; Convention on the Rights of the Child, 212–214;

debts, 68, 79–80; defense spending, 17–18; distribution of power, 78–91; global role, 124; global warming, 85–86, 266–274; health, 224–227, 229, 233; immigration to, 161–163, 173–176, 180–182; National Geospatial-Intelligence Agency, 83; National Intelligence Council, 85; National Oceanic and Atmospheric Administration, 274; oil consumption, 272, 282, 302–305; ozone depletion, 274–279, 314; poverty threshold, 139; trade deficits, 98–101, 104–106; US-India Civil Nuclear Cooperation Agreement, 34; WMD, 20–24. *See also* Iraq War
United States Agency for International Development (USAID), 216
Universal Declaration of Human Rights (1948), 58–65, 72, 75, 212, 221, 234
Urbanization, 161, 169, 173–174, 182, 186
Uruguay Round, 100, 252

Vaccination, 137, 152
Venezuela, 19, 130
Vienna Convention for the Protection of the Ozone Layer, 277
Vietnam Veteran's Memorial, 210–211
Vietnam War, 16
Voluntary export restraints, 101

Wan xi shao, 179
War: capital flows, 121–126; children, 205–210; conduct, 58; costs, 17–18, 81; fish, 294–298; human rights, 71–73
War crimes, 194. *See also* Crimes against humanity; Human rights, abuses and violations
War on terror, 35, 76, 86, 91; human rights, 71–72
Warsaw Pact, 124
Washington Consensus, 127, 131
Water, 298–301
Wealth, unequal distribution of, 315
Weapons, conventional: proliferation of, 16–18, 28, 33–34; sales of, 18–19
Weapons of mass destruction (WMD), 15, 83, 87; proliferation, 19–26; radiological, 27; reasons for building, 21; research, 23–26; social costs, 17–18; spread of technology, 29, 34; terrorism, 15, 20, 26–27, 88–90; types, 15, 20
Weatherhead, Betsy, 274
Weight-to-height ratio, 227
Williams, Jody, 83
Wilson, Woodrow, 120
WMD. *See* Weapons of mass destruction
Women: activism, 193–194; agriculture, 185–189, 198; education of, 150, 192; environment, 195; mortality rate, 222, 236, 256, 261; rights of, 72–74. *See also* Patriarchal society
Women's Decade, 191, 198
Women's World Banking, 198
WomenWatch, 192
Woods, Dorothea, 206
Work. *See* Labor
World Bank, 107, 122–123, 127–134, 198; development problems, 188; poverty reduction, 192; poverty thresholds, 139
World Commission on Environment and Development, 151, 244, 247
World Conference on Women: First 190; Fourth, 192
World Conservation Union (IUCN), 244
World Day Against Child Labour, 204
World Development Movement, 255
World fertility rates, 168
World Food Conference, 190
World Food Programme (WFP), 216
World government, 67, 311–312
World Health Organization (WHO), 216, 221, 224, 228, 232, 234, 238
World Meteorological Organization, 267
World Plan of Action, 190
World population, 166, 170–172
World Population Conference (Bucharest, 1974), 178, 190
World Population Plan of Action, 178
World Resources Institute, 259
World Summit for Children, 218
World Summit on Sustainable Development (WSSD), 256–257, 260
World Trade Organization (WTO), 64, 107–110, 131–133, 252–253; purpose of, 131
World Wide Fund for Nature (WWF), 244, 292, 313

WSSD. *See* World Summit on Sustainable Development
WTO. *See* World Trade Organization

Yugoslavia, 6, 67–69, 124, 194; breakup of, 40
Yukon River Salmon Agreement, 297
Yunus, Muhammed, 197

Zero population growth, 162
Zhdanov, Victor M., 228
Zionism, 45–46, 52

About the Book

ISSUES RANGING FROM CONFLICT AND SECURITY, TO THE ECONomy and economic development, to the environment are explored in this fully revised and updated edition of *Introducing Global Issues*.

Increased attention is given in the new edition to the historical conditions that have exacerbated the gap between North and South. Also notable are discussions of international efforts to deal with such challenges as terrorism, human rights violations, the threat of nuclear proliferation, global warming, and the seemingly intractable Israel-Palestine conflict.

The material has been successfully designed for readers with little or no prior knowledge of the topics covered. Each chapter provides an analytical overview of the issues addressed, identifies the central actors and perspectives, and outlines past progress and future prospects. Discussion questions are posed to enhance students' appreciation of the complexities involved, and suggestions for further reading additionally enrich the text.

Michael T. Snarr is associate professor of social and political studies at Wilmington College. **D. Neil Snarr** is professor of sociology at Wilmington College.